Advances in Ironmaking and Steelmaking Processes

Advances in Ironmaking and Steelmaking Processes

Editor

Pasquale Cavaliere

MDPI • Basel • Beijing • Wuhan • Barcelona • Belgrade • Manchester • Tokyo • Cluj • Tianjin

Editor
Pasquale Cavaliere
University of Salento
Italy

Editorial Office
MDPI
St. Alban-Anlage 66
4052 Basel, Switzerland

This is a reprint of articles from the Special Issue published online in the open access journal *Metals* (ISSN 2075-4701) (available at: https://www.mdpi.com/journal/metals/special_issues/Ironmaking_Steelmaking_Processes).

For citation purposes, cite each article independently as indicated on the article page online and as indicated below:

LastName, A.A.; LastName, B.B.; LastName, C.C. Article Title. *Journal Name* **Year**, *Volume Number*, Page Range.

ISBN 978-3-0365-7548-3 (Hbk)
ISBN 978-3-0365-7549-0 (PDF)

Contents

About the Editor

Pasquale Cavaliere

Professor of Metallurgy at the Department of Innovation Engineering of the University of Salento, Italy. Author of more than 10 books. Author of over 400 papers in international journals and international conferences: https://scholar.google.com/citations?user=5J0hKn4AAAAJ&hl=it. Member of the Editorial Board of over 20 international journals.

metals

Editorial

Advances in Primary Ironmaking and Steelmaking Processes

Pasquale Cavaliere

Department of Innovation Engineering, University of Salento, 73100 Lecce, Italy; pasquale.cavaliere@unisalento.it

1. Introduction

In the recent past, ironmaking and steelmaking saw the incorporation of various new processes and technologies that can be operated and organized in different combinations depending on the properties of raw materials and the required quality of the final products. Different raw materials, energy requirements, and investments can vary as a function of different plant configurations and the advanced technologies employed for to reduce emissions. Due to the low level of restrictions and international protocols active during the last 15–20 years, innovation has been growing so fast that knowledge of the best available technologies is fundamental for scientists and industrial operators.

Indications from the steel industry and local and global government institutions are that the breakthrough technologies for decarbonization will be based on hydrogen reduction. The employment of hydrogen in the ironmaking and steelmaking industries will push forward the global transformation of hard-to-abate industries. Given that the two main routes for primary steel production decarbonization will almost certainly be CCS (carbon capture and storage) and hydrogen-based reduction, public and private R&D spending, as well as investment in pilot plants, should focus on driving down their cost and increasing the efficiency of equipment and piloting and driving down the cost of hydrogen-based reduction.

For CO_2-lean process routes, three major solutions have been identified: decarbonizing, whereby coal would be replaced by hydrogen or electricity in the hydrogen reduction or electrolysis of iron ore processes; the introduction of CCS technology; and the use of sustainable biomass.

Through a hydrogen-based steelmaking route, CO_2 emissions would be reduced by more than 80%. Hydrogen steelmaking will depend profoundly on the availability of green hydrogen, which can be generated from natural gas by steam reforming, or from water by electrolysis. Today, hydrogen-based steelmaking is a potential low-carbon and economically attractive route, especially in countries where natural gas is cheap. In considering systems for increasing energy efficiency and reducing the environmental impact of steel production, CO_2 emissions may be greatly reduced by hydrogen-based steel production if the hydrogen is generated by means of carbon-free and renewable sources. Currently, the development of the hydrogen economy has received a great deal of attention in that H_2 is considered a promising alternative to replace fossil fuels. If H_2 is utilized as an alternative fuel, not only can the problem of the progressively exhausted fossil fuel reserves be solved, but the atmospheric greenhouse effect can also be mitigated. Based on hydrogen, the "hydrogen economy" is a promising clean energy carrier for decarbonized energy systems if the hydrogen used is produced from renewable energy sources or coupled with carbon capture and storage (CCS) or nuclear energy.

2. Contributions

Chen et al. [1] deeply analysed the basic thermodynamic conditions for the reduction of molten wüstite by hydrogen, with great significance for optimizing the ironmaking process, energy saving and reducing emissions. The results of this study provided fundamental data to support new hydrogen metallurgy technologies in the future. In Hauser et al. [2], a

Citation: Cavaliere, P. Advances in Primary Ironmaking and Steelmaking Processes. *Metals* **2023**, *13*, 781. https://doi.org/10.3390/met13040781

Received: 15 March 2023
Revised: 28 March 2023
Accepted: 4 April 2023
Published: 16 April 2023

first experimental demonstration is presented of an approach to the utilization of process off-gases generated in a steelworks that produced methane and methanol in hydrogen-intensified syntheses. Specifically, the integration of two methane synthesis reactors and one methanol synthesis reactor into a steel plant is experimentally simulated. The online tests carried out show that the calculation, optimization, and control architecture can lead to promising and satisfying results. In Cavaliere et al. [3], an innovative steel processing route developed by employing direct hydrogen reduced pellets and an open slag bath furnace is illustrated. The paper illustrates the direct reduction reactor employing hydrogen as a reductant on an industrial scale. The solution allows for the production of steel from blast furnace pellets transformed in the direct reduction reactor. The reduced pellets are then melted in open slag bath furnaces, allowing for carburization for further refining. The proposed solution is clean, contributing to the decarbonization of the steel industry. In Liu et al. [4], the effects of the width of the mould on the surface velocity, flow field pattern, turbulent kinetic energy distribution, and surface-level fluctuation in the mould were studied using the rod deflection method to measure the flow velocity near the surface of the mould at a high temperature, as well as numerical calculation. In actual production, the argon gas flow rate and the immersion depth of the submerged entry nozzle should be adjusted reasonably to optimize the flow field in the mould with different widths under the same fixed steel throughput. In El-Tawil et al. [5], coking coal blends with 5% and 10% additions of bio-coals (pre-treated biomass) of different origins and degrees of pre-treatment were carbonized at laboratory scale and at a technical scale using a 5% bio-coal addition, aiming to understand the impact on the bio-coal's properties (ash amount and composition, volatile matter content) and the addition of bio-coke reactivity. Types of bio-coke produced at the technical scale showed low reactivity in the thermogravimetric analysis but also showed a higher quality in standard tests for reactivity, strength after reaction, and mechanical strength. This indicates that the coking coal blend with a 5% high-temperature torrefied bio-coal could be suitable for industrial use. Steer et al. [6] conducted research comparing the laboratory-measured properties of injection coals that were used over a two-month production period compared to the process variables and measurements of the blast furnace during that study period. In Sun et al. [7], the effect of dephosphorization endpoint temperature on the dephosphorization of hot metal was studied for the double slag converter steelmaking process under the conditions of low temperature and low basicity using industrial experiments. Considering the experimental results and thermodynamic calculation results of the industrial experiments of the double slag dephosphorization process, the optimal temperature range for intermediate deslagging is about 1400–1420 °C. In Diniz et al. [8], a new algorithm is proposed to predict the silicon content time series up to 8 h ahead, immediately after the molten iron chemical analysis is delivered by the laboratory. The proposed model proved to be a promising tool for the prediction of pig iron silicon content, and it can be extended to other blast furnaces, with appropriate changes. In Hu et al. [9], the numerical simulation software ANSYS Fluent 18.2 was used to simulate the velocity field of molten steel under the condition of bottom-blowing stirring in different stages in the EAF steelmaking process. The properties of bottom-blowing and the kinetic conditions of the steel–slag interface were investigated. The physical model's verification confirmed the results; that is, the viscosity of molten steel decreased as the smelting progressed, and as the flow velocity of the molten steel caused by the agitation of bottom-blowing decreased, the effect of bottom-blowing also decreased. Based on these results, a theoretical basis was provided for the development of the bottom-blowing process. Finally, in the review from Cavaliere et al. [10], the main available technologies for water electrolysis for hydrogen production were described. The fundamentals of water electrolysis and the problems related to purification and/or desalinization of water before electrolysis were discussed. As a matter of fact, the authors underlined the energy efficiency issues, with particular attention given to the potential application of water electrolysis in the steel industry. The fundamental aspects related to the choice of high-temperature or low-temperature technologies are reviewed.

3. Conclusions and Outlook

This Special Issue aimed to focus on those traditional or innovative routes capable of reducing energy consumption and harmful greenhouse emissions. Obviously, discussions of energy took into account the direct and indirect energy consumption for each analysed technology. The methods to improve energy efficiency are energy consumption optimization, online monitoring, and energy audits.

This Special Issue describes some of the main approaches to produce and synthesize iron and steel through hydrogen-based technologies. For this Special Issue, we accepted contributions from both universities and the industry that evaluated the industrial feasibility of each selected technology. The main goal was to describe the most efficient solutions being applied by ironmaking and steelmaking factories all around the world.

I would like to thank all the authors for their contribution and the managing office of *Metals* (MDPI) for their fundamental and constant support in the development of this Special Issue.

Conflicts of Interest: The author declares no conflict of interest.

References

1. Chen, S.; Zhang, J.; Wang, Y.; Wang, T.; Li, Y.; Liu, Z. Thermodynamic Study of H_2-FeO Based on the Principle of Minimum Gibbs Free Energy. *Metals* **2023**, *13*, 225. [CrossRef]
2. Hauser, A.; Wolf-Zoellner, P.; Haag, S.; Dettori, S.; Tang, X.; Mighani, M.; Matino, I.; Mocci, C.; Colla, V.; Kolb, S.; et al. Valorizing Steelworks Gases by Coupling Novel Methane and Methanol Synthesis Reactors with an Economic Hybrid Model Predictive Controller. *Metals* **2022**, *12*, 1023. [CrossRef]
3. Cavaliere, P.; Perrone, A.; Silvello, A.; Stagnoli, P.; Duarte, P. Integration of Open Slag Bath Furnace with Direct Reduction Reactors for New-Generation Steelmaking. *Metals* **2022**, *12*, 203. [CrossRef]
4. Liu, J.; Yang, J.; Ma, C.; Guo, Y.; He, W.; Zhao, C.; Jiang, R.; Guo, Y. Effect of Mold Width on the Flow Field in a Slab Continuous-Casting Mold with High-Temperature Velocity Measurement and Numerical Simulation. *Metals* **2021**, *11*, 1943. [CrossRef]
5. El-Tawil, A.; Björkman, B.; Lundgren, M.; Robles, A.; Sundqvist Ökvist, L. Influence of Bio-Coal Properties on Carbonization and Bio-Coke Reactivity. *Metals* **2021**, *11*, 1752. [CrossRef]
6. Steer, J.; Greenslade, M.; Marsh, R. A Comparison of Laboratory Coal Testing with the Blast Furnace Process and Coal Injection. *Metals* **2021**, *11*, 1476. [CrossRef]
7. Sun, H.; Yang, J.; Lu, X.; Liu, W.; Ye, G.; Zhang, R.; Yang, W. Dephosphorization in Double Slag Converter Steelmaking Process at Different Temperatures by Industrial Experiments. *Metals* **2021**, *11*, 1030. [CrossRef]
8. Diniz, A.; Côco, K.; Gomes, F.; Salles, J. Forecasting Model of Silicon Content in Molten Iron Using Wavelet Decomposition and Artificial Neural Networks. *Metals* **2021**, *11*, 1001. [CrossRef]
9. Hu, H.; Yang, L.; Guo, Y.; Chen, F.; Wang, S.; Zheng, F.; Li, B. Numerical Simulation of Bottom-Blowing Stirring in Different Smelting Stages of Electric Arc Furnace Steelmaking. *Metals* **2021**, *11*, 799. [CrossRef]
10. Cavaliere, P.; Perrone, A.; Silvello, A. Water Electrolysis for the Production of Hydrogen to Be Employed in the Ironmaking and Steelmaking Industry. *Metals* **2021**, *11*, 1816. [CrossRef]

metals

MDPI

Article

Thermodynamic Study of H_2-FeO Based on the Principle of Minimum Gibbs Free Energy

Shuyue Chen [1], Jianliang Zhang [1], Yaozu Wang [2,*], Tengfei Wang [1], Yang Li [3] and Zhengjian Liu [1,*]

1 School of Metallurgical and Ecological Engineering, University of Science and Technology Beijing, 30th Xueyuan Road, Haidian District, Beijing 100083, China
2 School of Intelligence Science and Technology, University of Science and Technology Beijing, 30th Xueyuan Road, Haidian District, Beijing 100083, China
3 Ironmaking Department of Beijing Shougang Co., Ltd., Tangshan 064404, China
* Correspondence: yaozuwang@ustb.edu.cn (Y.W.); liuzhengjian@ustb.edu.cn (Z.L.); Tel.: +86-010-62332550 (Y.W. & Z.L.)

Abstract: Studies have shown that the reduction of wustite is the limiting factor in the ironmaking process, whether in hydrogen-based shaft furnaces, hydrogen-rich blast furnaces or smelting reduction vessels. The study of the thermodynamic conditions for the reduction of molten wustite by hydrogen is of great significance for the optimization of the ironmaking process, energy saving and emission reduction. Previous studies have mostly focused on the thermodynamic study of the reduction at a lower temperature, but the data at high temperatures are different, which makes the calculation of thermodynamics difficult. Moreover, it is difficult to obtain experimental evidence for the data at high temperature, so calculation is needed to verify its feasibility. In this paper, a thermodynamic calculation model for the reduction of molten wustite by hydrogen based on the principle of minimum Gibbs free energy is developed. The enthalpy changes of the reaction at different temperatures and the partial pressure of hydrogen required for the reaction to occur are calculated, and the energy change during the reaction is analyzed. The results show that the partial pressure of hydrogen for the reduction of molten wustite by hydrogen at high temperatures decreases from 0.67 at 1650 K to about 0.65 at 2000 K. The enthalpy changes of reaction at 1873 K are only 1/4 to 1/3 of that at 1173 K compared with that at the corresponding temperature between hydrogen for the reduction of molten wustite (1873 K) and hydrogen-based shaft furnace reaction (1173 K). Interestingly, the thermodynamic calculations show that the effect of energy absorption in the gas–liquid reaction of hydrogen with wustite at high temperatures is much lower than in the gas–solid reaction zone at low temperatures. These results indicate that the energy change of the reduction of molten wustite by hydrogen at high temperatures is better than that of hydrogen reduction at low temperatures, and the thermodynamic conditions are more favorable, with slightly different results from different thermodynamic databases, but the general trend is the same. The results of this study will provide fundamental data to support new hydrogen metallurgy technologies in the future. If its correctness can be verified experimentally in the future, this result will be promoted to the development of a new alternative ironmaking technology, hydrogen-based smelting reduction.

Keywords: hydrogen metallurgy; hydrogen reduction of iron oxides; alternative ironmaking; smelting reduction; thermodynamic

Citation: Chen, S.; Zhang, J.; Wang, Y.; Wang, T.; Li, Y.; Liu, Z. Thermodynamic Study of H_2-FeO Based on the Principle of Minimum Gibbs Free Energy. *Metals* **2023**, *13*, 225. https://doi.org/10.3390/met13020225

Academic Editor: Pasquale Cavaliere

Received: 21 December 2022
Revised: 20 January 2023
Accepted: 21 January 2023
Published: 25 January 2023

1. Introduction

The reduction of wustite by hydrogen is the most important part of the stepwise reduction process of iron oxides, which is widely present in the cohesive zone reduction of blast furnace ironmaking, direct reduction in hydrogen-based shaft furnaces and hydrogen-rich smelting reduction processes [1]. The studies of the reduction of wustite contribute to the understanding of different ironmaking processes and the development and optimization of new ironmaking processes. In a hydrogen-rich blast furnace, the blast furnaces are often operated with full oxygen blast to achieve the hot air temperature required for the reduction of wustite by hydrogen. The development of the reduction of wustite by hydrogen in gas-based shaft furnaces was reviewed by scholars, such as Zhang et al. [2], who pointed out that the commonly used reducing gases at this stage are generally H_2 and CO obtained from the cracking of natural gas, affirming the superior kinetic conditions of hydrogen in the stepwise reduction of iron oxides. In the hydrogen-based shaft furnace, the hydrogen content is higher than that of the hydrogen-rich blast furnace, and the heat load of the reduction reaction is higher, which requires a higher reduction gas feed temperature. Hayashi and Iguchi [3] determined the rate controlling step of smelting reduction by hydrogen through experiments, which reduces in the form of the progressive formation of the metallic iron spheres in the center of ferrous oxide droplets. The one-dimensional mass balance equation combined with the rate analysis shows that the chemical reaction itself is the rate controlling step in the early stages of reduction. In summary, the process of the hydrogen reduction of wustite contains gas–solid, gas–liquid and mixed gas-solid–liquid states, different reaction temperatures and states representing different smelting processes. The analysis of the reduction of wustite in different states is of great significance for the understanding and optimization of the ironmaking process.

Many pieces of work have been completed by domestic and international experts and scholars for the reduction of wustite by hydrogen. In terms of experimental research, at this stage, research on the reduction of molten iron oxides by hydrogen has mainly focused on the use of blast furnaces as experimental equipment in an attempt to simulate the mechanism by which the reduction reaction occurs. Katayama et al. [4] have studied the reduction of FeO in molten slag by hydrogen, based on the blast furnace process. The reduction of molten FeO by hydrogen has been studied by Ban-Ya et al. [5] using a thermal analysis balance, and the results show that the chemical reaction rate for the reduction of molten FeO by hydrogen is about 20 times the rate for the reduction of solid FeO by hydrogen. In terms of kinetic research, the reduction of wustite in molten slag by hydrogen has been studied by Nagasaka et al. [6]. The results show that the reduction rate of liquid wustite by hydrogen is much faster than that with solid carbon and CO gas. In terms of thermodynamic research, Fan X. et al. [7,8] have used thermodynamic calculations to simulate the effect of blowing different proportions of hydrogen in a hydrogen-rich blast furnace on the rate controlling step of the reaction; Luo's [9,10] study of the thermodynamics of CO reduction in a hydrogen-based shaft furnace concluded that there is a peak in the ability of CO to reduce FeO, with a decreasing trend after the peak is reached; Kashiwaya et al. [11] compared the activity of each component in the reaction between H_2 and CO reduction of FeO and concluded that the metal obtained by H_2 reduction had a high oxygen activity. Whether or not thermodynamic conditions are met will directly affect whether or not a chemical reaction can take place, and most current thermodynamic studies of iron oxides are concerned with gas–solid phase reactions. A great deal of work has been completed in the past on the experimental study and kinetic conditions for the hydrogen reduction of wustite, but there is a lack of fundamental data on the thermodynamics of wustite reduction, especially at high temperature, and the available data come from a wide range of sources, with differences between different data, making the understanding and calculation of thermodynamics difficult. At the same time, the thermodynamic calculation at high temperatures is often difficult to be proved by experiments, so it is necessary to verify whether it is meaningful to carry out by calculation. Finally, the research on hydrogen-based smelting reduction in alternative ironmaking is not perfect. The thermodynamic calculation results of hydrogen-based melting reduction of wustite at high temperatures are given in this paper.

To address this, this paper is based on the minimum Gibbs free energy model to research different wustite states at different temperatures. By collecting and comparing thermodynamic data from different sources and simulating the Gibbs free energy, the partial pressure of hydrogen required for reaction and enthalpy change of the reaction and the process of reducing wustite to obtain metallic iron was analyzed; the energy changes of the three processes, hydrogen-rich blast furnace reduction of FeO, hydrogen-based shaft furnace reduction of wustite and the reduction of molten wustite by hydrogen in the smelting reduction vessel were obtained and compared. The results provide the necessary theoretical basis for further research into the reduction reactions of molten iron oxides and are of some reference significance.

2. Model of Hydrogen Reduction of wustite

The minimum Gibbs free energy model is the classical model for the study of equilibrium in multiphase systems and was proposed by White in 1958. The basic principle is that for a multiphase system, the total free energy is minimized when the system reaches thermodynamic equilibrium. In a reaction system for reducing iron oxides by hydrogen, assuming that H_2 and Fe_2O_3 are present in the system at the beginning, there will be several possible groups of Fe_2O_3, Fe, Fe_xO, Fe_3O_4, H_2 and H_2O in the system at equilibrium. It is generally accepted that the process of reducing iron oxides by hydrogen consists of the following reactions:

$$3Fe_2O_3 + H_2 = 2Fe_3O_4 + H_2O, \quad (1)$$

$$\frac{x}{4x-3}Fe_3O_4 + H_2 = \frac{3}{4x-3}Fe_xO + H_2O, \quad (2)$$

$$Fe_xO + H_2 = xFe + H_2O, \quad (3)$$

$$\frac{1}{4}Fe_3O_4 + H_2 = \frac{3}{4}Fe + H_2O, \quad (4)$$

For the reaction of reducing molten wustite by hydrogen, four components of Fe, H_2, FeO and H_2O should be present in the equilibrium system. By fully considering the above four components, the result obtained from the calculation is the content of the system at equilibrium. The positive and negative values of the standard Gibbs free energy for the reduction of molten wustite by hydrogen allow a preliminary determination of the likelihood of the reaction proceeding and further analysis of how the reaction occurs. It is known empirically that reactions (1)–(4) can all occur within a certain temperature range. A ΔG^θ versus T curve is made from the ΔG^θ of the above reaction as a function of T, which is the iron oxide gas–solid reduction equilibrium diagram for this reaction. Due to the different thermodynamic reference data in different metallurgical textbooks and thermodynamic handbooks, the resulting gas–solid reduction equilibrium diagrams for iron oxides are different. This paper focuses on the special case of Reaction (3).

$$FeO + H_2 = Fe + H_2O, \quad (5)$$

This reaction is the rate controlling step in the reduction of iron oxides by hydrogen. Common thermodynamic handbooks were consulted to obtain the Gibbs free energy G of the reactants FeO, H_2 and the products Fe and H_2O at different temperatures, and the G versus T curves of the participating substances were obtained by fitting the (G, T) data points using MATLAB software (R2021b, MathWorks, Natick, MA, USA) [12–16]. It was again fitted using MATLAB software to obtain its ΔG^θ versus T curve.

$$\Delta G_T^\theta = G_{T,H2O}^\theta + G_{T,Fe}^\theta - G_{T,FeO}^\theta - G_{T,H2}^\theta \quad (6)$$

$$\Delta G^\theta = -RTlnK^\theta = -RTln\left(\frac{\varphi_{H_2O}}{\varphi_{H_2}}\right) \quad (7)$$

$$\varphi_{H_2} = \frac{1}{\left[1+\exp\left(-\frac{\Delta G^\theta}{RT}\right)\right]} \quad (8)$$

In addition, to examine Reaction (5) release or absorption of energy phenomena visually, the H^θ versus T curves of the substances involved in the reaction were obtained by fitting the *(H, T)* data points using MATLAB software according to the enthalpies of each substance involved in the reaction at different temperatures from the thermodynamic handbook.

Definition of Gibbs free energy:

$$G = H - TS \tag{9}$$

(*H* is the enthalpy of the substance and *S* is the entropy of the substance)
Isobaric thermal capacity C_p:

$$C_p = a + bT + cT^2 + dT^{-1} \tag{10}$$

$$H_T^\theta = \int C_p dT = H_0 + aT + \frac{b}{2}T^2 + \frac{c}{3}T^3 + dT^{-1} \tag{11}$$

$$\Delta H_T^\theta = H_{T,H2O}^\theta + H_{T,Fe}^\theta - H_{T,FeO}^\theta - H_{T,H2}^\theta \tag{12}$$

Gibbs–Helmholtz equation:

$$[\frac{\partial\left(\frac{\Delta G_T^\theta}{T}\right)}{\partial T}]_p = -\frac{\Delta H_T^\theta}{T^2} \tag{13}$$

$$d\left(\frac{\Delta G_T^\theta}{T}\right) = -\frac{\Delta H_T^\theta}{T^2}dT \tag{14}$$

For the indefinite integral of the above equation:

$$G_T^\theta = a_1 + a_2T + a_3T^2 + a_4T^4 + a_5T^{-1} + a_6TlnT \tag{15}$$

The temperature range 1650–2000 K is defined and calculated every 1 K to obtain φ_{H_2} versus *T* and ΔH^θ versus *T* , $T \in$ (1650–2000 K). The image is fitted to obtain the energy change regularity for Reaction (5).

3. Thermodynamic Calculations of Hydrogen Reduction of Wustite

3.1. Thermodynamic Data

For a more extensive and objective calculation of the reduction of molten wustite by hydrogen, this paper has consulted Thermochemical Properties of Inorganic Substances (1973) [17], Thermodynamic Data Manual of Practical Inorganic Matter (1981) [18], Thermodynamic Data Manual of Practical Inorganic Matter (2002) [19] and Thermodynamic Data Manual of Inorganic Matter (1993) [20], four common thermodynamic handbooks, referring to data given by the National Institute of Standards and Technology NIST website [21] and the thermodynamic calculation software FactSage [22,23] for metallurgical disciplines, some of which are selected for presentation purposes in Tables 1–6 below.

FactSage is a common data calculation software for metallurgical disciplines. FactSage holds thermodynamic data for thousands of pure substances and the data provided in FactSage version 8.2 (GTT-Technologies and Thermfact/CRCT, Aachen and Montreal, GER and CAN) are used in this paper. FactSage calculates the corresponding φ_{H_2} and ΔH^θ for every 1 K at the temperature range 298–2000 K. It has a large amount of data with high confidence and is used in common thermodynamic simulation studies. In the FactSage database, Reaction (5) has $\Delta H^\theta > 0$ in the 298–1371 K range, which turns to $\Delta H^\theta < 0$ after 1644 K and turns again after 1810 K with less energy absorption than at 1644–1810 K. The φ_{H_2} versus *T* and ΔH^θ versus *T* for Reaction (5) are obtained directly from the above numerical calculations. the specific values are given in Table 1.

Table 1. Results of φ_{H_2} and ΔH^θ calculations for Reaction (5) in the FactSage database.

Temperature/K	$\varphi(H_2)$	ΔH^θ/J·mol^{-1}
298	0.99995809	−20,001.1
300	0.99995346	−19,957.5
400	0.99815616	21,847.4
500	0.98510069	19,825.6
600	0.94573032	18,090.5
700	0.87919082	16,713.3
800	0.79792785	15,752.1
900	0.74496720	15,304.3
1000	0.70404393	15,584.0
1100	0.66642057	16,118.5
1200	0.63303482	16,620.8
1300	0.60399718	15,279.8
1400	0.58066874	13,927.7
1500	0.56207896	12,544.1
1600	0.54741453	11,111.6
1700	0.55429572	−20,636.4
1800	0.57477776	−21,767.6
1900	0.58325040	−8592.4
2000	0.59005277	−9137.9

NIST Chemical Database, a web-based database of physical properties from the National Institute of Standards and Technology. The thermodynamic data given in the database cover a wide range of temperatures and only the thermodynamic data for Fe, FeO, H_2 and H_2O in the high temperature range (1650–2000 K) and adjacent temperatures are listed here in Table 2 for subsequent calculations.

Table 2. NIST Standard Reference Database.

Temperature/K		1500	1600	1700	1800	1900	2000	2100
Fe (H_{298} = 0 KJ·mol^{-1})	H_T-H_{298}/KJ·mol^{-1}	45.643	49.338	54.066	58.263	76.641	81.24	85.845
	S_T/J·mol^{-1}·K	84.336	86.720	89.580	91.978	102.082	104.442	106.688
FeO	H_T-H_{298}/KJ·mol^{-1}	69.502	75.873	106.549	113.369	120.189	127.009	133.829
	H_{298}/KJ·mol^{-1}	−268.484	−267.642	−243.540	−242.774	−256.202	−255.866	−255.541
	S_T/J·mol^{-1}·K	151.323	155.435	174.026	177.924	181.611	185.109	188.437
H_2 (H_{298} = 0 KJ·mol^{-1})	H_T-H_{298}/KJ·mol^{-1}	36.290	39.541	42.835	46.169	49.541	52.951	56.397
	S_T/J·mol^{-1}·K	178.846	180.944	182.940	184.486	186.669	188.418	190.718
H_2O	H_T-H_{298}/KJ·mol^{-1}	48.151	52.908	57.758	62.693	67.706	72.790	77.941
	H_{298}/KJ·mol^{-1}	−250.265	−250.592	−250.881	−251.138	−251.368	−251.575	−251.762
	S_T/J·mol^{-1}·K	250.620	253.690	256.630	259.451	262.161	264.769	267.282

In addition to the two common databases mentioned above, the thermodynamic data for Fe, FeO, H_2 and H_2O from the following four thermodynamic manuals are presented in Tables 3–6.

Of the six sets of thermodynamic data, the data given by the FactSage database are special in that the software can directly calculate the values of φ_{H_2} and ΔH^θ at a given temperature. The remaining five sets of data give the enthalpy of the substances involved in the reaction; the Gibbs free energy; and, in some cases, the enthalpy changes and entropy values, but the units are not uniform. The normalization of the units resulted in the four thermodynamic handbooks giving essentially similar values, which differed significantly from the NIST website and FactSage, while the NIST website and FactSage database were not identical. However, it is worth noting that the thermodynamic handbooks give values at 100 K ranges, while FactSage is more precise, with each 1 K having its corresponding thermodynamic value.

Table 3. Thermochemical Properties of Inorganic Substances (1973).

Temperature/K		1500	1600	1700	1800	1900	2000	2100
Fe	H/kcal·mol^{-1}	11.302	12.208	13.393	14.443	18.795	19.851	20.911
	G/kcal·mol^{-1}	−19.473	−21.554	−23.109	−23.696	−25.908	−26.110	−28.345
FeO	H/kcal·mol^{-1}	−48.415	−46.885	−39.547	−37.917	−36.187	−34.657	−33.027
	G/kcal·mol^{-1}	−102.691	−106.359	−110.299	−114.508	−118.807	−123.192	−127.659
H_2	H/kcal·mol^{-1}	8.711	9.485	10.266	11.055	11.851	12.656	13.468
	G/kcal·mol^{-1}	−55.408	−59.708	−64.056	−68.451	−72.890	−77.370	−81.892
H_2O	H/kcal·mol^{-1}	−46.545	−45.431	−44.291	−43.126	−41.935	−40.719	−39.477
	G/kcal·mol^{-1}	−136.260	−142.277	−148.365	−154.520	−160.741	−167.026	−173.371

Table 4. Thermodynamic Data Manual of Practical Inorganic Matter (1981).

Temperature/K		1500	1600	1700	1800	1900	2000	2100
Fe(H_{298} = 0 kcal·mol^{-1})	H_T-H_{298}/cal·mol^{-1}	11,304	12,211	13,395	14,445	18,797	19,853	20,913
	S_T/cal·mol^{-1}·K	20.53	21.12	21.79	22.39	24.79	25.33	25.85
FeO(H_{298} = −65.02 kcal·mol^{-1})	H_T-H_{298}/cal·mol^{-1}	16,607	18,137	25,474	27,104	28,734	30,364	31,994
	S_T/cal·mol^{-1}·K	36.21	37.19	41.63	42.56	43.44	44.28	45.07
H_2(H_{298} = 0kcal·mol^{-1})	H_T-H_{298}/cal·mol^{-1}	8713	9487	10,268	11,057	11,854	12,658	13,470
	S_T/cal·mol^{-1}·K	42.75	43.25	43.73	44.18	44.61	45.02	45.42
H_2O(H_{298} = −57.98 kcal·mol^{-1})	H_T-H_{298}/cal·mol^{-1}	11,407	12,521	13,660	14,326	16,017	17,233	18,475
	S_T/cal·mol^{-1}·K	59.81	60.53	61.22	61.88	62.53	63.15	63.76

Table 5. Thermodynamic Data Manual of Practical Inorganic Matter (2002).

Temperature/K		1500	1600	1700	1800	1900	2000	2100
Fe(H_{298} = 0 J·mol^{-1})	H_T-H_{298}/J·mol^{-1}	46,131	49,826	54,555	58,751	77,129	81,731	86,334
	S_T/J·mol^{-1}·K	84.779	87.164	90.024	92.423	102.526	104.887	107.132
FeO(H_{298} = −272,044 J·mol^{-1})	H_T-H_{298}/J·mol^{-1}	69,479	75,881	106,583	113,402	120,222	127,042	133,862
	S_T/J·mol^{-1}·K	151.414	155.545	174.152	178.050	181.738	185.236	188.563
H_2(H_{298} = 0 J·mol^{-1})	H_T-H_{298}/J·mol^{-1}	36,453	39,689	42,957	46,258	49,591	52,957	56,355
	S_T/J·mol^{-1}·K	178.66	180.954	182.935	184.822	186.624	188.351	190.009
H_2O(H_{298} = −241,814 J·mol^{-1})	H_T-H_{298}/J·mol^{-1}	47,723	52,385	57,153	62,029	67,011	72,100	77,297
	S_T/J·mol^{-1}·K	250.262	253.270	256.161	258.947	261.641	264.251	266.786

Table 6. Thermodynamic Data Manual of Inorganic Matter.

Temperature/K		1600	1800	2000	2200
Fe	H/kJ·mol^{-1}	49.81	58.73	81.72	90.92
	G/kJ·mol^{-1}	−89.61	−107.58	−128.01	−149.43
FeO	H/kJ·mol^{-1}	−196.17	−158.64	−145.00	−131.36
	G/kJ·mol^{-1}	−445.01	−479.10	−515.44	−553.14
H_2	H/kJ·mol^{-1}	39.68	46.25	52.95	59.78
	G/kJ·mol^{-1}	180.94	184.87	188.33	191.59
H_2O	H/kJ·mol^{-1}	−189.43	−179.79	−169.72	−159.22
	G/kJ·mol^{-1}	−594.64	−645.86	−698.19	−751.54

3.2. Thermodynamic Equilibrium Diagram

The thermodynamic calculation of G_i^θ versus T for each component in Reaction (5) gives the φ_{H_2} versus T function for the reaction, as shown in Table 7. The above ΔG^θ versus T function was fitted using MATLAB software to obtain the ΔG^θ versus T

function for Reaction (5) at the temperature range 1650–2000 K in the above table reference. The image of φ_{H_2} versus T function is shown in Figure 1.

Table 7. ΔG^{θ} versus T calculations from the NIST website and four thermodynamic handbooks.

References	ΔG^{θ} versus T
NIST Standard Reference Database	$121467.46 - 2942.761T - 0.42514T^2 + 5.144 \times 10^{-5}T^4 + 1293.0311T^{-1} + 462.88303TlnT$
Thermochemical Properties of Inorganic Substances (1973)	$38148.142 - 308.0349T - 0.045298T^2 + 8.13 \times 10^{-6}T^4 + 442.08621T^{-1} + 46.348489TlnT$
Thermodynamic Data Manual of Practical Inorganic Matter (1981)	$25155.3608 + 32.5768065T + 0.01043708T^2 - 4.424 \times 10^{-7}T^4 + 286.502223T^{-1} - 7.797753TlnT$
Thermodynamic Data Manual of Practical Inorganic Matter (2002)	$41071.779 - 235.4074T - 0.034435T^2 + 7.255 \times 10^{-6}T^4 + 447.86741T^{-1} + 34.312306TlnT$
Thermodynamic Data Manual of Inorganic Matter	$35359.172 - 181.6009T - 0.019334T^2 + 3.876 \times 10^{-6}T^4 + 356.47473T^{-1} + 25.435745TlnT$

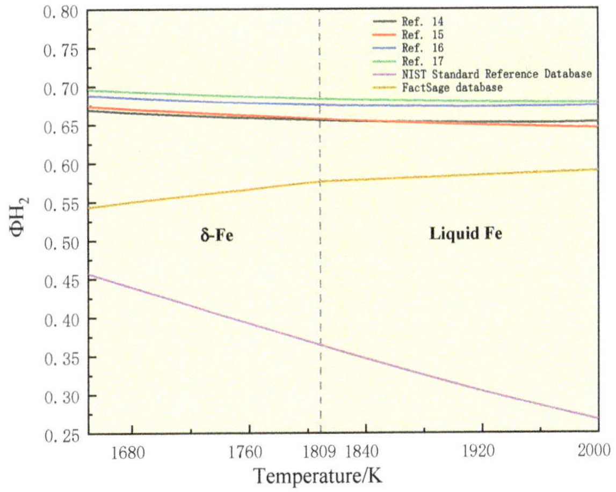

Figure 1. Relationship between φ_{H_2} and temperature.

Figure 1 shows the calculated and fitted φ_{H_2} versus T function image for Reaction (5) obtained from six sets of thermodynamic data. φ_{H_2} versus T function images are commonly known as fork curves in the ironmaking industry. φ_{H_2} is the partial pressure of hydrogen required for the reaction to occur. In all four thermodynamic handbooks, the φ_{H_2}versus T function images show a decreasing trend and similar magnitude, all in the range of 0.6–0.7. The NIST database gives values that also show a decreasing trend but to a slightly greater extent than the four thermodynamic handbooks mentioned above. In addition, FactSage is more unusual, giving a slight upward trend and a folding point at 1809 K. The analysis of the φ_{H_2} versus T function image shows that the trend of the equilibrium curve is related to the ΔH^{θ} of the reaction. If the reaction is endothermic ($\Delta H^{\theta} > 0$), the equilibrium curve shows a decreasing trend with increasing temperature. Therefore, the results obtained from the four thermodynamic handbooks and the NIST database all show that Reaction (5) is absorbing energy in the high temperature range (1650–2000 K) and the FactSage database shows a weakly releasing energy reaction.

In addition, φ_{H_2} characterises the partial pressure of hydrogen required for the reaction to occur, and the data given in References 14–17 all show a decreasing trend when fitted to data from the NIST database. Although the FactSage database gives the opposite result in this dimension, it does not show a significant upward trend until 1809 K and a much slower upward trend after 1809 K. It is considered that this fit does not affect subsequent conclusions. This suggests that the amount of hydrogen required for the reaction to occur is essentially constant or even less as the temperature increases.

3.3. Thermodynamic Enthalpy Change

The H_i^θ versus T of each group in Reaction (5) was calculated thermodynamically to obtain the ΔH^θ versus T function for the reaction, as shown in Table 8.

Table 8. ΔH^θ versus T calculations from the NIST website and four thermodynamic handbooks.

References	ΔH^θ versus T
NIST Standard Reference Database	$59.261072 + 98.954207T - 0.121208T^2 + 3.213 \times 10^{-5}T^3 + 10439.744T^{-1}$
Thermochemical Properties of Inorganic Substances (1973)	$19920.093 + 29.385873T - 0.035393T^2 + 8.207 \times 10^{-6}T^3 + 162.2773T^{-1}$
Thermodynamic Data Manual of Practical Inorganic Matter (1981)	$24749.9641 + 15.4796745T - 0.0238181T^2 + 5.2918 \times 10^{-6}T^3 + 242.831376T^{-1}$
Thermodynamic Data Manual of Practical Inorganic Matter (2002)	$21380.396 + 22.271131T - 0.028354T^2 + 6.24 \times 10^{-6}T^3 + 10418.891T^{-1}$
Thermodynamic Data Manual of Inorganic Matter	$20930.1194 + 14.8469727T - 0.0171551T^2 + 2.6461 \times 10^{-6}T^3 + 10383.6518T^{-1}$

The above ΔH^θ versus T function was fitted using MATLAB software to obtain the ΔH^θ versus T function in the above table reference for doing Reaction (5) at the 298–2000 K temperature range. The image of the ΔH^θ versus T function is shown in Figure 2.

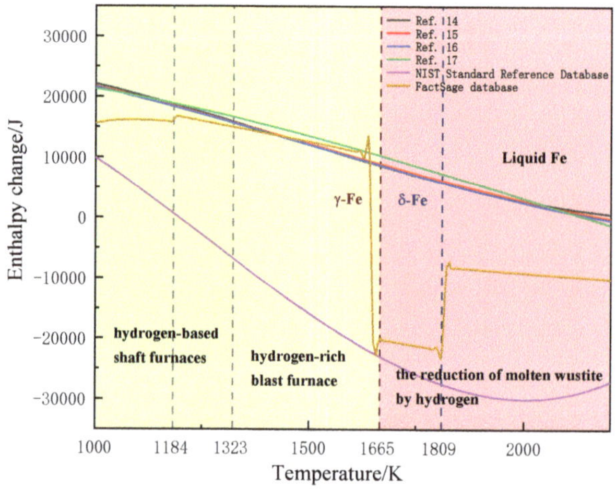

Figure 2. Relationship between enthalpy changes and temperature.

The image of the ΔH^θ versus T function provides a visual representation of the release or absorb energy of a reaction. If $\Delta H^\theta > 0$, the reaction is endothermic, and if $\Delta H^\theta < 0$, the reaction is exothermic. All six sets of thermodynamic data above show that Reaction (5) absorbs energy until 1184 K. Thereafter, the NIST database shows that the reaction gradually releases energy, and the energy released increases with increasing temperature until an inflection point at around 2000 K. For the FactSage database, there are abrupt changes around two temperature points, 1644 K and 1811 K. This is due to the fact that these points are the melting points of FeO and Fe, respectively. The remaining four thermodynamic handbooks all show a decreasing trend, i.e., the energy absorption becomes less as the temperature increases. The difference is that the calculations of the four thermodynamic manuals show a weakly absorption of energy at the high temperature range, while FactSage shows an energy released one. The fitting results show that Reaction (5) absorbs energy up to the temperature of 1642 K. Between the melting point of FeO (around 1642 K) and the melting point of Fe (around 1811 K), Reaction (5) absorbs energy strongly, while the release energy remains weak in the temperature range of the reduction of molten wustite by hydrogen after 1811 K, and the release of energy increases gradually with the increase in temperature. Furthermore, the data given on the NIST website show that Reaction (5) exhibits a release energy reaction after 1184 K. This data set differs considerably from the others, and the regularity it shows is questionable.

4. Discussion

It was shown that the rate of the reduction of the molten wustite by hydrogen at 1673 K is very fast and that the rate controlling step in this reaction is not the phase transfer but the chemical reaction itself. A synthesis of results from several papers found that the rate of the reduction of molten wustite by hydrogen was much faster than using other reducing agents, such as solid carbon, carbon dissolved in liquid iron and CO gas [24]. The rate controlling step, whether hydrogen-rich blast furnace, hydrogen-based shaft furnace or the reduction of molten wustite by hydrogen in the smelting reduction vessel is Reaction (5). The simulations of common thermodynamic handbooks' data reveal that Reaction (5) has different energy changes in different temperature ranges.

At low temperatures (1123–1323 K), i.e., the temperature range applicable to hydrogen-based shaft furnaces, the fits given by the four thermodynamic handbooks and the FactSage database all showed a strong absorption of energy. At 1320 K, the enthalpy changes for Reaction (5) were 15,918.1411 J/mol (Reference 14), 15,854.2651 J/mol (Reference 15), 15,733.8543 J/mol (Reference 16), 16,730.8318 J/mol (Reference 17) and 15,011 J/mol (FactSage database), which is thought to absorb more energy. The NIST database differs significantly from the above five data sets in this range and the regularity is questionable.

At the mid-temperatures (1323–1650 K), i.e., the temperature range applicable to hydrogen-rich blast furnace, the fits given in the four thermodynamic handbooks, the FactSage database and the NIST database also show the absorbed energy, but the amount of energy decreases to varying degrees with increasing temperature. The enthalpy changes for Reaction (5) at 1650 K were 8919.1 J/mol (Reference 14), 9219.39377 J/mol (Reference 15), 8971.07641 J/mol (Reference 16) and 10,615.5912 J/mol (Reference 17), respectively. It is worth noting that at 1644 K (which can be considered to be around 1642 K) the enthalpy changes in the FactSage database change abruptly to a negative value (−22,451.538 J/mol); it is thought that this is due to the phase change of FeO.

At the high temperatures (1650–2000 K), i.e., the temperature range for the reduction of molten wustite by hydrogen in the smelting reduction vessel, the enthalpy changes given in the four thermodynamic handbooks already show a weak absorption of energy, and at 2000K the enthalpy changes for Reaction (5) are 2780.72054 J/mol (Reference 14), 2773.43342 J/mol (Reference 15), 2431.86755 J/mol (Reference 16), 3177.25483 J/mol (Reference 17), 2431.86755J/mol (Reference 16) and 3177.25483J/mol (Reference 17). As the temperature increases, the amount of energy absorbed by the reaction decreases, gradually approaching thermal equilibrium. The thermodynamic data given in the FactSage database even show a release energy (−30,059.105 J/mol), which becomes more pronounced as the temperature increases. Furthermore, at 2000K, the partial pressure of hydrogen for Reaction (5) drops to 65.29% (Reference 14), 64.57% (Reference 15), 67.51% (Reference 16) and 67.92% (Reference 17), respectively. Although the FactSage database gives an increasing trend in the partial pressure of hydrogen, it is still only about 59.01% at 2000 K.

The energy absorption and release at around 1173 K (900 °C) is compared with that at around 1873 K (1600 °C) and the results are shown in Table 9. According to the FactSage database and the NIST database results, Reaction (5) is already release energy at a high temperature of 1873 K. The data given in four other thermodynamic handbooks show that the energy absorbed by Reaction (5) at 1873K is only 1/4–1/3 of that at 1173 K, which can be considered as weak absorption or release of energy.

Table 9. The 1173 K versus 1873 K energy absorption and release.

References	1173 K/J	1873K/J	Ratios
FactSage database	13,294.2	−9880.6	-
NIST Standard Reference Database	1176.1	−28,886.9	-
Thermochemical Properties of Inorganic Substances (1973)	18,938.5	4726.6	4.01
Thermodynamic Data Manual of Practical Inorganic Matter (1981)	18,677.1	4959.2	3.77
Thermodynamic Data Manual of Practical Inorganic Matter (2002)	18,571.3	4631.6	4.01
Thermodynamic Data Manual of Inorganic Matter	19,021.0	5948.2	3.20

In practice, comparing the energy absorption values at the two temperatures, the results are obvious. This means that at higher temperatures, the reduction of wustite to iron by hydrogen absorbs less energy. Although it costs more to provide heat with oxygen and hydrogen than to provide heat with oxygen and carbon, this result shows that a reaction with less energy absorption is obviously more energy efficient when it comes to heating with oxygen and hydrogen. That is, the reduction of smelting wustite by hydrogen at high temperature is more energy saving.

In another similar paper, a reduction method based on a hydrogen-based shaft furnace is proposed for solid-state direct reduction combined with hydrogen plasma reduction, a method that can also have the effect of saving hydrogen resources and which is set at an experimental temperature of 700 °C [25]. If the equipment allows, the reduction of molten wustite by hydrogen should also be experimented frequently. The Technical Research Laboratories in Korea Pohang, which attaches great importance to hydrogen metallurgy, points out that the problem facing the reduction of wustite by hydrogen at this stage is the heat absorption of the reaction and states that the application of this technology must be accompanied by a variety of processes and equipment changes [26].

5. Conclusions

Based on the above six sets of data, the variation in Gibbs free energy in the range of 298–2000 K was obtained using the principle of minimum Gibbs free energy, and the relationship between φ_{H_2}versus T and ΔH^{θ}versus T was further deduced. The different data were found to be consistent in the calculation of enthalpy changes below 1665 K and similar at the temperature range 1809–2000 K; the four thermodynamic handbooks were similarly consistent in the calculation of φ_{H_2}. In the temperature range of 1665–1809 K, the calculation results of enthalpy changes of different data are different. The calculation results of φ_{H_2} in the four thermodynamic handbooks and the FactSage database are different, and further verification by experiment is required subsequently.

At low temperatures of 1123 K to 1323 K, i.e., in the hydrogen-based shaft furnace, the energy absorption is all around 15,000 J/mol, and the reaction energy absorption is violent.

At medium temperatures of 1323 K to 1650 K, i.e., in the hydrogen-rich blast furnace, the energy absorption drops to around 9000 J/mol, and the reaction is still in an energy-absorbing state, but the energy absorption is significantly reduced.

At high temperatures of 1650 K to 2000 K, i.e., in the reduction of molten wustite by hydrogen, the energy absorption drops to around 2500 J/mol and can be considered as weak absorption or release of energy. Comparing this result with the energy variation of the hydrogen-based shaft furnace, the difference is obvious. The energy absorption of the hydrogen-based shaft furnace is about three to four times greater than the energy absorption of the reduction of molten wustite by hydrogen at high temperatures. The partial pressure of hydrogen required for the reaction to occur at this time is similarly at a low level of about 60–65%.

In summary, it can be concluded that the reduction of molten wustite by hydrogen has better thermodynamic conditions, absorbs less energy and requires a lower partial pressure of hydrogen for the reaction to occur compared to the other two processes. Alternative ironmaking is gradually developing and Corex, Finex and HIsmelt processes are emerging; at this stage, it is still the direct reduction process that is more dominant, but with the rise in coke prices, smelting reduction will be the direction of future development of alternative ironmaking [27]. This paper provides support for the thermodynamic calculation of the reduction of molten wustite by hydrogen at high temperatures, which will be beneficial for the development of the molten reduction process when supplemented by experimental verification in the future.

Author Contributions: Z.L., Y.W. and T.W. conceived and designed the calculations; S.C. and T.W. performed the calculations; J.Z., Z.L., S.C. and T.W. analyzed the data; Y.W., S.C., T.W. and Y.L. wrote the paper. All authors have read and agreed to the published version of the manuscript.

Funding: This research was funded by the National Natural Science Foundation of China, (52174291), the Beijing New-star Plan of Science and Technology (Z211100002121115), the Central Universities Foundation of China (06500170), the Guangdong Basic & Applied Basic Research Fund Joint Regional Funds-Youth Foundation Projects (2020A1515111008) and the China Postdoctoral Science Foundation (2021M690369).

Data Availability Statement: All the data that support the findings of this study are included within the article.

Conflicts of Interest: The authors declare no conflict of interest.

References

1. Zhang, J.; Li, Y.; Liu, Z.; Wang, T.; Wang, Y.; Li, K.; Wang, G.; Xu, T.; Zhang, Y. Isothermal kinetic analysis on reduction of solid/liquid wustite by hydrogen. *Int. J. Miner. Metall. Mater.* **2022**, *29*, 1830–1838.
2. Zhang, F.; Cao, Z.; Xu, H. Development Status and Prospect of Direct Reduction Technology in Gas-based Shaft Furnace. *Iron Steel* **2014**, *49*, 1–10.
3. Hayashi, S.; Iguchi, Y. Influence of gangue species on hydrogen reduction rate of liquid wustite in gas-conveyed systems. *ISIJ Int.* **1995**, *35*, 242–249. [CrossRef]
4. Katayama, H.; Taguchi, S.; Tsuchiya, N. Reduction of Iron Oxide in Molten Slag with H2 Gas. *Tetsu-to-Hagané* **1982**, *68*, 2279–2286. [CrossRef] [PubMed]
5. Ban-ya, S.; Iguchi, Y.; Nagasaka, T. Rate of reduction of liquid wustite with hydrogen. *Tetsu-to-Hagané* **1984**, *70*, 1689–1696. [CrossRef] [PubMed]
6. Nagasaka, T.; Hino, M.; Ban-Ya, S. Interfacial kinetics of hydrogen with liquid slag containing iron oxide. *Metall. Mater. Trans. B* **2000**, *31*, 945–955. [CrossRef]
7. Fan, X.; Li, C.; Wang, M. Effects of Adding Different Proportions of H2 to the Simulated Hydrogen Rich Blast Furnace. *Chem. Eng. Technol.* **2022**, *45*, 2284–2291. [CrossRef]
8. Iqbal, S.; Bahadur, A.; Ali, S.; Ahmad, Z.; Javed, M.; Irfan, R.M.; Ahmad, N.; Qamar, M.A.; Liu, G.; Akbar, M.B.; et al. Critical role of the heterojunction interface of silver decorated ZnO nanocomposite with sulfurized graphitic carbon nitride heterostructure materials for photocatalytic applications. *J. Alloy.Compd.* **2021**, *858*, 158338.
9. Luo, W. Review and Thermodynamic Calculation of Gas-based Shaft Furnace Direct Reduction Ironmaking. *EDP Sci.* **2020**, *218*, 01032. [CrossRef]
10. Sher, M.; Javed, M.; Shahid, S.; Iqbal, S.; Qamar, M.A.; Bahadur, A.; Qayyum, M.A. The controlled synthesis of g-C3N4/Cd-doped ZnO nanocomposites as potential photocatalysts for the disinfection and degradation of organic pollutants under visible light irradiation. *RSC Adv.* **2021**, *11*, 2025–2039. [CrossRef] [PubMed]
11. Kashiwaya, Y.; Hasegawa, M. Thermodynamics of impurities in pure iron obtained by hydrogen reduction. *ISIJ Int.* **2012**, *52*, 1513–1522. [CrossRef]
12. Tang, J. Nonlinear curve fitting based on matlab. *Comput. Mod.* **2008**, 15. [CrossRef]
13. Zheng, L.; Li, L.; Chen, Y. The comparison of Matlab and Mathematica in the application of nonlinear fitting. *China Sci. Technol. Inf.* **2012**, *13*, 185.
14. Guo, H. *The Course of Metallurgical Physical Chemistry*, 2nd ed.; Metallurgical Industry Press: Beijing, China, 2012.
15. Wang, F.; Wang, X. *Mathematica and Its Applications to Mathematics*; China Agriculture Press: Beijing, China, 2013.
16. *MATLAB*, version 9.11; Software for technical computation; Cleve Moler Research, MathWorks, Inc.: Natick, MA, USA, 1984.
17. Knacke, O.; Hesselmann, K. *Thermochemical Properties of Inorganic Substances*; Springer: Berlin, Germany, 1973.
18. Ye, D. *Thermodynamic Data Manual of Practical Inorganic Matter*; Metallurgical Industry Press: Beijing, China, 1981.

19. Ye, D. *Thermodynamic Data Manual of Practical Inorganic Matter*; Metallurgical Industry Press: Beijing, China, 2002.
20. Liang, Y.; Che, Y.; Liu, X. *Thermodynamic Data Manual of Inorganic Matter*; Northeastern University Press: Shenyang, China, 1993.
21. U.S. Department of Commerce. Available online: https://janaf.nist.gov/ (accessed on 13 November 2022).
22. *FactSage*, version 8.2; Software for chemical thermodynamics, calculation; Thermfact/CRCT: Montreal, Canada; GTT-Technologies: Aachen, Germany, 2001.
23. Cao, Z.; Song, X.; Qiao, Z. Thermodynamic Simulation Software FactSage and its application. *Rare Met.* **2008**, *32*, 216–219.
24. Li, B.; Guo, H.; Guo, J.; Sun, G. Thermodynamic study of gas-solid reduction of iron oxides based on minimum Gibbs Free energy principle. *J. Eng. Sci.* **2017**, *39*, 1653–1660.
25. Filho, I.; Springer, H.; Ma, Y.; Mahajan, A.; Silva, C.; Kulse, M. Green steel at its crossroads: Hybrid hydrogen-based reduction of iron ores. *J. Clean. Prod.* **2022**, *340*, 130805. [CrossRef]
26. Yi, S.; Lee, W.; Lee, Y.; Kim, W. Hydrogen-based reduction ironmaking process and conversion technology. *J. Korean Inst. Met. Mater.* **2021**, *59*, 41–53. [CrossRef]
27. Mameshin, V.; Zhuravlova, S. Modern condition and development prospects of alternative processes of the ferrous metallurgy. *Theory Pract. Metall.* **2018**, *6*, 71–75. [CrossRef]

metals

MDPI

Article

Valorizing Steelworks Gases by Coupling Novel Methane and Methanol Synthesis Reactors with an Economic Hybrid Model Predictive Controller

Alexander Hauser [1,*], Philipp Wolf-Zoellner [2], Stéphane Haag [3], Stefano Dettori [4], Xiaoliang Tang [3], Moein Mighani [3], Ismael Matino [4], Claudio Mocci [4], Valentina Colla [4], Sebastian Kolb [1], Michael Bampaou [5], Kyriakos Panopoulos [5], Nina Kieberger [6], Katharina Rechberger [7] and Juergen Karl [1]

[1] Chair of Energy Process Engineering, Friedrich-Alexander-Universität Erlangen-Nürnberg, Fürther Straße 244f, 90429 Nürnberg, Germany; sebastian.kolb@fau.de (S.K.); juergen.karl@fau.de (J.K.)
[2] Chair of Process Technology and Industrial Environmental Protection, Montanuniversität Leoben, Franz-Josef-Str. 18, 8700 Leoben, Austria; philipp.wolf-zoellner@unileoben.ac.at
[3] AIR LIQUIDE Forschung und Entwicklung GmbH, Gwinnerstrasse 27-33, 60388 Frankfurt am Main, Germany; stephane.haag@airliquide.com (S.H.); xiaoliang.tang@airliquide.com (X.T.); moein.mighani@airliquide.com (M.M.)
[4] TeCIP Institute, School of Advanced Studies Sant'Anna, Via Moruzzi 1, 56124 Pisa, Italy; s.dettori@santannapisa.it (S.D.); i.matino@santannapisa.it (I.M.); claudio.mocci@santannapisa.it (C.M.); valentina.colla@santannapisa.it (V.C.)
[5] Chemical Process & Energy Resources Institute, Centre for Research and Technology Hellas, 6th km Charilaou-Thermi Road, 57001 Thessaloniki, Greece; bampaou@certh.gr (M.B.); panopoulos@certh.gr (K.P.)
[6] Voestalpine Stahl GmbH, voestalpine-Straße 3, 4020 Linz, Austria; nina.kieberger@voestalpine.com
[7] K1-MET GmbH, Stahlstraße 14, 4020 Linz, Austria; katharina.rechberger@k1-met.com
* Correspondence: alexander.hauser@fau.de

Citation: Hauser, A.; Wolf-Zoellner, P.; Haag, S.; Dettori, S.; Tang, X.; Mighani, M.; Matino, I.; Mocci, C.; Colla, V.; Kolb, S.; et al. Valorizing Steelworks Gases by Coupling Novel Methane and Methanol Synthesis Reactors with an Economic Hybrid Model Predictive Controller. *Metals* **2022**, *12*, 1023. https://doi.org/10.3390/met12061023

Academic Editor: Pasquale Cavaliere

Received: 3 May 2022
Accepted: 11 June 2022
Published: 16 June 2022

Abstract: To achieve the greenhouse gas reduction targets formulated in the European Green Deal, energy- and resource-intensive industries such as the steel industry will have to adapt or convert their production. In the long term, new technologies are promising. However, carbon capture storage and utilization solutions could be considered as short-term retrofitting solutions for existing steelworks. In this context, this paper presents a first experimental demonstration of an approach to the utilization of process off-gases generated in a steelworks by producing methane and methanol in hydrogen-intensified syntheses. Specifically, the integration of two methane synthesis reactors and one methanol synthesis reactor into a steel plant is experimentally simulated. An innovative monitoring and control tool, namely, a dispatch controller, simulates the process off-gas production using a digital twin of the steel plant and optimizes its distribution to existing and new consumers. The operating states/modes of the three reactors resulting from the optimization problem to be solved by the dispatch controller are distributed in real time via an online OPC UA connection to the corresponding experimental plants or their operators and applied there in a decentralized manner. The live coupling test showed that operating values for the different systems can be distributed in parallel from the dispatch controller to the test rigs via the established communication structure without loss. The calculation of a suitable control strategy is performed with a time resolution of one minute, taking into account the three reactors and the relevant steelworks components. Two of each of the methane/methanol synthesis reactors were operated error-free at one time for 10 and 7 h, respectively, with datasets provided by the dispatch controller. All three reactor systems were able to react quickly and stably to dynamic changes in the load or feed gas composition. Consistently high conversions and yields were achieved with low by-product formation.

Keywords: steelworks gas valorization; methane synthesis; methanol synthesis; predictive control; carbon capture and usage; hydrogen enrichment

1. Introduction

The European process industry is committed to align its objectives with the expectation of the European Union (EU), which plans a sustainable growth through the European Green Deal [1] and the "Fit for 55" legislative package [2] to achieve the ambitious goal of the reduction in greenhouse gases (GHG) by 55% compared to 1990 and pave the way for carbon neutrality in 2050. Consequently, energy- and resource-intensive industries such as steelworks are transforming their production routes to lower their GHG emissions.

The steel sector, which is responsible for about 6% of the total EU emissions [3], designed the Clean Steel Partnership (CSP) to tackle the twofold challenge of countering climate change and supporting sustainable growth in the EU [4] and is investing in the development of technologies to progressively reduce CO_2 emissions.

The work reported in the present paper fits this context by addressing the valorization of process off-gases (POGs) generated in integrated steelworks and providing valuable products, i.e., methane and methanol.

POGs derive from different steps of steel production from virgin raw materials: coke oven gas (COG) is generated by the ovens producing carbon coke, blast furnace gas (BFG) derives from materials melting to generate pig iron, and basic oxygen furnace gas (BOFG) is output by the converter transforming pig iron into steel through carbon removal. The POG composition depends strongly on the feedstock used and the process conditions of the individual sub-processes. Table 1 reports the main average features of the POGs relevant for this paper's case study.

Table 1. Main process off-gas features (averages) [5].

Compound/Feature	Unit	COG [b,c]	BFG [c]	BOFG [c]
N_2	mol.%	2.9	48.3	27.6
CO_2	mol.%	1.2	23.1	20.0
CO	mol.%	5.8	24.9	51.8
H_2	mol.%	65.7 [c]	3.7	0.6
CH_4	mol.%	21.8	traces	-
C_nH_m	mol.%	2.5	traces	-
O_2	mol.%	0.1	traces	traces
NCV [a]	kWh/Nm^3	5.9	1.0	2.4
Other features	-	Significant content of minor compounds (i.e., potential catalyst poisons)	Continuously produced	Discontinuously produced
Main users (in decreasing order)	-	Power plant, hot rolling mill, mixing and enrichment station, coke plant area, plate annealing, blast furnace area	Mixing and enrichment station [d], blast furnace area	Mixing and enrichment station [d]

[a] CH_4 NCV = 10.5 kWh/Nm^3. [b] COG composition measurement takes place after a gas cleaning unit (i.e., filter or desulphurization). [c] POG compositions are in good agreement with the values given in the BAT document [6] and in further literature papers [7–9]. [d] Mixed enriched gas is mainly used in power plant, coke ovens, and blast furnaces.

BFG and BOFG are rich in CO_x, while COG holds the highest amount of H_2. Usually, POGs are used internally, and their fate is mainly linked to their net calorific value (NCV): COG is highly used in the steelworks' power plant (PP) as it is, while other POGs are often mixed and/or enriched with natural gas (NG) to increase their NCV.

POG production and usage is often discontinuous and/or not synchronized. Thus, two limiting conditions can arise: a lack of gas, compensated by NG purchase, and excess gas, compensated by flaring, with the consequent release of CO_2 and energy waste. Internal electricity production is not always the best solution to exploit POGs if one considers CO_2 emissions: With respect to NG (or methane), BFG and BOFG hold significantly lower

NCVs and show potentially negative effects on CO_2 emissions. For this reason, Zhang et al. suggest reducing BFG consumption in the power plant [10].

Considering this background, two kinds of solutions can be followed to improve the valorization of POGs:

- Optimal management avoiding limiting conditions;
- Alternative POG routes exploiting carbon capture storage and usage (CCS and CCU) solutions.

Both approaches are relevant in the transition period towards C-lean production processes, such as the ones using hydrogen for the direct reduction of iron ore [11–13], as during the transition POGs will still be produced.

On POG network management, relevant work was carried out by Porzio et al. [14,15] concerning POG distribution and by Maddaloni et al. [16] on gas network layout optimization. More recently, an online tool for the optimal real-time management of POGs was proposed in [17,18], which also exploits machine learning (ML) [19–21].

CCS and CCU are considered in the smart carbon usage (SCU) pathway provided by the CSP [22]. CCS solutions (e.g., standard and innovative chemical absorption) can lead to significant decreases in global warming potential (between 48 and 76%) [23], especially if applied to BFG [24,25]. Important achievements can also be obtained through CCU technologies for the production of chemicals or fuels from POGs due to their similarity to syngas. For instance, the Fischer–Tropsch process can exploit POGs as feedstock [26] and ammonia can also be produced with them [27].

The present paper addresses methane and methanol production in an innovative way, as suggested in [28], by establishing a synergy between different sectors (i.e., steel, chemical, energetic, and automation) to maximize the advantages conveyed by CCS and CCU. In particular, experimental campaigns are presented in which the potential integration of synthesis reactors into steelworks is experimentally simulated. For this purpose, a previously developed control tool (dispatch controller) calculates the optimized operating control trends for three synthesis reactors (methane/methanol) based on hydrogen-enriched POGs from the steel industry. These data are made available to the decentralized reactors in real time via an online connection and the specified load and concentration gradients are applied live.

The paper is organized as follows: the state-of-the-art method of methane and methanol production from steelworks POGs is depicted in Section 2, together with a brief overview of the exploited control methodology. Section 3 presents the reactors and controller exploited in the experiments. Section 4 summarizes the main results, while Section 5 provides some concluding remarks and hints for future work.

2. State-of-the-Art Analysis

The idea of methanol production in steelworks is not new: the first investigations date back to the 1980s [29,30]. However, only in the last decade, the technology became mature enough to be considered as a valid CCU solution [31–34]. Nevertheless, the approaches are not optimized for today's challenges related to smaller capacities, integration into other processes (e.g., steelworks,) and fluctuating operation (e.g., related to POG availability and/or H_2 availability from renewables). Recently, these challenges have increasingly become the focus of research [7,35–39]. In addition, the techno-economic evaluation of these processes is of great interest [40–42].

Conversely, methane production from CO_2-rich gases is quite recent. The main studies on methanation from steelworks POGs refer to use of COG [43,44], which, however, decreases the availability of the most suitable POG for internal heating and power production. Furthermore, only preliminary investigations can be found on the use of BFG and BOFG [45,46], including some works of the authors that were the basis of the investigation presented in this work [47,48].

Methane and methanol synthesis involve the following hydrogenation reactions ((1) and (2) for methane and (3) and (4) for methanol) and the water gas shift reaction (5):

$$CO + 3H_2 \rightleftharpoons CH_4 + H_2O \tag{1}$$

$$CO_2 + 4H_2 \rightleftharpoons CH_4 + 2H_2O \tag{2}$$

$$CO + 2H_2 \rightleftharpoons CH_3OH \tag{3}$$

$$CO_2 + 3H_2 \rightleftharpoons CH_3OH + H_2O \tag{4}$$

$$CO_2 + H_2 \rightleftharpoons CO + H_2O \tag{5}$$

Suitable ratios among reagents are needed as well as ad hoc catalysts to allow high reagent conversions and the related significant product yields. In particular, the stoichiometric number (*SN*), defined as follows, is commonly used to express the ratios among the reagents:

$$SN_{CH_4} = \frac{[H_2]}{3[CO] + 4[CO_2]} \in [1 \quad 1.1] \tag{6}$$

$$SN_{CH_3OH} = \frac{[H_2] - [CO_2]}{[CO] + [CO_2]} \in [1.5 \quad 2.1] \tag{7}$$

The required SN ranges and the POG compositions (see Table 1) highlight that BFG and BOFG must be enriched in hydrogen before the syntheses. The amount of required hydrogen is significant, and green hydrogen production (e.g., through polymer electrolyte membrane (PEM) electrolysis powered by green electricity) is fundamental to achieve GHG emission reduction [49–51].

To sum up, valorizing POGs through methane and methanol synthesis requires addressing a series of issues. Some of them are specific to steelworks, such as the management of synthesis processes with uncommon feedstocks and in uncommon fluctuating operations, hydrogen production through green processes, and POG distribution towards new users. In addition, standard POG management and usage cannot be neglected.

Therefore, reactors for methanation and methanol production can be included in integrated steelworks only by coupling them with an ad hoc developed control system that considers the different constraints related to all the involved sub-processes and accounts for the volatility of energy commodity markets.

An evolution of model predictive control (MPC) was adopted here (see Section 3.4). MPC is a control approach developed in industrial contexts and deepened by numerous academic studies that have defined its main characteristics and amplified its effectiveness [52].

The previously mentioned coupling was tested and is reported in the present paper targeting the following objectives:

- The synchronization and control of different processes belonging to the steelmaking (i.e., internal gas users), energetic (i.e., power plant), chemical (i.e., synthesis reactors), and electrochemical (i.e., H_2 production process) domains;
- The suitable distribution of POGs between standard and novel users;
- POG mixing (both pure and mixed POGs can be fed to the reactors) and enrichment with hydrogen for optimal usage in synthesis reactors;
- The correct operation of the communication structure for the transmission of the control values to the real synthesis plants;
- The smooth operation of synthesis reactors with dynamics in feed gas composition and load;
- Optimized methane and methanol production, usage, and sale.

3. Materials and Methods

Two concepts of methanation reactors and a novel pilot plant for methanol production were exploited and coupled to an ad hoc developed supervision and control system. Their behavior under dynamic conditions was tested by reproducing the flows and compositions of POG feedstock to the synthesis reactors and related hydrogen enrichment as indicated by the control system by mixing bottled dry gases. For this reason, no additional purification processes of the synthetic POGs were required; these treatment steps would have been necessary in the case of real POGs, as described in [50].

These concepts are described in the following subchapters.

3.1. Lab-Scale Methanation Unit at Friedrich-Alexander-Universität Erlangen-Nürnberg (FAU)

A lab-scale methanation test rig (FAU_CH4_TR) placed at FAU in Nürnberg, Germany was used for the experiments. Figure 1 shows its simplified flowchart [48]. The test rig was designed for a maximum syngas power (P_{syn}) of 5 kW (related to the NCV of the reactants) and can be operated at pressures up to 5 bar. It comprised two methanation stages that were operated in series to obtain a higher carbon conversion. Between these stages, a condenser separated the water formed during the reaction, and the gas was pre-heated before entering the second reactor. Gas sampling points after the first and second reactor stage enabled the analysis of intermediate and product gas using an AO2000 analyzer (ABB Automation GmbH, Mannheim, Germany).

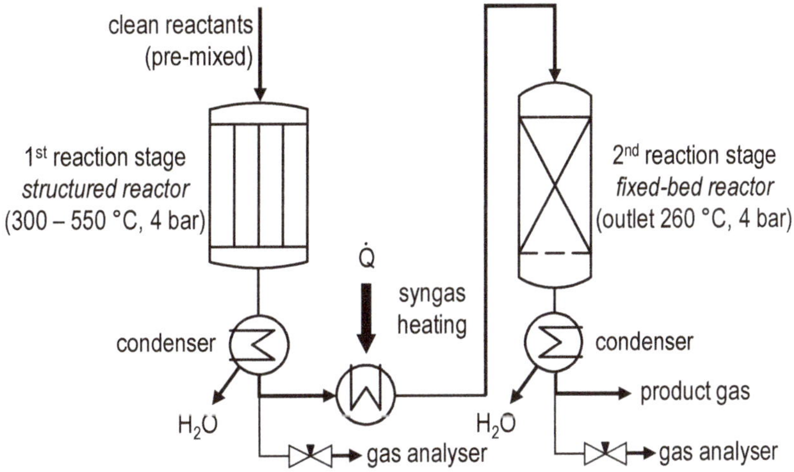

Figure 1. Simplified flowchart of FAU_CH4_TR [48].

The first reactor stage was designed as a heat-pipe cooled, structured fixed-bed reactor. The design conditions are presented in [53], and its main aim was the advanced temperature control of the highly exothermic methanation reaction.

The structured fixed-bed reactor (see Figure 2) had a total of nine reaction channels with diameters of 8 mm each, which were incorporated in a steel reactor body and contained semi-commercial catalyst pellets (Ni/Al_2O_3). To remove the reaction heat from the main reaction zone and the reactor, the reactor had 16 drillings in which 16 heat pipes fit exactly. Furthermore, drillings for internal gas preheating, reactant distribution, gas redirection, and product gas collection were incorporated in the reactor body. Figure 2 shows an exemplary gas pathway. To cool the heat pipes, defined volume flows of compressed air could be applied to their condenser zones. Heating cartridges allowed reactor preheating at start-up and trace heating during experiments. Several thermocouples recorded the temperature conditions of the reactor; in particular, three of them measured the temperatures at different axial positions at the wall of a reaction channel. A thermowell was inserted into the central reaction channel, where a thermocouple could be automatically moved along the vertical direction by means of a linear motor. Due to the slow travel speed, quasi-stationary temperature profiles could be recorded along the reactor axis.

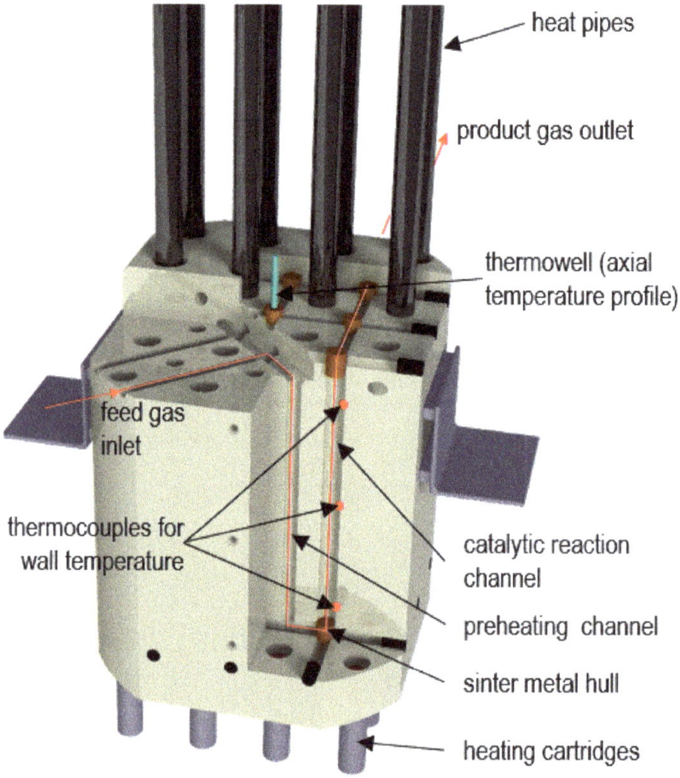

Figure 2. Cutaway scheme of the FAU structured reactor and exemplary gas pathway (reprint from [53] with permission from Elsevier).

The second reactor stage was designed as a classic fixed bed with an inner diameter of 27.6 mm. It could be cooled by compressed air through a double jacket and heated electrically. This reactor was oversized and was used to achieve full conversion.

3.2. *Lab-Scale Methanation Unit at Montanuniversität Leoben (MUL)*

The second methanation test rig (MUL_CH4_TR) was located at MUL in Leoben, Austria [47,54] and consisted of three cylindrical reactors (DN80, height 300 mm) operated in series (for a simplified flowchart see Figure 3). Each reactor was either filled with commercial Ni-based bulk catalyst or with wash-coated ceramic honeycomb monoliths [54]. In both cases, a volume of 0.25 L per reactor was covered with catalyst, and the remaining part was filled with an inert material (stoneware balls). The dimensions and design specifications of the reactors met the following operating parameters:

- Flow rate range: 0.3–3 Nm^3/h (≈5–50 NL/min);
- Gas hourly space velocity (*GHSV*) range: 1200–12,000 h^{-1};
- Pressure rating: up to 20 bar;
- Maximum reactor temperature: 700 °C.

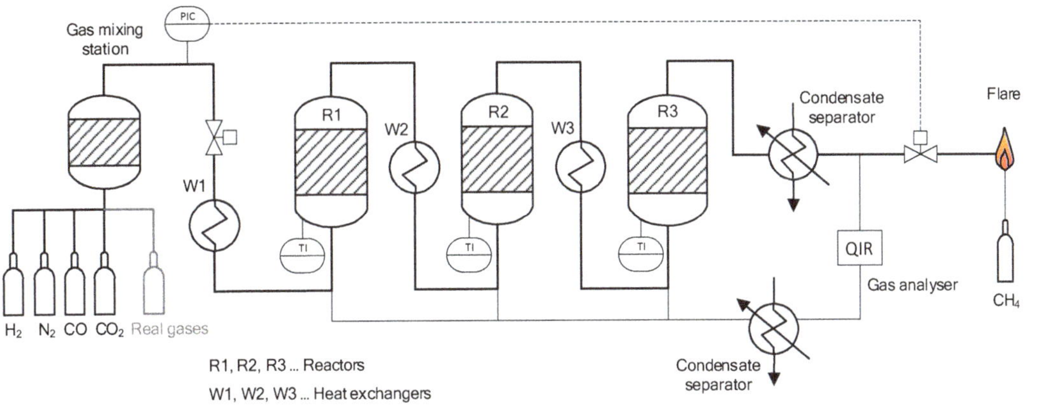

Figure 3. Simplified flowchart of MUL_CH4_TR [47].

Furthermore, four gas sampling stations were installed, allowing the composition analysis of both inlet and produced stream after each reactor stage. A URAS 26 infrared photometer was used for measuring CO, CO_2, and CH_4, while a CALDOS 27 thermal conductivity analyzer measured H_2 (both from ABB AG, Wiener Neudorf, Austria). The final product gas stream was cooled and guided through a condensate separator to extract any H_2O formed during synthesis. The product gases were combusted in a flare connected to the aspiration system. The reactor temperatures were measured with multi-thermocouples, providing an axial profile of the temperature distribution inside each reactor. In addition, the MUL_CH4_TR included heat exchangers up- and downstream of the reactors, a water dosing pump at the reactant gas side, a cooling unit, and a water separator as well as heating sleeves coating the outside of each reactor.

3.3. Pilot Plant for Methanol Synthesis at Air Liquide Forschung und Entwicklung (ALFE)

The novel pilot plant for methanol synthesis (ALFE_CH3OH_PLP) was located at Air Liquide Forschung und Entwicklung (ALFE) in Frankfurt am Main, Germany. It was designed to optimize the transformation of the CO_x contained in POGs into methanol and was successfully started in 2020.

The configuration of the new multi-stage setup allowed the adaption of the reaction section as desired to optimize the conversions of gases with different compositions (max. of four stages with inter-stage condensation). Therefore, this plant could deal with unconventional gas compositions, such as those of POGs, under once-through conditions or with low recycling at very fast response times regarding changes in loads and/or in gas compositions.

The key features of this pilot plant can be described as follows:

- Several reactor stages;
- Various possible flow schemes;
- Heat transfer to steam system;
- Temperature profile measurement in the different process stages through multi-point thermocouples;
- Throughput:
  - Feed gas up to 35 Nm^3/h;
  - Raw methanol product up to 20 kg/h.

Figure 4 presents pictures of the outdoor and indoor parts of ALFE_CH3OH_PLP.

(**a**) (**b**)

Figure 4. (**a**) Outdoor and (**b**) indoor sections of ALFE_CH3OH_PLP.

For the test campaign, four stages were filled with a commercial methanol catalyst. The reaction section consisted of several stages set up for the optimal conversions of CO, CO_2, and H_2 (X_{CO}, X_{CO2}, and X_{H2}, respectively) combined with fast adaption to changing conditions. Between each stage there was the possibility to remove the produced raw methanol. The collected methanol/water mixture was then analyzed to determine the amounts of by-products. The new unit allowed high overall methanol production and extended the lifetime of the catalyst. The operation in the ALFE_CH3OH_PLP using four stages filled with catalyst under once-through operation can be described as follows:

- Fresh feed gas was fed to stage 1 of the reaction section;
- Unconverted gases from stage 1 were fed to stage 2;
- Unconverted gases from stage 2 were fed to stage 3;
- Unconverted gases from stage 3 were fed to stage 4;
- Raw methanol product was removed after each stage and analyzed accordingly;
- By-product amounts were evaluated in the raw methanol removed between the stages;
- The final raw methanol product gathered all the contributions coming from each stage.

One major advantage to using a once-through multi-stage concept for fluctuating loads and fluctuating off-gas compositions is that the time of response is significantly lower than for a setup with high recycling. Therefore, only a few minutes were necessary without recycling to have the desired feed composition and load at the reactor (or stage) inlet in the methanol plant, which was not the case when traditional recycling was used. Typically, by using classical recycling, a few hours would be required to provide the desired conditions at the reactor inlet in terms of feed composition and load.

Moreover, the multi-stage concept with once-through operation is much more flexible regarding the amount of inerts in the feed (nitrogen and/or methane). These inerts can rapidly lead to accumulation at the reactor inlet when recycling. The multi-stage concept can be operated regardless of the composition of the feed and can still produce a raw methanol that is able to be treated in a classical distillation section. This concept fits perfectly for a coupling with an advanced control system, as the desired conditions at the reactor/stage inlet are fixed quickly. This allows the adaption of all the parameters to optimize the operating conditions in a very short time in order to maximize methanol production and to lower by-product formation to ensure a proper distillation.

The online analytical concept is based on process gas chromatography. The system was developed and implemented for the analysis of syngas composition and methanol as well as some main by-products in a single gas phase with a broad concentration range and a detection limit of 0.01 mol.% or better.

In particular, hot gas streams from reactor outlets, unconverted gases, methanol, water, and some oxygenates under high pressure and temperature can be online analyzed in the gas phase using a dedicated online sample conditioning and analysis system. Moreover, a gas stream dilution system can be applied for hot gas streams once the chromatography signal is saturated. Additionally, a continuous evaporation mixture system is integrated as a validation system to monitor the conditioning and analysis system with methanol and water vapor mixture during operation in a certain time interval. The online analytical system, including a heated conditioning system, a four-channel micro-GC system, and a continuous evaporation mixture system is shown in Figure 5.

Figure 5. Online conditioning and analysis system of ALFE_CH3OH_PLP.

The short response time of the online analyzer enables the real-time optimization of process variables and the adjustment of feedstock composition to control the side reactions and achieve the desired product yield.

3.4. Dispatch Controller at ICT-COISP Center of TeCIP Institute of Scuola Superiore Sant'Anna

The supervision and control system exploited in the test campaign, from now on named the dispatch controller (DC), was developed at Scuola Superiore Sant'Anna (SSSA) in Pisa, Italy to investigate the integration of the synthesis reactors into steelworks and to online optimize POG distribution to the synthesis reactors while considering all the issues related to POG management and the interactions between gas networks and gas consumers. A schematic overview of the concept of the POG-based methane and methanol production and control system is depicted in Figure 6, where the coupling between the POG networks, synthesis reactors, power plant, hydrogen production system, and DC is highlighted.

To optimize the production of methane and methanol from POGs, several factors must be considered. First, POG production, standard usage, and gasholder dynamics must be taken into account. POG excess is stored in gasholders, which hold minimum and maximum level limits and, in general, show quite fast dynamics, as they tend to fill and empty in about 30–60 min, depending on the steelworks size. POG availability is varied by leading synthesis reactors to work under variable load conditions, which are characterized by highly non-linear dynamics that make their control challenging.

POG-based methane and methanol production requires gas enrichment by large amounts of hydrogen, the production of which is very energy-intensive. To ensure economic and environmental sustainability, green electricity sources must be exploited for hydrogen production, especially in the hours when energy is cheaper [49,50].

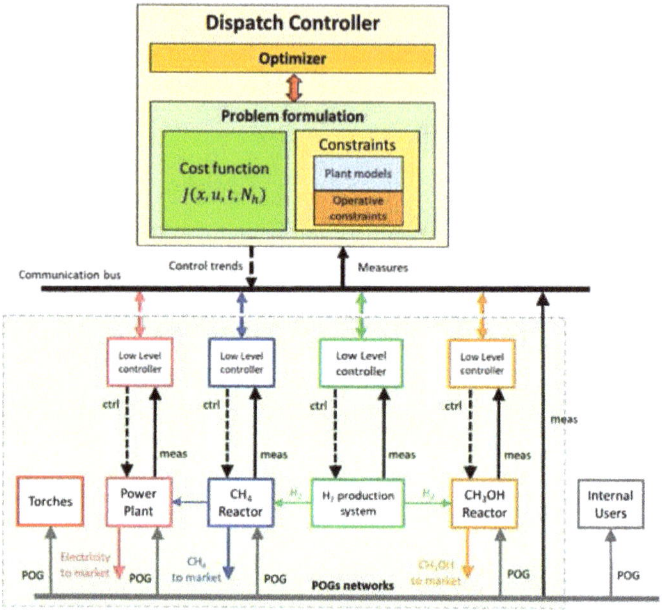

Figure 6. Conceptual coupling between DC and POG-based methane and methanol production units.

The mentioned plants and processes can be mathematically modelled through input/output correlations that often require the use of discrete control variables. For instance, synthesis reactors can operate in different discrete modes (i.e., OFF, standby, only BFG, only BOFG, and mixed gas), power plant groups can be switched on/off according to the amount of power to be produced, and the hydrogen production system (e.g., PEM electrolyzers) is composed of different stacks that can be switched on/off, depending on the required amount of hydrogen. Furthermore, the DC must be able to calculate an appropriate control action in real time to deal with variable POG availability.

The DC was developed through a methodology based on the economic hybrid model predictive control (EHMPC) [55], which maximizes revenues and minimizes environmental impact while complying with the constraints of each plant and process. Details on the design strategy, mathematical implementation and preliminary results were presented in the paper of some of the authors [56], where the main characteristics were reported. The optimization problem is formulated as a mixed integer linear programming (MILP) for a receding horizon, where the cost function to be maximized is calculated as follows:

$$J(x,u,t,N_h) = \sum_{k=t}^{t+N_h} \alpha^k \Big(R_{ES}(k) + R_{MEOH}(k) + R_{CH4}(k) + R_{CO_2}(k) - C_{PEM}(k) - C_T(k) - C_{fOPEX}(k) \Big) \quad (8)$$

where x and u are, respectively, the state of the system and the manipulated variables, t is the current time, N_h the prediction horizon, k is the time along the prediction horizon, R_{ES} is the revenue related to the electricity sold to the external grid, and R_{MEOH} and R_{CH4} are, respectively, the revenues related to the methanol and methane sold. C_{PEM} is the price of the green electricity consumed by the PEM electrolyzer system, C_T is the cost of wasting POG excess in the torches, and C_{fOPEX} are the fictitious operative costs that penalize the continuous switching of reactors and PEM stacks, the variation in gas volume flow at the reactor inlet, and the decreasing of gasholder levels below a security threshold. The economic costs related to the environmental impact of the CO_2 emissions in integrated steelworks typically depend on the amount of virgin coal-based raw materials exploited and on the purchased natural gas. For this reason and considering that the CO_2

emissions related to virgin coal-based raw materials are not affected by POG management, R_{CO_2} is calculated here as the revenue related to CO_2 cost savings due to internal methane production. At the present stage of the research, the economic costs do not take into account the eventual operative costs of methane and methanol clean-up equipment and treatments.

The constraints of the optimization problem mathematically describe the dynamics of each reactor, power plant group, PEM electrolyzer, and gasholder and the operative limits of each piece of equipment and unit. Each unit and piece of equipment was modelled by a linear state space model whose parameters were tuned using real data. The reactors' models were designed in MATLAB/Simulink (R2020b, MathWorks®, Natick, MA, USA) by implementing the physical/chemical equations of synthesis reactions and thermal transmission [57]; these equations were tuned using real data of the reactors and of preliminary experimental tests. Each synthesis reactor model did not take into account the clean-up stages and related operative costs.

Furthermore, the DC included a set of advanced models forecasting future POG excess (i.e., digital twins of steelworks units involved in POG production and storage) and the costs of energy media (electricity, methane, and methanol), both developed through ML techniques. The models forecasting POG excess were developed using real data of voestalpine steelworks located in Linz (Austria), which were used also in this work for simulating a possible real application of methane and methanol synthesis processes [58].

The DC calculated the control strategy for a time window of 2 h ahead with a control period of 1 min. Every control period, the controller linearized the reactor models around the last operative point, allowing the solution of a linear optimization problem within a fixed maximum time of 30 s.

3.5. Data Communication Infrastructure

Figure 7 shows the data communication infrastructure established for the live tests, which was required due to the different locations of the reactors and dispatch controller.

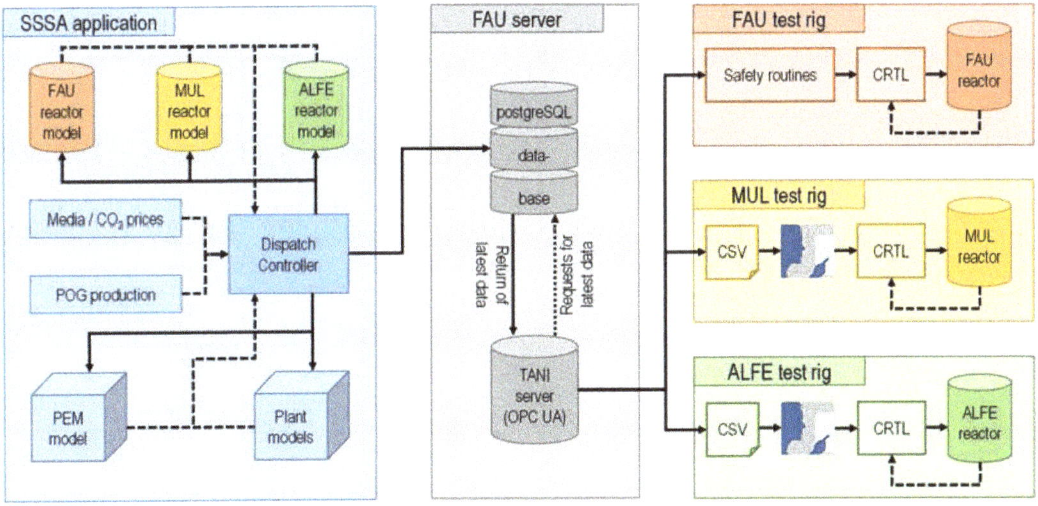

Figure 7. Data communication structure between DC and test rigs for online test campaign.

The DC is highlighted in blue on the left side with all related models and input parameters. This routine maximized the objective function (Equation (8)) by iterating the feedback generated from the individual models for the PEM electrolysis, the three synthesis reactors, as well as eleven further models mapping the various steelworks units. It performed calculations every minute and wrote the best solution for the following POG operating parameters into a PostgreSQL database (grey section, middle): total volumetric flow rate, gas composition, pressure, and temperature. These data included the solution for the current timestamp and a look ahead for the upcoming two hours, which was available for FAU_CH4_TR (orange, top right), MUL_CH4_TR (yellow, middle right), and ALFE_CH3OH_PLP (green, bottom right) to be retrieved. Therefore, an open platform communications unified architecture (OPC UA) server was implemented, which sent queries to the database in selectable regular intervals and requested the last valid dataset. Three query routines were included, each of them requested the operating values for one test rig. Once the data exchange took place successfully, human operators checked the suggested parameters at MUL_CH4_TR and ALFE_CH3OH_PLP to ensure safe plant operations. For FAU_CH4_TR, automated safety routines took this task and checked, e.g., whether the stoichiometry (Equation (6)) and syngas power matched the reactor limits. Therefore, the datasets were handed over directly to the plant's programmable logic controller (PLC) via the OPC UA. When an incoming dataset did not meet the specifications, it was ignored, and the test rig was further operated with the last valid dataset.

With respect to the conceptual idea (described in Section 3.4), the DC interacted with digital twins of the integrated steelworks and the synthesis reactors in the live test campaign presented in this paper. For safety reasons, at the current stage of development, the coupling between the DC and reactors was simulated through the models without direct feedback of the real state of each piece of involved equipment. Since the DC delivered the control strategy to the synthesis reactors without receiving any direct feedback from the real plants, the overall system operated in an open loop.

4. Results and Discussion

Different scenarios were designed and then live tested where, generally, two reactors were operated in parallel. The following subchapters describe the experimental online coupling campaign and the main results.

4.1. Definition of Scenarios

To evaluate the validity of the approach to synthesize methane and methanol from POGs and the integration of this solution in integrated steelworks, two essential aspects need to be assessed: (i) the ability of the controller to calculate optimal control trends in real time and (ii) the ability of the synthesis reactors and test rigs to work, even under particularly variable dynamic conditions. To these aims, a set of scenarios was defined through a wide campaign of offline simulations in which the advanced models of the reactors and the digital twin of the considered integrated steelworks was used. In order to obtain indications suitable for both the laboratory setup and for the possible development of future industrial plants, the DC interacted with virtual reactors that were linearly scaled up with respect to the pilot/bench-scale ones. In more detail, the industrial scales of the reactors were evaluated to produce a methane amount sufficient to satisfy the steelworks' needs and a methanol amount sufficient for commercial purposes.

The first phase of the simulation campaign was devoted to a fine tuning of the controller and optimizer parameters to balance the aggressiveness of the control action on the plants and the computational time; some simulated control trends were reported in [56,59] where the results of a preliminary investigation on economic cost improvement were also presented. In [59], the authors reported a first attempt at sensitivity analysis, varying the price of renewable sources of electricity for producing green hydrogen, and an estimate of CO_2 emission reductions.

Different scenarios for the online test campaign were tested, among which the most significant ones are:

- Scenario 1 (SC1): two methane reactors (FAU_CH4_TR and MUL_CH4_TR) running in parallel in variable load conditions with strong disturbances in the BFG network, where several sequential BFG shortages are simulated;
- Scenario 2 (SC2): one methane (FAU_CH4_TR) and one methanol (ALFE_CH3OH_PLP) reactor running in parallel with variable operating conditions due to a BFG shortage.

This scenario is of particular interest, as it allows the simulation of a context in which it is possible to access the product market (i.e., methane, methanol, and possibly electricity) in which the flexibility and diversification of production could be an economically winning factor. In particular, considering the current variability in the prices of both electricity and products such as methane, it is considered fundamental to test the validity of the developed control approach.

To obtain sufficient variability in the control trends and to verify the behavior of the reactors at different operating points, each scenario was designed to simulate the operation of reactors characterized by different industrial scales and different disturbances in the POG networks. The main idea was testing the DC and the plants in extreme conditions characterized by fast dynamics in a wide range and variability in the inlet gas both in terms of flow rate and composition. This implies a different POG distribution to the plants and the possibility, for instance, of making the reactors work within the maximum operating points (by decreasing the industrial scale of the reactors, a greater relative availability of POGs was observed) or with greater variability within the operating ranges (with reactors characterized by larger industrial scales or the simulation of a temporary shortage of POGs). In order to better understand the nature of the simulated disturbances, the starting scenario data refer, in terms of POG production, to the data of the industrial partner of the project, which are related to a "constant and standard" steel production period. Regarding the internal POG consumption, on the other hand, we referred to an average consumption baseline within the integrated steelworks plant and some artificial disturbances simulated on top of them. Artificial additional disturbances had the purpose of testing the behavior of both control systems and reactors in extreme conditions.

Furthermore, all the scenarios were characterized by a set of constant boundary conditions:

- POGs distributed to the power plant and synthesis reactors;
- Produced methanol sold to external users;
- Produced methane distributed to external users and internal consumers based on the economic gain;
- A warmup phase followed by about 10–12 h of methane and/or methanol production;
- The methane reactors exploited BFG and BOFG with the constant stoichiometric number $SN_{CH_4} = 1.04$;
- The methanol reactor exploited BFG, BOFG, and COG with $SN_{CH_3OH} \in [1.5 \quad 2.1]$;
- The price of electricity sold to the external grid was equal to EUR 80/MWh;
- The prices of methane and methanol sold to external users were equal to EUR 25/MWh and EUR 400/ton, respectively;
- The price of CO_2 emissions was equal to EUR 30/ton;
- The price of green electricity for the PEM electrolyzer was equal to EUR 5/MWh;
- The number of available PEM electrolyzer stacks was equal to 70, each characterized by a nominal power of 17.5 MW and a resulting hydrogen production of 340 kg/h at full capacity.

Exemplary results obtained in the two mentioned scenarios are depicted in Figures 8–13, which were calculated online during the live test campaign and exploited as control references for the involved reactors. In particular, Figure 8a depicts SC1 in terms of POG production in blue and internal consumption in orange (from top to bottom, respectively, BFG, BOFG, and COG). Figure 8b shows the POG distribution to the PP (blue) and reactors (orange), respectively, for BFG and BOFG, while the third diagram shows the methane produced and distributed to the PP (blue) and to external users (orange). Figure 9 depicts the bench-scale control action calculated by the DC for FAU_CH4_TR. In more detail, Figure 9a shows the POG volume flow and hydrogen mixture at the inlet of the reactor and its limits in the function of the modality (only BFG, only BOFG, or a mixture of both), while Figure 9b shows the control action in terms of inlet gas composition. Figure 10 depicts the same control action in the case of MUL_CH4_TR running in parallel.

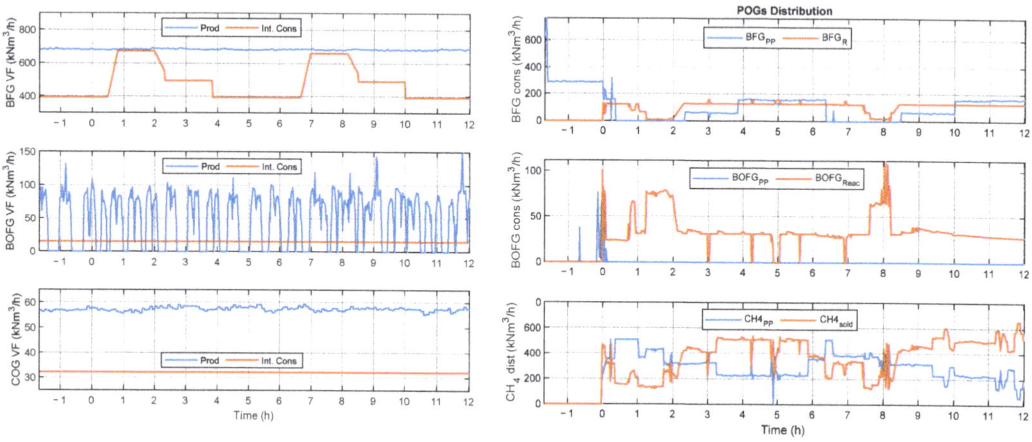

(**a**) (**b**)

Figure 8. SC1. (**a**) POG production and consumption by internal users; (**b**) POG distribution to PP and methane reactors and methane consumption in PP and sent to market.

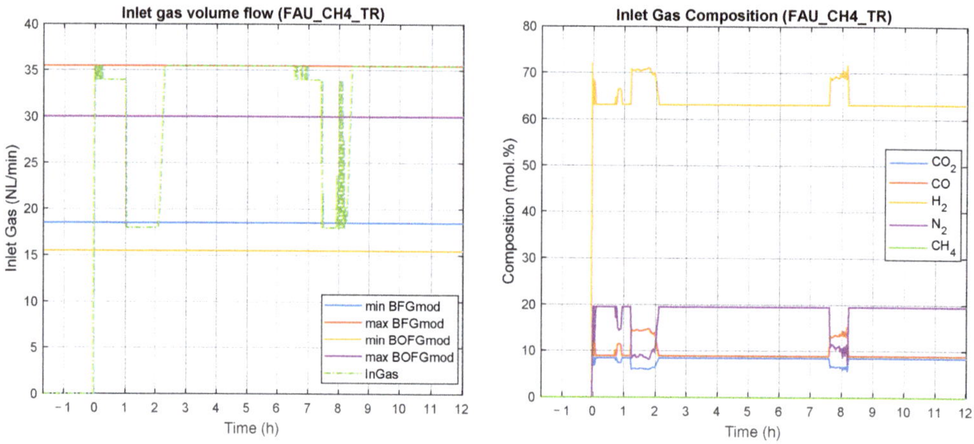

(**a**) (**b**)

Figure 9. SC1. Control strategy for FAU_CH4_TR. (**a**) Inlet gas volume flow; (**b**) inlet gas composition.

The representation of Figure 11 reports the same information as Figure 8 but for SC2 and extended by the methanol sent to market (yellow) in Figure 11b. Figure 12 represents the control action calculated by the DC for FAU_CH4_TR in bench scale during SC2, while Figure 13 depicts the control action for the methanol plant ALFE_CH3OH_PLP running in parallel.

An in-depth analysis of the results of the scenarios revealed several interesting behaviors. They are described below together with some preliminary results regarding potential reductions of CO_2 emissions.

In SC1, the objective was to achieve a sufficient variability in load conditions. Figure 8a shows two sequential disturbances in BFG availability: two simulated shortages of BFG (between 0.5–2.5 h and between 6.5–8.5 h), during which the main POG in terms of quantity was less available for the methanation process. The BFG shortages drove reactors to work in several dynamic conditions and led the DC to calculate a control action that decreased the

BFG consumption in both the reactors and PP (see Figure 9b) with a consequent switching of FAU_CH4_TR to BOFG/mixed gas modalities with suitable hydrogen enrichment to reach the required SN_{CH4}. Concerning MUL_CH4_TR, it mainly ran through BOFG enriched with hydrogen with few switches to the BFG modality. The COG was exploited only in the PP, while the produced methane was distributed to the PP and market with a certain variability. Regarding the CO_2 reduction in the SC1 period, an approximate evaluation was conducted by starting from the rough estimate provided in [59]. The reduction in CO_2 in the SC1 period was about 1151 t, corresponding to a reduction of about 62% with respect to the CO_2 emissions that would have been obtained by feeding the same amount of POGs exploited in the synthesis reactors directly to a power plant. At the same time, the produced CH_4 usage in power plant in replacement of the POGs allowed the production of an amount of electrical energy 1.65 times higher.

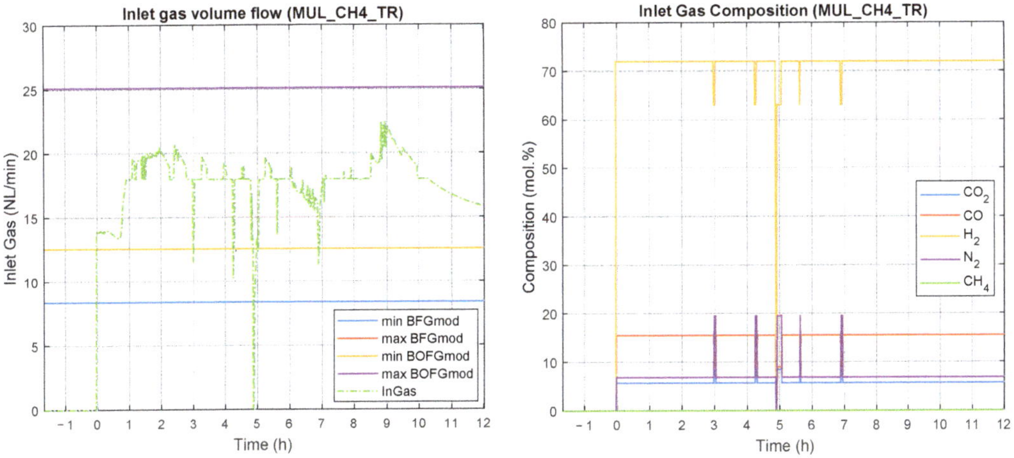

(a) (b)

Figure 10. SC1. Control strategy for MUL_CH4_TR. (**a**) Inlet gas volume flow; (**b**) inlet gas composition.

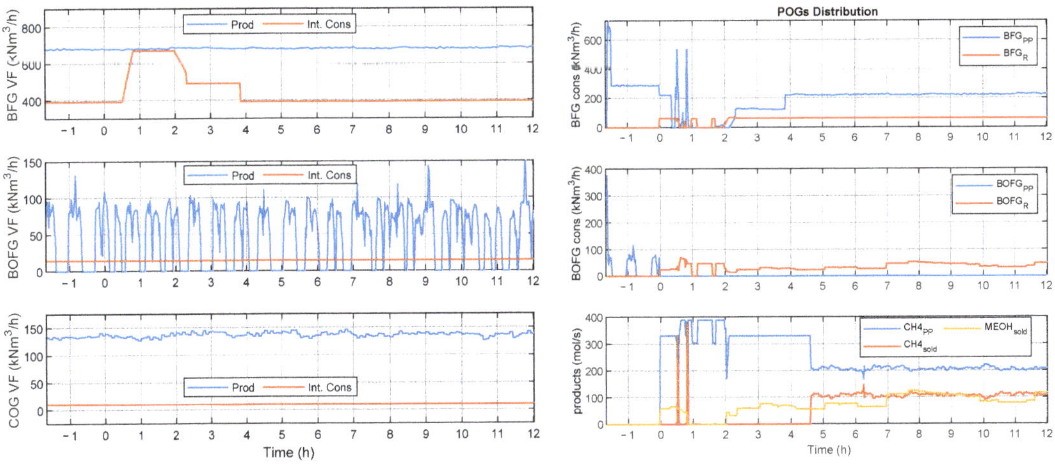

(a) (b)

Figure 11. SC2. (**a**) POG production and consumption by internal users; (**b**) POG distribution to PP and methane reactors, methane consumption in PP and sent to market, and methanol sold.

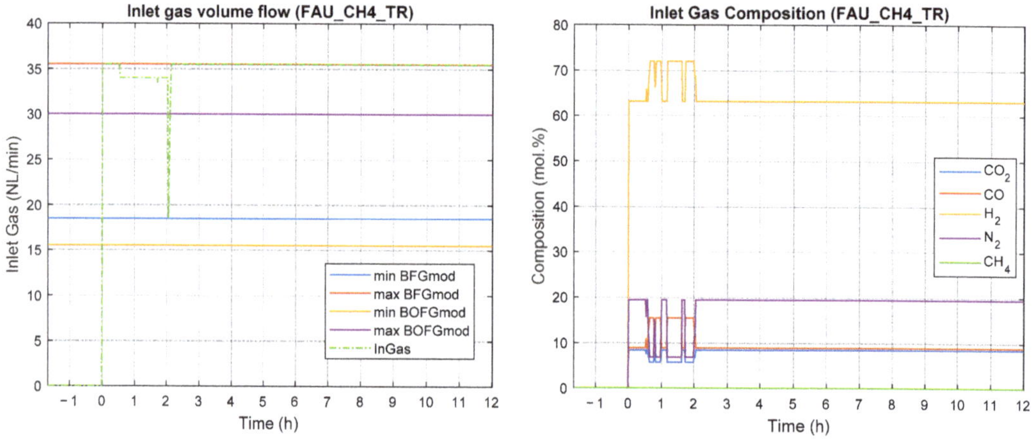

(a) (b)

Figure 12. SC2. Control strategy for FAU_CH4_TR. (**a**) Inlet gas volume flow; (**b**) inlet gas composition.

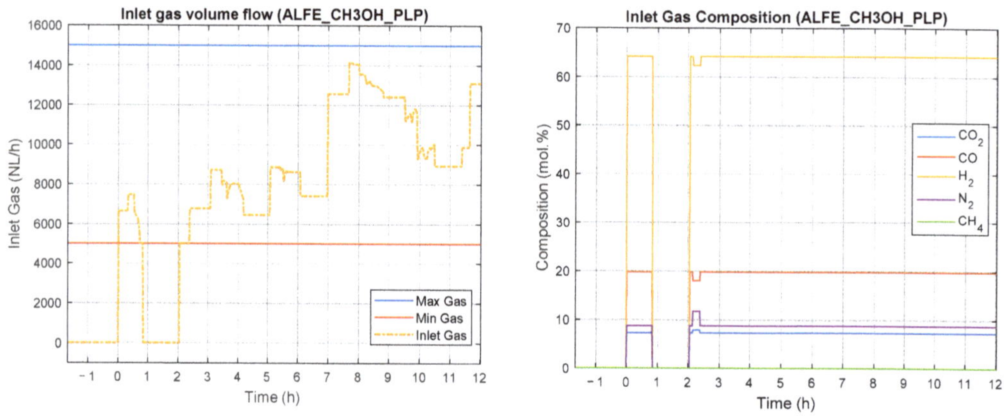

(a) (b)

Figure 13. SC2. Control strategy for ALFE_CH3OH_PLP. (**a**) Inlet gas volume flow; (**b**) inlet gas composition.

In SC2, FAU_CH4_TR and ALFE_CH3OH_PLP worked in parallel. Moreover, in this case, the control action calculated by the DC forced the reactors to work in a complementary way. FAU_CH4_TR operated with an inlet gas composed of a mixture of BFG and hydrogen, while during the disturbance period, the reactor exploited mixed gas modalities (a mixture of BFG/BOFG enriched with hydrogen) to produce methane. On the other hand, the ALFE_CH3OH_PLP exploited BOFG and hydrogen as a unique inlet gas source. For very short periods during the disturbance in BFG availability, the methanol reactor exploited a mixture of BOFG and hydrogen with a small addition of BFG. Moreover, in this case, an estimate was calculated for the CO_2 reduction by summing the reduction contribution related to the use in the power plant of produced CH_4 in place of POGs and by considering the avoided emissions due to the production of CH_3OH. The CO_2 reduction results of about 774 t, corresponding to about 61% of those emitted in the case of burning the same amount of POGs exploited in the synthesis reactors directly in the power plant. An electrical energy production about 1.52 times higher was also obtained.

To sum up, the DC showed the capability to successfully react to disturbances in the main POG network while still allowing the correct operations of reactors. In addition, it also ensured significant benefits in terms of CO_2 emission reductions.

4.2. Results of Experimental Campaign with FAU_CH4_TR

As described in Section 4.1, the two methanation plants FAU_CH4_TR and MUL_CH4_TR were operated in parallel in SC1. The two phases of dynamic operation specified by the DC (approx. 0.5–2.5 h and 6.5–8.5 h without considering the warm-up phase) could be observed in the operation of FAU_CH4_TR (see Figure 14, solid lines). The dynamics involved transients in the volume flow rate ($\dot{V}_{tot}$) and gas composition. Five operating points (OP) for the evaluation were defined (see Table 2). OP1.1, OP1.3, and OP1.5 represent stoichiometrically adjusted BFG while OP1.2 and OP1.4 represent stoichiometrically adjusted mixtures of BOFG (main part) and BFG.

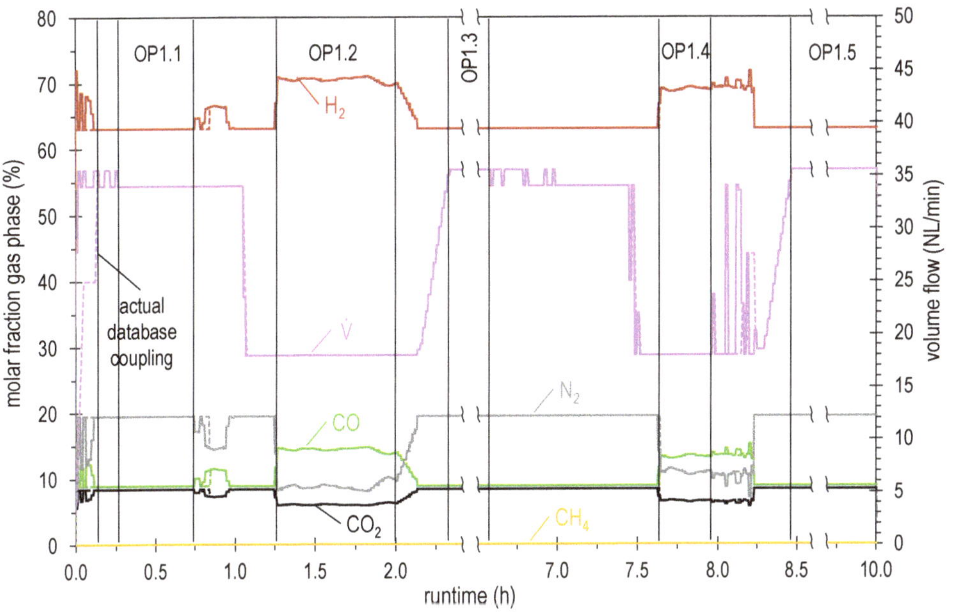

Figure 14. SC1. Handover values received from DC (solid lines) and set values of MFC (dashed lines) for FAU_CH4_TR.

Table 2. FAU_CH4_TR operating points defined for the evaluation; mean set values specified by the DC. p = 4 bar for all operating points.

Operating Point	$\dot{V}_{tot}$	H_2	CO_2	CO	CH_4	N_2	SN_{CH4}	P_{syn}
	NL/min	mol.%	mol.%	mol.%	mol.%	mol.%	-	kW
OP1.1	34.0	63.1	8.5	8.9	0.0	19.5	1.04	4.39
OP1.2	18.0	70.5	6.2	14.4	0.0	8.9	1.04	2.71
OP1.3	35.5	63.1	8.5	8.9	0.0	19.5	1.04	4.59
OP1.4	18.0	69.0	6.6	13.3	0.0	11.0	1.04	2.65
OP1.5	35.5	63.1	8.5	8.9	0.0	19.5	1.04	4.59
OP2.1	35.5	63.1	8.5	8.9	0.0	19.5	1.04	4.59

SC2 included the parallel operation of FAU_CH4_TR and ALFE_CH3OH_PLP. For FAU_CH4_TR, it comprised only one phase of dynamic operation (approx. 0.5–2.25 h, without considering the warm-up phase). Since the gain in knowledge was rather small compared to SC1, only one operating point is defined here for evaluation (see Table 2) OP2.1 represents stoichiometrically adjusted BFG.

Figure 14 compares the data received from DC to the set values of the mass flow controllers (MFC) for SC1. The latter (dashed lines) follow the specified values of the DC (solid lines) at almost every point during the 10 h of the experiment. Both gas volume flow and composition changes were automatically adopted by the PLC. However, there are exceptions where the graphs are not congruent. These deviations resulted from the implemented security query regarding the permitted stoichiometric limits, as some datasets were not forwarded to the MFC. This was caused by rounding errors in data processing resulting in calculated stoichiometric ratios smaller than 1.04. As a short-term corrective measure during the experiment, the corresponding values were typed in manually. After the experiment, the number of significant digits of rounding was increased, and the lower limit of the safety function checking the stoichiometry was slightly lowered.

Furthermore, the comparison of the data stored in the database by the DC and the received values recorded by the test rig's PLC showed that all datasets were transferred completely and without loss.

The temperature trends over the runtime for selected measurements in the structured reactor during SC1 are depicted in Figure 15. At the beginning of OP1.1, the influence of the reactor start-up phase can be clearly seen. The temperatures stabilized during this operating point, as the syngas power was relatively high (P_{syn} = 4.39 kW) and active cooling was applied. Between OP1.1 and OP1.2, the influence of the short switch in gas composition and the following decreasing volume flow ramp (see Figure 14) can be noticed. The temperature level did not stabilize during OP1.2 (CO-rich feed gas and low volume flow) despite stopping active cooling. For a longer operation of the structured reactor at this operating point, additional trace heating would have to be applied. During the start of OP1.3 (stoichiometrically adjusted BFG and high syngas power), the temperature rose again and was stabilized by active cooling at a desirable overall temperature level. The temperature over runtime for OP1.4 showed a behavior similar to OP1.2, as the conditions were almost the same. OP1.5 was identical to OP1.3, but the temperature stabilization this time did not succeed as fast as before.

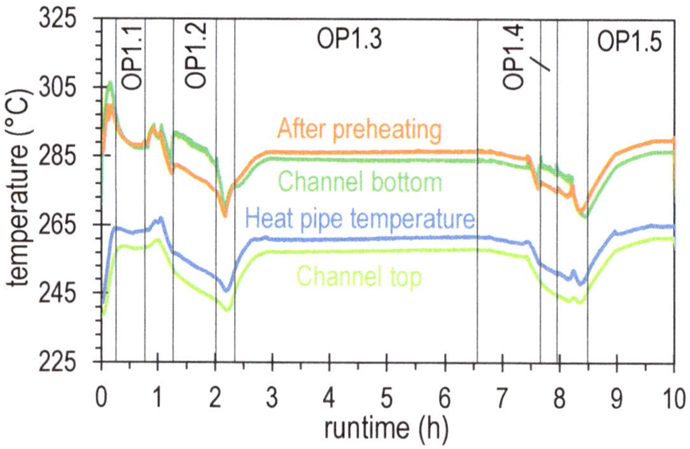

Figure 15. SC1. Temperature over runtime for selected measurements in the structured reactor of FAU_CH4_TR.

The curves in Figure 15 and the above-described dependencies clearly show that permanent monitoring and, if necessary, adjustment of the heat management (cooling and trace heating) is essential for the stable operation of the methanation system. These adjustments were made manually, despite the automated value transfer for the feed. In further development for the demonstration of a possible application in an industrial environment, a fully automated control concept will be required to take over this task.

Figure 16 illustrates the averaged axial temperature profiles of the structured reactor for the operating points OP1.3 and OP1.5. These showed a pronounced polytropic pattern with a temperature maximum of ≈500 °C about 48 mm after the start of the reaction channel. The maximum temperature was, thus, ≈200 °C lower than the expected adiabatic synthesis temperature and 50 °C lower than the maximum catalyst temperature. After the hotspot, the temperature declined to 265 °C at the reactor outlet, which increased the methane yield, $Y_{CH4,COx}$. The two requirements for an active cooling concept formulated in [53] (the limitation of the hotspot temperature to a high but materially acceptable level and a lowered outlet temperature) are, thus, met. The additionally plotted temperatures at the wall of a reaction channel (squares in Figure 16) clearly illustrate the large radial temperature gradient of up to 75 °C/mm. As the profiles of OP1.3 and OP1.5 are almost identical, it can be stated that same reaction and temperature conditions were reached during these operating points. Furthermore, the 12 individual profiles recorded at OP1.3 were completely congruent. The absence of a progressive change in temperature conditions in the structured reactor indicates that there was no change in the catalyst activity and, thus, no on-going catalyst deactivation.

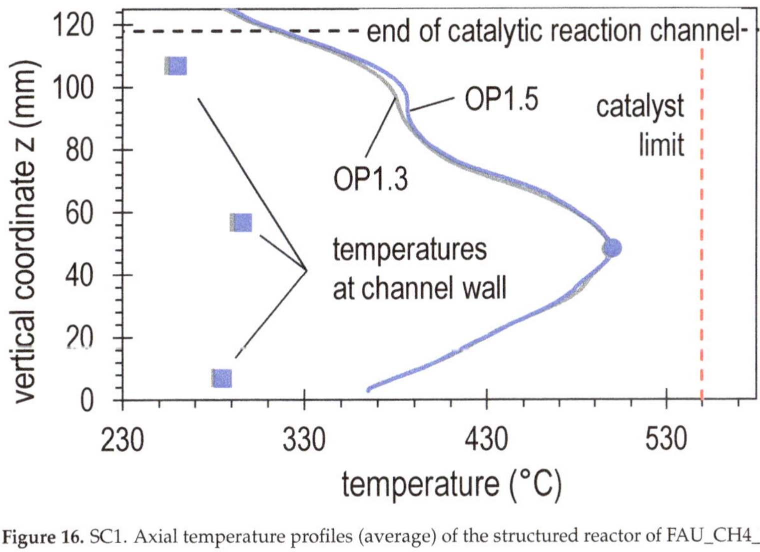

Figure 16. SC1. Axial temperature profiles (average) of the structured reactor of FAU_CH4_TR.

A summary of the evaluation of the hydrogen conversion, X_{H2}, and methane yield, $Y_{CH4,COx}$, is reported in Figure 17 for the five operating points of SC1. At the three comparable operating points, OP1.1, OP1.3, and OP1.5, full methane yield was achieved after the two-stage process with intermediate water separation. The hydrogen conversion was around 94.5% (over-stoichiometric feed gas composition). For OP1.2 and OP1.4, similar values for the final product gas are to be expected, but no product gas measurement was made after the second reactor stage at these points. The X_{H2} and $Y_{CH4,COx}$ of OP1.2 and OP1.4 (CO-rich gases and lower syngas power) after the first stage were higher (≈6.5%) than the values at the operating points with equal amounts of CO and CO_2 and higher volume flow rates. A methane yield of around 91% was achieved in just one reactor stage.

The lower volume flow allowed a longer residence time in the catalyst bed. In addition, the reactants had a higher reactivity due to the increased CO content. These dependencies indicate a kinetic limitation in the structured reactor under the given conditions.

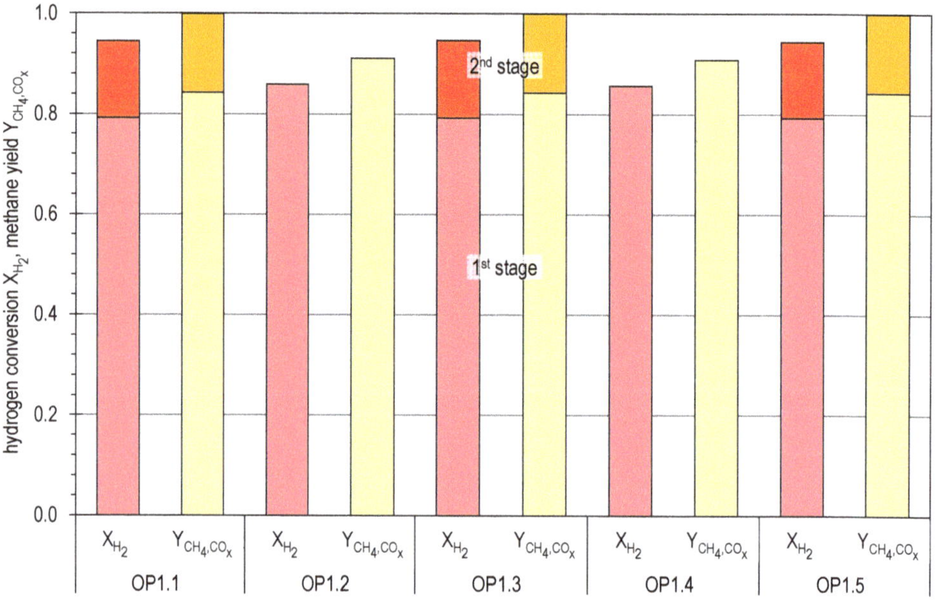

Figure 17. SC1. Hydrogen conversion, X_{H2}, and methane yield, $Y_{CH4,COx}$, after first and part of second methanation stage for the different operating points of FAU_CH4_TR.

Concerning SC2, Figure 18 shows the graphs of the data received from the DC in comparison to the set values of the MFC. Analogously to SC1, all datasets were completely and correctly transferred from the DC to the control system of the test rig. Nevertheless, noticeably in the dynamic part, some datasets did not comply with the reactor operating limits. The syngas power was too high at these points, and the programmed safety mechanisms skipped them. This greatly reduced the dynamics. The optimization tool should only calculate operating points that can also be run with the test rig. However, the maximum volume flows for BFG and BOFG, equivalent to meet the maximum syngas power with these gases, were stored in the DC models. Since the variation from 0.5 to 2.0 h in Figure 18 involved a mixture of both of these gases, there were probably deviations in calculating the maximum allowed volume flow of this mixture and thus the allowable syngas power was exceeded.

During the subsequent operating point, OP2.1, steady-state temperature conditions were achieved in the structured reactor and stationary gas compositions in both the intermediate and final gas. After the two-stage process, a full methane yield was achieved again with a hydrogen conversion of about 93.5%. This is in agreement with the values of comparable operating points from SC1 (see Figure 17).

4.3. Results of Experimental Campaign with MUL_CH4_TR

As described in Section 4.1 the operating parameters provided by the DC for MUL_CH4_TR in SC1 were mainly characterized by BOFG with highly dynamic fluctuations through jumps and impulses, ramps, and smooth changes. Nevertheless, the input gas composition was kept constant for each gas type, including a consistent hydrogen surplus of 4% (SN_{CH4} = 1.04). Such an excess in hydrogen compared to the reaction stoichiometry

(Equation (6)) met the optimum parameter for a methanation setup as used during the live experiments [60].

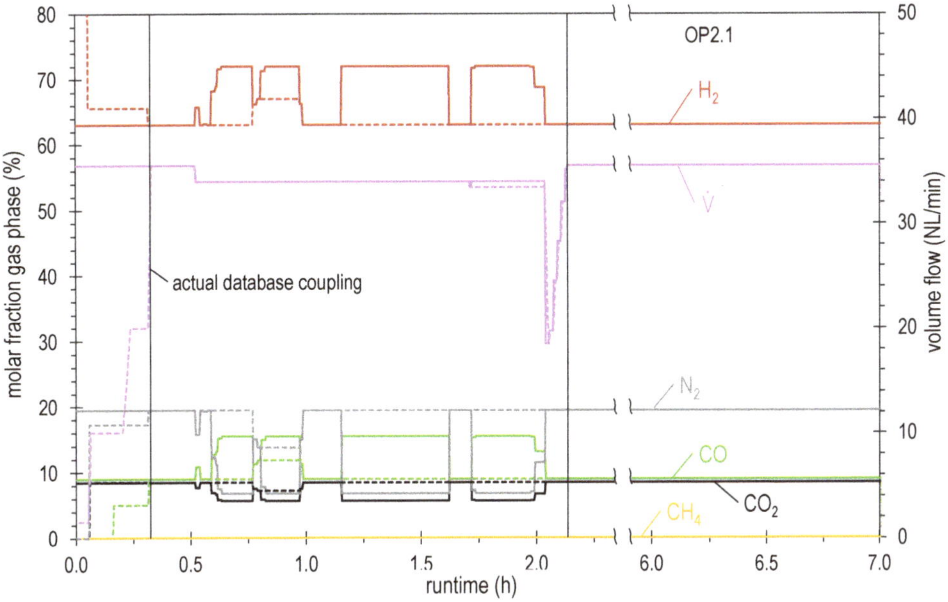

Figure 18. SC2. Handover values received from DC (solid lines) and set values of the MFC (dashed lines) for FAU_CH4_TR.

Prior to the start of methanation, the three reactors were heated up following the design specifications given by the catalyst manufacturer. The methanation of BOFG started with very smooth changes in the total volumetric flow rate, varying between 13.5 and 14 NL/min ($\approx$3200–3350 h^{-1} *GHSV*, 0 h). After 45 min, the flow rate increased to 18 NL/min ($\approx$4300 h^{-1} *GHSV*) over a period of 15 min and stayed at this elevated level for 5 h. During this period, multiple step increases, jumps in the total flow rate in the range of 2–3 NL/min were observed from one data point to the next one, representing a $\pm$16.7% increase in gas input power within 1 min. Furthermore, four changes to the BFG composition occurred instantly. These disturbances lasted for 4, 5, 14, and 2 min and resulted in a decrease in gas input power of up to 59.1%. Again, a consistent hydrogen surplus of 4% was guaranteed during the BFG disturbances. The second half of the live experiment was again characterized by a high usage of BOFG at about 18 NL/min, including step changes in the total volume flow as well as one disturbance with BFG. After a period of 1.5 h without any changes, the maximum flow rate of the live experiments of 22.5 NL/min was reached. The increase was achieved through imminent changes in the range of 1–2 min. During the whole duration of the experiment, a constant pressure of 4 bar was kept, meeting the specifications of the steelworks plant.

Figure 19 shows the time-based data for the following parameters: *GHSV* and product gas compositions for CH_4, H_2, CO, and CO_2 measured after the third reactor (N_2 = rest to 100%) as well as the CO_x conversion rate. Sixteen representative operating points are highlighted with black dots flagging dynamic changes in critical parameters.

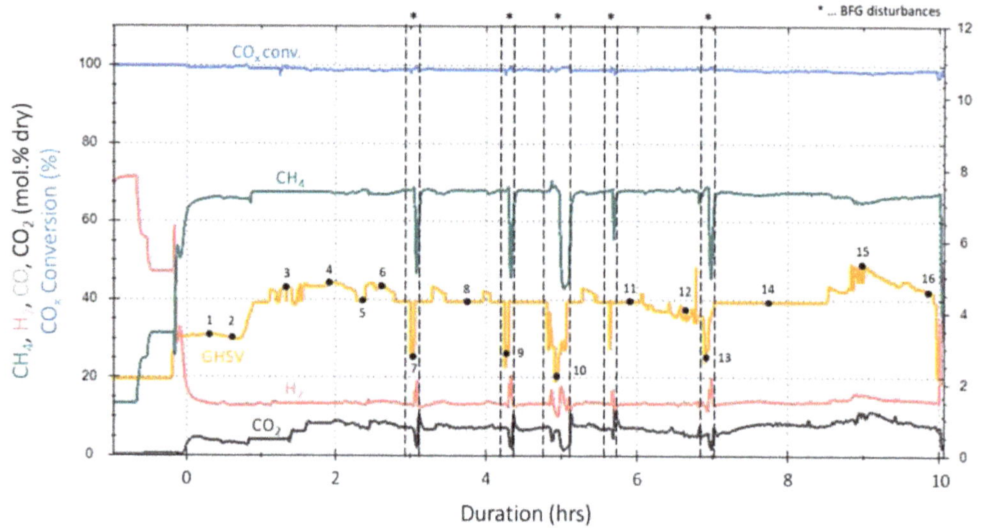

Figure 19. SC1. Time-based data for live experiments with MUL_CH4_TR: gas composition in mol.% (dry), CO_x conversion in %, and *GHSV* in 10^3 h^{-1}; sixteen representative operating points are marked with black dots.

Although highly dynamic variations in BOFG input power up to 45% were performed, the product gas composition measured after the third reactor stage remained nearly constant. The individual gas components, H_2, CO_2, CO, CH_4, and N_2, varied within a ±2.8% pt. range around their averages. Figure 20 gives an overview of the detailed gas composition for the 16 reference points, each representing a relevant dynamically changing operating condition. Hydrogen varied between 12.7 and 15.4 mol.% (dry), and methane stayed between 65.2 and 70.1 mol.% (dry). The concentration of CO_2 remained at a very low level of about 0.7 mol.% (dry) on average. As no CO was measured in the product gas, the CO_x conversion constantly remained, on average, above 99.0% (see Table 3).

The results of the instant changes from BOFG to BFG were noticeable. The CO_x conversion increased during these changes (#7, 9, 10, and 13), as the blue curve in Figure 19 and the values in Table 3 indicate; this was due to the lower flow rates and related *GHSV* and the resulting longer residence times. With an optimized temperature control of the reactors, conversion rates of up to 99.8% were achieved. Furthermore, the product gas composition changed to ≈47% for CH_4, ≈17% for H_2, and ≈36% for N_2 due to the different injected gas compositions. Verification experiments carried out after the live experiment did not indicate any signs of catalyst deactivation due to the highly dynamic changes in operating conditions.

4.4. Results of Experimental Campaign with ALFE_CH3OH_PLP

The methanol synthesis experiments during SC2 were conducted in ALFE_CH3OH_PLP with a commercial methanol catalyst filled in the four stages of the multi-stage setup. The use of the control tool described in Section 3.4. in real time to provide set points for ALFE_CH3OH_PLP considering the use of POGs was demonstrated. Methanol was successfully produced under fluctuating conditions, and the test served as a proof of concept for advanced process control with synthetic gas mixtures and load profiles provided by the DC through an online connection.

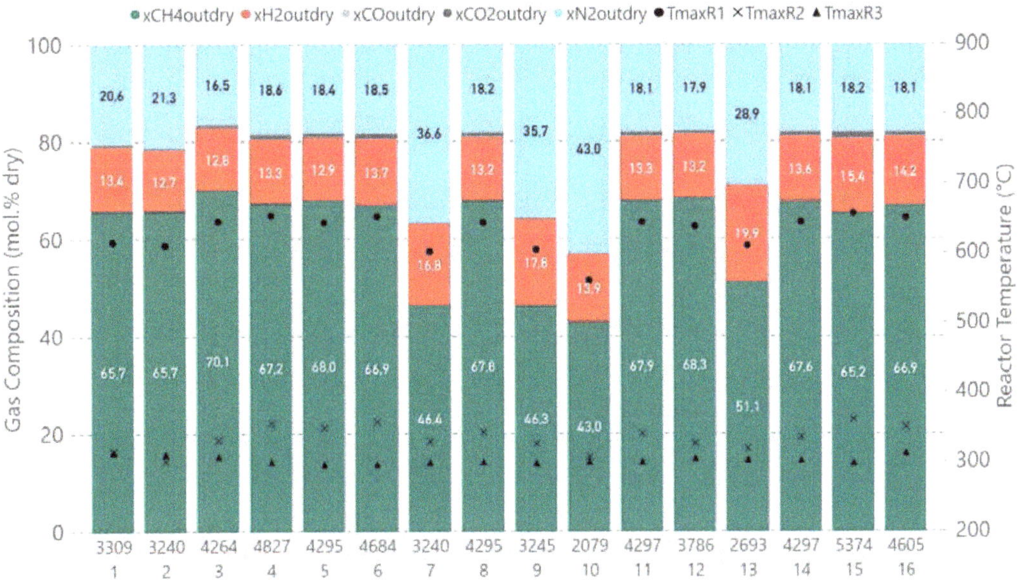

Figure 20. SC1. Product gas compositions and max. reactor temperatures for 16 representative points in time (values in mol.% (dry) and °C) for MUL_CH4_TR.

Table 3. SC1. Operating parameters, product gas compositions, and CO_x conversion (X_{COx}) rates for 16 representative dynamic operating points for MUL_CH4_TR. BFG disturbances are highlighted with *. SN_{CH4} = 1.04 for all reference points.

Operating Point	$\dot{V}_{tot}$	*GHSV*	H_2	CO_2	CO	CH_4	N_2	X_{COx}
	NL/min	h^{-1}	mol.%	mol.%	mol.%	mol.%	mol.%	%
1	13.9	3300	13.4	0.3	0.0	65.7	20.6	99.6
2	13.6	3200	12.7	0.3	0.0	65.7	21.3	99.6
3	17.9	4300	12.8	0.6	0.0	70.1	16.5	99.2
4	20.2	4800	13.3	1.0	0.0	67.2	18.6	98.7
5	18.0	4300	12.9	0.8	0.0	68.0	18.4	98.9
6	19.6	4700	13.7	0.9	0.0	66.9	18.5	98.7
7 *	18.0	3200	16.8	0.3	0.0	46.4	36.6	99.5
8	18.0	4300	13.2	0.8	0.0	67.8	18.2	98.9
9 *	18.0	3200	17.8	0.2	0.0	46.3	35.7	99.6
10 *	12.3	2100	13.9	0.2	0.0	43.0	43.0	99.7
11	18.0	4300	13.3	0.8	0.0	67.9	18.1	98.9
12	15.9	3800	13.2	0.6	0.0	68.3	17.9	99.2
13 *	16.3	2700	19.9	0.2	0.0	51.1	28.9	99.8
14	18.0	4300	13.6	0.7	0.0	67.6	18.1	99.0
15	22.5	5400	15.4	1.2	0.0	65.2	18.2	98.3
16	19.3	4600	14.2	0.8	0.0	66.9	18.1	98.8

As described in Section 4.1., FAU_CH4_TR and ALFE_CH3OH_PLP simultaneously received information from the DC and ran in parallel to produce methane and methanol. Hence, 10 points were tested at ALFE under fluctuating conditions to successfully and effectively produce methanol as set points were received through the online connection in real time from the DC. The high flexibility of the plant and the dedicated analytics allowed the plant to respond very quickly and effectively to load variations.

A typical BOFG composition was used together with additional H_2 to adjust the SN_{CH3OH} to 2.1. A once-through operation was chosen to lower the plant response time regarding new parameters for the synthesis. This allowed quick adaptation to load changes. The composition of the feedstock for the methanol synthesis was:

CO_2: 7.3 mol.%;

CO: 19.8 mol.%;

H_2: 64.2 mol.%;

N_2: 8.7 mol.%.

The temperatures and pressures during the synthesis were adapted to the changes in load, and the online analysis gave rapid information on conversions and stream compositions at the reactor outlet. Particular attention was given to combine high conversions with moderate temperature profiles to limit by-product formation.

Figure 21 describes the different loads provided in real time by the DC (black) and demonstrates very clearly the fast response of the unit to these variations (red). The time between the decision on a new set point and the process value of the flow controller was very fast, and the desired set point was reached in a few minutes.

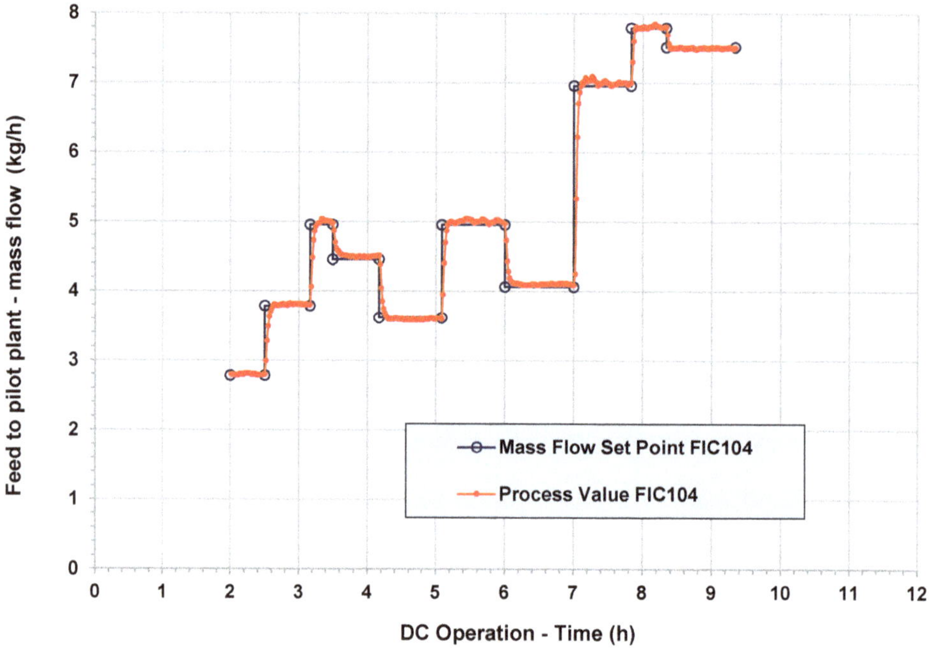

Figure 21. SC2. Mass flow set points versus time during test campaign of ALFE_CH3OH_PLP with DC.

In a deviation from the inlet conditions for the pilot plant shown in Figure 13, the operators did not start the methanol production from the very first values suggested by the DC (time 0 in Figures 13 and 21) but only after two hours of DC operation. The stop-and-go operation was not part of this test campaign at ALFE. Moreover, a certain smoothing of the load profile was performed (see comparison Figures 13a and 21), and only one gas composition (as mentioned above) was used.

The results in terms of catalytic performance were good and in accordance with expectations. The CO conversion (X_{CO}) for each load and without any recycling was always higher than 99% and in most of the cases overcame 99.5%. The CO_2 conversion (X_{CO2}) was also high but still lower than for CO. In more detail, CO_2 conversions were between 70% and 78% during the campaign depending on the amount of feed entering the synthesis section. This was expected, as CO_2 is well-known to be more difficult to be converted in one pass compared to CO at mild conditions for methanol synthesis. Nevertheless, CO_2 conversion can be increased by optimizing the operating conditions, especially acting on temperatures and pressures.

By increasing loads, the amount of methanol produced was also increased accordingly, even if slightly lower CO and CO_2 conversions were observed, as shown in Table 4. The flexibility of ALFE_CH3OH_PLP and the dedicated analytics ensured a quick unit adaptation, even when the load was rapidly increased or decreased. The plant flexibility was also guaranteed by the deep control of temperature (see Section 3.3) that provided valuable real-time information to the operators.

Table 4. SC2. Catalytic performance in the four-stage reactor system ALFE_CH3OH_PLP under once-through conditions.

Time of Sampling	p	T	T_{max}	Feed	X_{CO2}	X_{CO}	X_{H2}	Raw CH_3OH
hours	bar	°C	°C	NL/h (kg/h)	%	%	%	kg/h
2.00 (start of operation)	80	220	278	4927 (2.80)	77.8	99.6	82.5	2.06
2.80	80	220	272	6530 (3.71)	77.3	99.6	79.7	2.72
3.30	80	220	269	8441 (4.79)	77.7	99.6	80.2	3.52
3.80	80	220	269	8000 (4.54)	76.4	99.6	79.2	3.32
4.50	80	220	272	6459 (3.66)	76.3	99.6	81.1	2.67
5.30	80	220	269	8663 (4.91)	75.0	99.5	78.8	3.56
6.30	80	230	280	7323 (4.15)	74.3	99.4	79.2	2.99
7.30	80	230	273	11,945 (6.78)	72.6	99.2	77.6	4.85
8.00	80	230	271	13,618 (7.73)	70.5	99.1	77.0	5.47
9.50 (end of operation)	80	230	271	13,265 (7.52)	70.4	99.1	78.3	5.32

The amount of by-products was still lower than 5000 wt.ppm despite the high CO concentration and the high maximum temperature in the first stage (see Table 5). The by-product concentration in raw methanol decreased when the feed amount increased for a given gas composition. The by-product amount lied in a range that can be covered by a traditional distillation section to produce a commercial methanol suitable for the market without additional treatment.

Table 5. SC2. By-product amount in raw methanol from ALFE_CH3OH_PLP at different feed amounts.

Time of Sampling	p	T	T_{max}	Feed	Raw CH_3OH	Water in Raw CH_3OH	Total By-Products
hours	bar	°C	°C	NL/h (kg/h)	kg/h	%	wt.ppm
2.00 (start of operation)	80	220	278	4927 (2.80)	2.06	11.8	4680
9.50 (end of operation)	80	230	271	13,265 (7.52)	5.32	10.9	4460

In both cases, low load and high load, it can be observed in Tables 6 and 7 that the high conversion of CO in the first stage also led to higher maximum temperatures in the catalytic bed. This caused the formation of a high amount of by-products in the first stage. The CO_2 conversion was much lower in the first stage than the conversion of CO and, as a consequence, the concentration of water in the raw methanol produced in the first stage was low. The amount of by-products produced in the first stage could be compensated by the lower production of by-products at later stages, and this delivered a collected and gathered final raw methanol product that was able to be purified in the distillation section.

Table 6. Conversion of CO_2, CO, and H_2 in each stage and for the overall plant at low load (4927 NL/h (2.80 kg/h)) in the four-stage methanol plant (T = 220 °C, p = 80 bar); time of sampling: 2.00 h (start of operation).

Stage	Feed Inlet	T_{inlet}	T_{max}	X_{CO2}	X_{CO}	X_{H2}	By-Products in Raw MeOH	Water in Raw MeOH
-	NL/h	°C	°C	%	%	%	wt.ppm	wt.%
1	4927	220	278	16.9	93.0	56.9	6450	3.2
2	2132	220	226	40.3	84.4	29.5	671	27.5
3	1557	220	224	50.9	52.1	24.1	396	35.6
4	1236	220	224	60.0	41.1	21.6	339	35.7
Overall Plant	4927	220	278	77.8	99.6	82.5	4680	11.8

Table 7. Conversion of CO_2, CO, and H_2 in each stage and for the overall plant at high load (13,265 NL/h (7.52 kg/h)) in the four-stage methanol plant (T = 230 °C, p = 80 bar); time of sampling: 9.50 h (end of operation).

Stage	Feed Inlet	T_{inlet}	T_{max}	X_{CO2}	X_{CO}	X_{H2}	By-Products in Raw MeOH	Water in Raw MeOH
	NL/h	°C	°C	%	%	%	wt.ppm	wt.%
1	13,265	230	271	13.1	82.3	49.7	6463	2.5
2	6656	230	243	26.9	82.1	29.3	1258	17.2
3	4802	230	236	38.9	58.0	22.4	519	32.8
4	3854	230	236	47.0	44.7	20.0	387	35.4
Overall Plant	13,265	230	271	70.4	99.1	78.3	4460	10.9

This means that CO_2-rich gases can be used as well as CO-rich gases in this type of equipment, which is a perfect complement to the dispatch controller, which distributes in real time the different loads and different compositions of steelworks off-gases to convert them into methanol and/or methane.

The amount of feed and the gas composition entering the first stage was provided by the dispatch controller as described in Tables 6 and 7. The outlet of the first stage was then separated into a liquid phase that represented the raw methanol and a gas phase that was sent as a feed to stage 2. This process was then repeated for all the stages, as already explained in Section 3.3 of this publication. Therefore, the gas composition as well as the amount of feed at the stage inlet were different for all the stages. As a consequence, the amount of raw methanol produced also differed. Hence, the raw methanol produced after each stage was removed, gathered, and sent to a low-pressure separator, which separated the solved gases from the final raw methanol product.

The overall conversions of CO for the different loads were above 99% but were slightly better for the lower amount of feed and lower inlet temperatures. The conversion of CO_2 was more sensible to the amount of feed entering the reactor, with an overall CO_2 conversion of 70.4% for the high-load test point in comparison to 77.8% for the low-load test point, even if the temperature was 10 °C lower in this case (220 °C for low-load test versus 230 °C for high-load test). The control of the maximum temperature in the different stages in real time was very important to avoid severe degradation of the catalyst.

The same exercise can be conducted with different compositions of BFG, BOG, COG, and/or mixture of these gases, and the plant can perform well, producing a suitable amount of methanol with limited by-product volume. The control of gas compositions and available hydrogen are key parameters. Nevertheless, ALFE_CH3OH_PLP can adapt to a low hydrogen amount and to a high amount of inerts in the feed, ensuring suitable methanol production.

Moreover, ALFE_CH3OH_PLP can be put in a stand-by condition and started and stopped without any significant issue. No loss of performance and no unusual deterioration of the catalyst were observed during a further extended test campaign of about 1800 h on a stream with different gas compositions representing POGs. The 1800 h on the stream included the commissioning and validation phase of the pilot plant with well-known gas compositions and operating conditions as well as several test points at a steady state with synthetic steel gas mixtures as the feed for the methanol synthesis. In this publication, exclusively the results linked to the test under dynamic conditions with a direct coupling to the developed dispatch controller are described.

5. Summary, Conclusions, and Future Works

The work and tests reported in this paper concern the design and construction of a complex, virtually coupled test rig consisting of several synthesis reactors for the production of methane and methanol located in three different locations (one in Austria and two in Germany), a remote computing, supervision, and control system located in Italy, and a data processing system that allows the distribution of measurements, simulated data, and control strategies between the systems of interest.

The carried out online tests have shown that the calculation, optimization, and control architecture can lead to promising and satisfying results. First of all, the data exchange architecture based on an OPC UA infrastructure allowed the distribution of measures and control strategies to several recipients in parallel without transmission/reception issues. The control and supervision system, namely, the dispatch controller (DC), allowed the calculation of control strategies in real time with a temporal resolution of one minute and sufficiently smooth and optimal/sub-optimal control trends in the pre-established time. The optimization problem formulated as a non-linear MILP, linearized in real time, allowed the description of the dynamic behavior of the reactors and of all the equipment connected to the methane/methanol production system with sufficient accuracy.

The operation of the three synthesis reactors with steelworks process off-gases ran without errors for the two scenarios presented. Two plants each were operated in parallel over periods of 10 and 7 h, respectively, with values that were specified by the DC and distributed to the plants or their operators via an OPC UA.

FAU's bench-scale methanation test rig received the DC's specified values directly via the OPC UA and processed them automatically. The safety routines implemented for this purpose worked reliably, albeit in part with very strict limit specifications. The two-stage methanation with a structured fixed bed reactor as the first stage responded quickly to specified load or feed gas composition changes with a stable operating behavior and a high reproducibility of results. A full methane yield was obtained after the two-stage methanation with intermediate water separation, of which up to 91% could already be achieved in the first reactor stage. No evidence of progressive catalyst deactivation was observed.

MUL's bench-scale methanation plant also showed resilient behavior to the load fluctuations specified by the DC. Even with quick and large changes in input power of up to 45%, the product gas composition remained almost constant after the three-stage process. Consistently high CO_x conversion rates of over 99% on average were achieved. Moreover, no additional catalyst deactivation was noticeable.

ALFE's innovative multi-stage pilot plant for methanol synthesis showed high flexibility in once-through operation in terms of inlet gas composition and variable operating conditions. The plant reacted to a load change within minutes and always showed a stable operating behavior. Both the handling of a high content of inerts and unusual reactant gas mixtures (POGs, high N_2 content, and a mixture of CO and CO_2) were mastered, even under fluctuating conditions. CO and CO_2 conversions were over 99% and between 70–78%, respectively. The amount of by-products was low for the prevailing reaction conditions (<5000 wt.ppm).

The results of the live coupling tests via online connection suggest that it is possible to integrate methane and/or methanol generation plants in combination with an intelligent supervision and control system into existing steelworks as a CCU concept while also obtaining significant advantages in terms of CO_2 emission reductions. However, further testing, for example, the live feedback of the real plants to the dispatch controller, the direct coupling of the plants to an electrolyzer, and a higher level of automation (automated value transfer and processing to/at all plants and an automated stabilization of operating points) are necessary before an on-site demonstration at a steelworks can take place.

Furthermore, future works will be aimed at investigating different topics, including a complete sensitivity analysis of the economic costs as a function of energy media and CO_2 prices. Further studies will focus on the more detailed computation of CO_2 reductions through the production of methane and methanol with the plants presented on an industrial scale and considering all the sources of avoided CO_2 (e.g., also those related to the extraction, preparation, and transportation of primary NG). In addition, an improvement in the dispatch controller will be studied through the implementation of more accurate models, including product treatment and cleaning equipment. In this sense, future works will probably also make it possible to refine the control strategies and/or the design of the plants to maximize their effectiveness.

Author Contributions: Conceptualization, A.H., P.W.-Z., S.H., S.D., M.M., I.M., C.M., V.C., S.K., M.B., K.P. and J.K.; methodology, A.H., P.W.-Z., S.D., X.T., M.M., I.M., V.C., S.K., M.B. and J.K.; software, A.H., P.W.-Z., S.D., M.M., I.M., C.M. and S.K.; validation, A.H., P.W.-Z., S.D., X.T., M.M., I.M., V.C., S.K., M.B. and K.P.; formal analysis, A.H., P.W.-Z., S.H., S.D., X.T., M.M., I.M. and V.C.; investigation, A.H., P.W.-Z., S.H., S.D., X.T., M.M. and I.M.; resources, S.H., V.C., N.K., K.R. and J.K.; data curation, A.H., P.W.-Z., S.D., X.T., M.M., I.M. and C.M.; writing—original draft preparation, A.H., P.W.-Z., S.H., S.D., X.T., M.M. and I.M.; writing—review and editing, A.H., P.W.-Z., S.H., S.D., I.M., C.M., V.C., S.K., M.B., K.P., N.K., K.R. and J.K.; visualization, A.H., P.W.-Z., S.D. and M.M.; supervision, S.H., V.C., K.P., N.K., K.R. and J.K.; project administration, A.H., S.H., V.C., K.P., N.K., K.R. and J.K.; funding acquisition, S.H., V.C., K.P. and J.K. All authors have read and agreed to the published version of the manuscript.

Funding: The work carried out was funded by the European Union through the Research Fund for Coal and Steel (RFCS) within the project entitled "i3upgrade: Integrated and intelligent upgrade of carbon sources through hydrogen addition for the steel industry", Grant Agreement No. 800659. This paper reflects only the author's view and the European Commission is not responsible for any use that may be made of the information contained therein.

Institutional Review Board Statement: Not applicable.

Informed Consent Statement: Not applicable.

Data Availability Statement: Not applicable.

Conflicts of Interest: The authors declare no conflict of interest.

References

1. European Commission. *The European Green Deal*; European Commission: Brussels, Belgium, 2019.
2. European Commission. *"Fit for 55": Delivering the EU's 2030 Climate Target on the Way to Climate Neutrality*; European Commission: Brussels, Belgium, 2021.
3. European Commission. *Towards Competitive and Clean European Steel*; European Commission: Brussels, Belgium, 2021.
4. Estep Clean Steel Partnership. *Strategic Research and Innovation Agenda (SRIA)*; Estep Clean Steel Partnership: Brussels, Belgium, 2021.

5. Voestalpine Stahl GmbH. *Analysenergebnisse Der Umwelt—Und Betriebsanalytik (Durchgeführt Für i3 Upgrade)*; Technical Report; Voestalpine Stahl GmbH: Linz, Austria, 2019.
6. Remus, R.; Roudier, S.; Aguado Monsonet, M.A.; Sancho, L.D. *Best Available Techniques (BAT) Reference Document for Iron and Steel Production*; Publications Office of the European Union: Luxembourg, 2013.
7. Schlüter, S.; Hennig, T. Modeling the Catalytic Conversion of Steel Mill Gases Using the Example of Methanol Synthesis. *Chem. Ing. Tech.* **2018**, *90*, 1541–1558. [CrossRef]
8. Razzaq, R.; Li, C.; Zhang, S. Coke oven gas: Availability, properties, purification, and utilization in China. *Fuel* **2013**, *113*, 287–299. [CrossRef]
9. Moral, G.; Ortiz-Imedio, R.; Ortiz, A.; Gorri, D.; Ortiz, I. Hydrogen Recovery from Coke Oven Gas. Comparative Analysis of Technical Alternatives. *Ind. Eng. Chem. Res.* **2022**, *61*, 6106–6124. [CrossRef] [PubMed]
10. Zhang, Q.; Li, Y.; Xu, J.; Jia, G. Carbon element flow analysis and CO2 emission reduction in iron and steel works. *J. Clean. Prod.* **2018**, *172*, 709–723. [CrossRef]
11. Bhaskar, A.; Assadi, M.; Somehsaraei, H.N. Decarbonization of the Iron and Steel Industry with Direct Reduction of Iron Ore with Green Hydrogen. *Energies* **2020**, *13*, 758. [CrossRef]
12. Wang, R.; Zhao, Y.; Babich, A.; Senk, D.; Fan, X. Hydrogen direct reduction (H-DR) in steel industry—An overview of challenges and opportunities. *J. Clean. Prod.* **2021**, *329*, 129797. [CrossRef]
13. Patisson, F.; Mirgaux, O. Hydrogen Ironmaking: How it Works. *Metals* **2020**, *10*, 922. [CrossRef]
14. Porzio, G.F.; Fornai, B.; Amato, A.; Matarese, N.; Vannucci, M.; Chiappelli, L.; Colla, V. Reducing the energy consumption and CO_2 emissions of energy intensive industries through decision support systems—An example of application to the steel industry. *Appl. Energy* **2013**, *112*, 818–833. [CrossRef]
15. Porzio, G.F.; Nastasi, G.; Colla, V.; Vannucci, M.; Branca, T.A. Comparison of multi-objective optimization techniques applied to off-gas management within an integrated steelwork. *Appl. Energy* **2014**, *136*, 1085–1097. [CrossRef]
16. Maddaloni, A.; Matino, R.; Matino, I.; Dettori, S.; Zaccara, A.; Colla, V. A quadratic programming model for the optimization of off-gas networks in integrated steelworks. *Matér. Tech.* **2019**, *107*, 502. [CrossRef]
17. Colla, V.; Matino, I.; Dettori, S.; Petrucciani, A.; Zaccara, A.; Weber, V.; Salame, S.; Zapata, N.; Bastida, S.; Wolff, A.; et al. Assessing the efficiency of the off-gas network management in integrated steelworks. *Matér. Tech.* **2019**, *107*, 104. [CrossRef]
18. Matino, I.; Dettori, S.; Castellano, A.; Matino, R.; Mocci, C.; Vannocci, M.; Maddaloni, A.; Colla, V.; Wolff, A. Machine Learning-Based Models for Supporting Optimal Exploitation of Process Off-Gases in Integrated Steelworks. In Proceedings of the Cybersecurity Workshop by European Steel Technology Platform, ESTEP 2020: Impact and Opportunities of Artificial Intelligence Techniques in the Steel Industry, Pisa, Italy, 15–16 October 2020; Springer: Cham, Switzerland, 2020; pp. 104–118.
19. Dettori, S.; Matino, I.; Colla, V.; Speets, R. A Deep Learning-Based Approach for Forecasting off-Gas Production and Consumption in the Blast Furnace. *Neural Comput. Appl.* **2022**, *34*, 911–923. [CrossRef] [PubMed]
20. Dettori, S.; Matino, I.; Colla, V.; Weber, V.; Salame, S. Neural Network-based modeling methodologies for energy transformation equipment in integrated steelworks processes. *Energy Procedia* **2019**, *158*, 4061–4066. [CrossRef]
21. Matino, I.; Dettori, S.; Colla, V.; Weber, V.; Salame, S. Forecasting blast furnace gas production and demand through echo state neural network-based models: Pave the way to off-gas optimized management. *Appl. Energy* **2019**, *253*, 113578. [CrossRef]
22. Rieger, J.; Colla, V.; Matino, I.; Branca, T.; Stubbe, G.; Panizza, A.; Brondi, C.; Falsafi, M.; Hage, J.; Wang, X.; et al. Residue Valorization in the Iron and Steel Industries: Sustainable Solutions for a Cleaner and More Competitive Future Europe. *Metals* **2021**, *11*, 1202. [CrossRef]
23. Chisalita, D.-A.; Petrescu, L.; Cobden, P.; van Dijk, H.; Cormos, A.-M.; Cormos, C.-C. Assessing the environmental impact of an integrated steel mill with post-combustion CO2 capture and storage using the LCA methodology. *J. Clean. Prod.* **2018**, *211*, 1015–1025. [CrossRef]
24. Huang, Z.; Ding, X.; Sun, H.; Liu, S. Identification of main influencing factors of life cycle CO2 emissions from the integrated steelworks using sensitivity analysis. *J. Clean. Prod.* **2010**, *18*, 1052–1058. [CrossRef]
25. Saima, W.H.; Mogi, Y.; Haraoka, T. Development of PSA System for the Recovery of Carbon Dioxide and Carbon Monoxide from Blast Furnace Gas in Steel Works. *Energy Procedia* **2013**, *37*, 7152–7159. [CrossRef]
26. Steynberg, A. Chapter 1—Introduction to Fischer-Tropsch Technology. In *Studies in Surface Science and Catalysis*; Steynberg, A., Dry, M., Eds.; Elsevier: Amsterdam, The Netherlands, 2004; Volume 152, pp. 1–63.
27. Valera-Medina, A.; Roldan, A. Ammonia from Steelworks. In *Sustainable Ammonia Production. Green Energy and Technology*; Inamuddin, Boddula, R., Asiri, A., Eds.; Springer: Cham, Switzerland, 2020; pp. 69–80.
28. De Ras, K.; Van De Vijver, R.; Galvita, V.V.; Marin, G.B.; Van Geem, K.M. Carbon capture and utilization in the steel industry: Challenges and opportunities for chemical engineering. *Curr. Opin. Chem. Eng.* **2019**, *26*, 81–87. [CrossRef]
29. Lyke, S.E.; Moore, R.H. *Chemical Production from Industrial By-Product Gases: Final Report*; Battelle Pacific Northwest Labs: Richland, WA, USA, 1981.
30. Cordier, J.; Dussart, B. Ammonia and Methanol Production—How Savings Can Be Made. *Pet. Technol.* **1984**, *307*, 38–45.
31. Kim, S.; Kim, J. The optimal carbon and hydrogen balance for methanol production from coke oven gas and Linz-Donawitz gas: Process development and techno-economic analysis. *Fuel* **2020**, *266*, 117093. [CrossRef]
32. Deng, L.; Ii, T.A.A. Techno-economic analysis of coke oven gas and blast furnace gas to methanol process with carbon dioxide capture and utilization. *Energy Convers. Manag.* **2019**, *204*, 112315. [CrossRef]

33. Kim, D.; Han, J. Techno-economic and climate impact analysis of carbon utilization process for methanol production from blast furnace gas over Cu/ZnO/Al_2O_3 catalyst. *Energy* **2020**, *198*, 117355. [CrossRef]
34. Lundgren, J.; Ekbom, T.; Hulteberg, C.; Larsson, M.; Grip, C.-E.; Nilsson, L.; Tunå, P. Methanol production from steel-work off-gases and biomass based synthesis gas. *Appl. Energy* **2013**, *112*, 431–439. [CrossRef]
35. Haag, S.; Castillo-Welter, F.; Schuhmann, T.; Williams, B.A.; Oelmann, T.; Günther, A.; Gorny, M. How to Convert CO_2 to Green Methanol. In Proceedings of the Challenges for Petrochemicals and Fuels: Integration of Value Chains and Energy Transition (DGMK Conference), Berlin, Germany, 10–12 October 2018.
36. Oelmann, T.; Gorny, M.; Schuhmann, T.; Strozyk, M.; Castillo-Welter, F.; Drosdzol, C.; Haag, S. A New Reactor Concept for Conversion of CO_2 to Methanol. *Oil Gas-Eur. Mag.* **2021**, *47*, 28–32.
37. Girod, K.; Lohmann, H.; Schlüter, S.; Kaluza, S. Methanol Synthesis with Steel-Mill Gases: Simulation and Practical Testing of Selected Gas Utilization Scenarios. *Processes* **2020**, *8*, 1673. [CrossRef]
38. Girod, K.; Breitkreuz, K.; Hennig, T.; Lohmann, H.; Kaluza, S.; Schlüter, S. Steel Mills as Syngas Source for Methanol Synthesis: Simulation and Practical Performance Investigations. *Chem. Eng. Trans.* **2019**, *74*, 475–480.
39. Periodic Reporting for Period 3—FReSMe (From Residual Steel Gasses to Methanol) | H2020 | CORDIS | European Com-Mission. Available online: https://cordis.europa.eu/project/id/727504/reporting (accessed on 9 June 2022).
40. Lee, J.-K.; Lee, I.-B.; Han, J. Techno-economic analysis of methanol production from joint feedstock of coke oven gas and basic oxygen furnace gas from steel-making. *J. Ind. Eng. Chem.* **2019**, *75*, 77–85. [CrossRef]
41. Thonemann, N.; Maga, D. Life Cycle Assessment of Steel Mill Gas-Based Methanol Production within the Carbon2Chem®Project. *Chem. Ing. Tech.* **2020**, *92*, 1425–1430. [CrossRef]
42. Rigamonti, L.; Brivio, E. Life cycle assessment of methanol production by a carbon capture and utilization technology applied to steel mill gases. *Int. J. Greenh. Gas Control* **2022**, *115*, 103616. [CrossRef]
43. Müller, K.; Rachow, F.; Günther, V.; Schmeisser, D. Methanation of Coke Oven Gas with Nickel-Based Catalysts. *J. Environ. Sci.* **2019**, *4*, 73–79.
44. Razzaq, R.; Zhu, H.; Jiang, L.; Muhammad, U.; Li, C.; Zhang, S. Catalytic Methanation of CO and CO_2 in Coke Oven Gas over Ni–Co/ZrO_2–CeO_2. *Ind. Eng. Chem. Res.* **2013**, *52*, 2247–2256. [CrossRef]
45. Rosenfeld, D.C.; Böhm, H.; Lindorfer, J.; Lehner, M. Scenario analysis of implementing a power-to-gas and biomass gasification system in an integrated steel plant: A techno-economic and environmental study. *Renew. Energy* **2019**, *147*, 1511–1524. [CrossRef]
46. Perpiñán, J.; Bailera, M.; Romeo, L.M.; Peña, B.; Eveloy, V. CO_2 Recycling in the Iron and Steel Industry via Power-to-Gas and Oxy-Fuel Combustion. *Energies* **2021**, *14*, 7090. [CrossRef]
47. Wolf-Zoellner, P.; Medved, A.R.; Lehner, M.; Kieberger, N.; Rechberger, K. In Situ Catalytic Methanation of Real Steelworks Gases. *Energies* **2021**, *14*, 8131. [CrossRef]
48. Hauser, A.; Weitzer, M.; Gunsch, S.; Neubert, M.; Karl, J. Dynamic hydrogen-intensified methanation of synthetic by-product gases from steelworks. *Fuel Process. Technol.* **2021**, *217*, 106701. [CrossRef]
49. Zaccara, A.; Petrucciani, A.; Matino, I.; Branca, T.A.; Dettori, S.; Iannino, V.; Colla, V.; Bampaou, M.; Panopoulos, K. Renewable Hydrogen Production Processes for the Off-Gas Valorization in Integrated Steelworks through Hydrogen Intensified Methane and Methanol Syntheses. *Metals* **2020**, *10*, 1535. [CrossRef]
50. Bampaou, M.; Panopoulos, K.; Seferlis, P.; Voutetakis, S.; Matino, I.; Petrucciani, A.; Zaccara, A.; Colla, V.; Dettori, S.; Branca, T.A.; et al. Integration of Renewable Hydrogen Production in Steelworks Off-Gases for the Synthesis of Methanol and Methane. *Energies* **2021**, *14*, 2904. [CrossRef]
51. Kolb, S.; Plankenbühler, T.; Hofmann, K.; Bergerson, J.; Karl, J. Life cycle greenhouse gas emissions of renewable gas technologies: A comparative review. *Renew. Sustain. Energy Rev.* **2021**, *146*, 111147. [CrossRef]
52. Lee, J.H. Model predictive control: Review of the three decades of development. *Int. J. Control Autom. Syst.* **2011**, *9*, 415–424. [CrossRef]
53. Neubert, M.; Hauser, A.; Pourhossein, B.; Dillig, M.; Karl, J. Experimental evaluation of a heat pipe cooled structured reactor as part of a two-stage catalytic methanation process in power-to-gas applications. *Appl. Energy* **2018**, *229*, 289–298. [CrossRef]
54. Biegger, P.; Kirchbacher, F.; Medved, A.R.; Miltner, M.; Lehner, M.; Harasek, M. Development of Honeycomb Methanation Catalyst and Its Application in Power to Gas Systems. *Energies* **2018**, *11*, 1679. [CrossRef]
55. e Silva, D.P.; Salles, J.L.F.; Fardin, J.F.; Pereira, M.M.R. Management of an island and grid-connected microgrid using hybrid economic model predictive control with weather data. *Appl. Energy* **2020**, *278*, 115581. [CrossRef]
56. Dettori, S.; Matino, I.; Iannino, V.; Colla, V.; Hauser, A.; Wolf-Zöllner, P.; Haag, S. Optimizing methane and methanol production from integrated steelworks process off-gases through economic hybrid model predictive control. *IFAC-PapersOnLine* **2022**, *55*, 66–71. [CrossRef]
57. Bampaou, M.; Kyriakides, A.S.; Panopoulos, K.; Seferlis, P.; Voutetakis, S. Modelling of Methanol Synthesis: Improving Hydrogen Utilisation. *Chem. Eng. Trans.* **2021**, *88*, 931–936. [CrossRef]
58. Matino, I.; Dettori, S.; Colla, V.; Rechberger, K.; Kieberger, N. Echo-state neural networks forecasting steelworks off-gases for their dispatching in CH_4 and CH_3OH syntheses reactors. In Proceedings of the 29th European Symposium on Artificial Neural Networks, Computational Intelligence and Machine Learning, Online Conference, 6–8 October 2021. [CrossRef]

59. Matino, I.; Dettori, S.; Zaccara, A.; Petrucciani, A.; Iannino, V.; Colla, V.; Bampaou, M.; Panopoulos, K.; Rechberger, K.; Kolb, S.; et al. Hydrogen role in the valorization of integrated steelworks process off-gases through methane and methanol syntheses. *Matériaux Tech.* **2021**, *109*, 308. [CrossRef]
60. Medved, A.R.; Lehner, M.; Rosenfeld, D.C.; Lindorfer, J.; Rechberger, K. Enrichment of Integrated Steel Plant Process Gases with Implementation of Renewable Energy: Integration of Power-to-Gas and Biomass Gasification System in Steel Production. *Johns. Matthey Technol. Rev.* **2021**, *65*, 453–465. [CrossRef]

metals

Article

Integration of Open Slag Bath Furnace with Direct Reduction Reactors for New-Generation Steelmaking

Pasquale Cavaliere [1,*], Angelo Perrone [1], Alessio Silvello [2], Paolo Stagnoli [3] and Pablo Duarte [4]

[1] Department of Innovation Engineering, University of Salento, Via per Arnesano, 73100 Lecce, Italy; angelo.perrone@unisalento.it
[2] Thermal Spray Center CPT, Universitat de Barcelona, 08007 Barcelona, Spain; asilvello@cptub.eu
[3] TENOVA S.p.A., Via Gerenzano 58, 21053 Castellanza, Italy; paolo.stagnoli@tenova.com
[4] TENOVA HYL, San Nicolás de los Garz 66450, Mexico; pablo.duarte@tenova.com
* Correspondence: pasquale.cavaliere@unisalento.it

Abstract: The present paper illustrates an innovative steel processing route developed by employing hydrogen direct reduced pellets and an open slag bath furnace. The paper illustrates the direct reduction reactor employing hydrogen as reductant on an industrial scale. The solution allows for the production of steel from blast furnace pellets transformed in the direct reduction reactor. The reduced pellets are then melted in open slag bath furnaces, allowing carburization for further refining. The proposed solution is clean for the decarbonization of the steel industry. The kinetic, chemical and thermodynamic issues are detailed with particular attention paid to the slag conditions. The proposed solution is also supported by the economic evaluation compared to traditional routes.

Keywords: direct reduced pellets; open slag bath furnace; slag; blast furnace pellets; hydrogen; decarbonization

Citation: Cavaliere, P.; Perrone, A.; Silvello, A.; Stagnoli, P.; Duarte, P. Integration of Open Slag Bath Furnace with Direct Reduction Reactors for New-Generation Steelmaking. *Metals* **2022**, *12*, 203. https://doi.org/10.3390/met12020203

Academic Editor: Alexander McLean

Received: 4 January 2022
Accepted: 20 January 2022
Published: 21 January 2022

1. Introduction

The direct reduction of iron oxide technologies is considered the best available technique able to greatly reduce the carbon dioxide emissions of steel plants [1]. Mainly, this is obtained by employing natural gas instead of coke or coal [2]. Many approaches have been followed in the development of these solutions; a summary of them is shown in Figure 1.

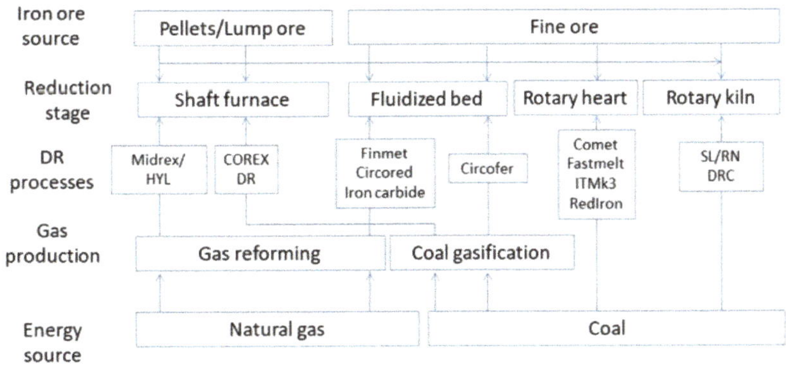

Figure 1. Main developed direct reduction processes.

Today, direct reduction processes account for over 70% of the total production of direct reduced iron (DRI) and hot briquetted iron (HBI) based on natural gas being used as the main reducing agent. The natural gas is converted to the reducing agents, mainly carbon monoxide and hydrogen, acting as reducing sources of iron oxides. The main industrially diffused technologies are Midrex and HYL-Energiron [3]. The energy consumption of the whole process is essentially dependent on metallurgical, chemical and physical properties of the raw materials [4]. This is amplified in the direct reduction processes because all the transformations act at the solid state. Taking into account the further operations of steel making, ash and sulfur quantities influence the electric arc furnace (EAF) operations [5], so natural gas needs pre-heating in order to be more active for sulfur removal. As a matter of fact, high-quality ores are required for the DRI-EAF steelmaking route [6]. Obviously, the direct reduction reactors are largely located in those regions characterized by large natural gas production or by NG (natural gas) low price availability. A more recently developed solution is the employment of hydrogen instead of natural gas for the reduction of iron ores. The hydrogen direct reduction of iron ores produces mainly iron and water vapor but also CO_2 [7,8]. This vapor is optimal for employment in high-temperature electrolyzers for further hydrogen production. Now, more than 90% of hydrogen is produced via fossil sources through various technologies generating carbon dioxide that needs to be treated, captured and stored [9]. In this way, the best means of producing iron with the lowest impact on the environment is the production of hydrogen through electrolysis [10,11]. By employing hydrogen produced via green energy sources as the reducing agent, carbon dioxide emissions can be reduced by 300 kg/t.

The integration of open slag bath furnaces (OSBFs) with direct reduction reactors is an innovative and interesting solution for emission reduction, energy saving and cost reduction. It is believed (and sometimes demonstrated in the literature) that OSBFs are very useful in treating those materials difficult to be melted and processed with an electric arc furnace. So, the OSBF is highly versatile for steel processing in a wide range of potential raw materials to be employed in the direct reduction reactors. In order to precisely describe the OSBFs, it is fundamental to underline those differences with respect to the submerged-arc furnaces. In this case, the furnace electrodes are submerged in the pre-reduced iron and carbon mixture. Electrodes provide the power in the order of 20 MW to the furnace, and the burden resistance allows for the transformation of power into heat in order to melt the material. As a consequence of the difference in the density of the slag and of the metals, the first flows toward the electrodes while the second flows to the bottom of the furnace (Figure 2).

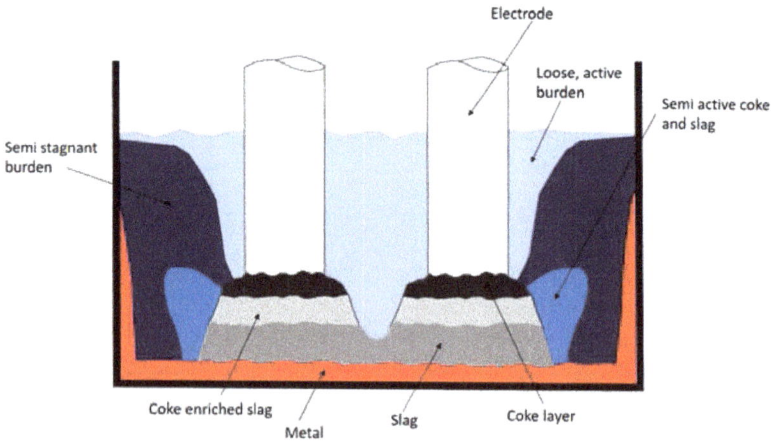

Figure 2. OSBF (open slag bath furnaces) schematic process behavior.

In the present paper, we present an innovative solution for steelmaking production based on hydrogen-assisted direct reduced iron and subsequent melting and refining in open slag bath furnaces. All the presented results belong to industrial experiences performed on real-scale plants.

2. Experimental Procedure

The employed direct reduction plant was an HYL (TENOVA) reactor (TENOVA Castellanza, Italy) with a production capacity of 900,000 t/year. The employed reducing gas was hydrogen produced via high-temperature water electrolysis; the water vapor from the DR (direct reduction) reactor is employed for electrolysis.

The system is designed in order to reduce blast furnace (BF)-grade pellets, and the composition of the employed raw material is listed in Table 1.

Table 1. BF-grade pellets employed in the present study.

Raw Materials Analyses (% by Mass)		
Fe_2O_3	93.09	min
FeO	0.1	max
MnO	0.22	max
SiO_2	1.93	max
Al_2O_3	1.76	max
S	0.012	max
P	0.01	max
TiO_2	0.55	max
CaO	1.72	-
MgO	0.39	-
Na_2O	0.08	-
V_2O_5	0.07	-
K_2O	0.08	-

The employed OSBF furnace was produced by TENOVA Pyromet (Figure 3).

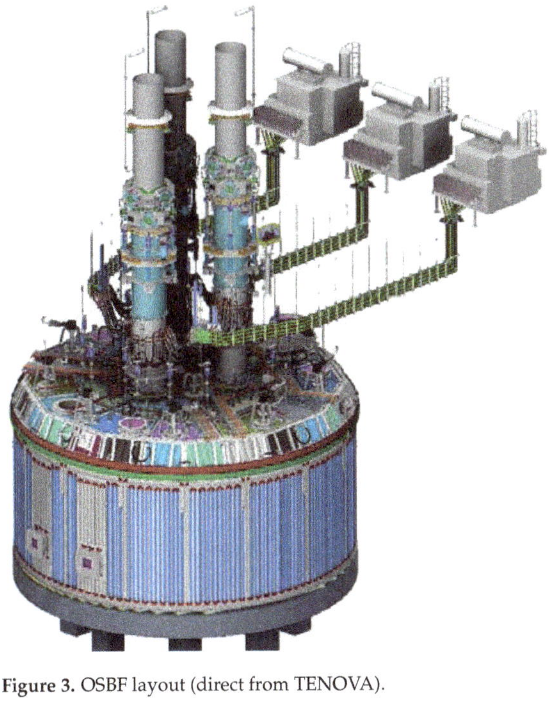

Figure 3. OSBF layout (direct from TENOVA).

The main characteristics were: internal shell diameter 20 m; +/−110 MVA AC furnace; 75 MW peak power input; 3 single phase transformers; 6 feed bins and 12 feed chutes; MgO conductive lining; 1700 mm Soderberg Electrode (TENOVA, Castellanza, Italy) (Figure 4).

Figure 4. Soderberg electrodes.

Autofurn™ furnace control system. The gas cleaning is performed through Wet gas scrubbing plant, twin venturi plant design. Chemical energy contained in off-gas is used as an energy source for the DRI reactor.

Figure 5 shows the TENOVA-HYL hydrogen-based direct reduction plant employed in the present study.

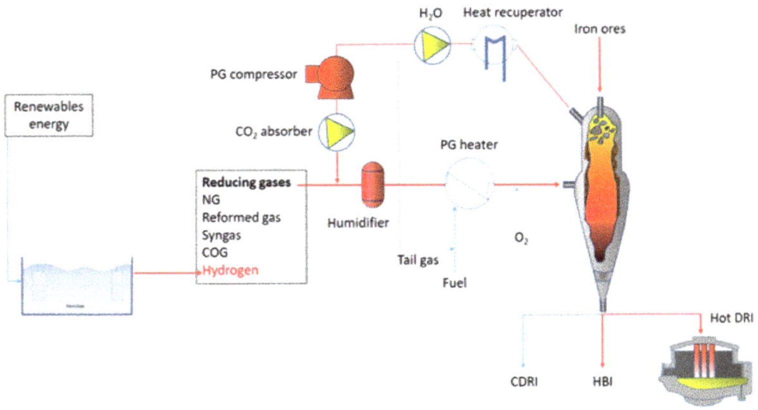

Figure 5. Schematic of TENOVA HYL-DRI plant.

Due to the high volatility of hydrogen, the pressure employed into the reactor chamber is fundamental for the whole process efficiency. As a matter of fact, the reactor described in Figure 5 has an operating pressure of 6–8 bar, reducing iron ores at a temperature of around 1050 °C. The high pressure of this configuration solves the problem of the gas volatility leading to a remarkable increase in the process efficiency with respect to room pressure configurations. In addition, the described configuration allows for the employment of different reducing gases with different percentages of hydrogen additions. This allows for high flexibility in the use of many energy sources. The iron ore dimensions are in the range 3.2–18 mm. As shown in Figure 5, the plant is equipped with a carbon capture and storage/utilization device in the case of natural gas addition to the hydrogen reducing agent. The schematic of the carbon dioxide removal and utilization is shown in Figure 6.

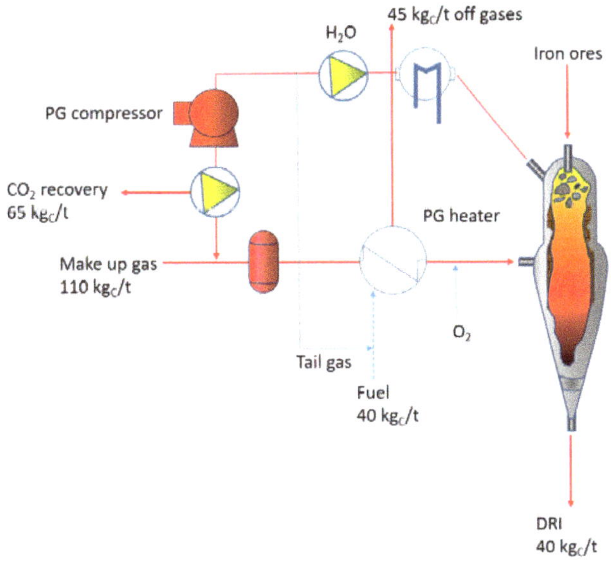

Figure 6. Carbon balance in the HYL-DRI plant; (PG is the pressurized gas).

In this configuration, carbon dioxide is captured at a rate of 45% and total emissions to the atmosphere reach 30%. The remaining carbon finishes in the direct reduced iron.

Figure 7 shows the comparison of carbon fluxes in the plant schematized in Figure 5 with the equivalent MIDREX plant.

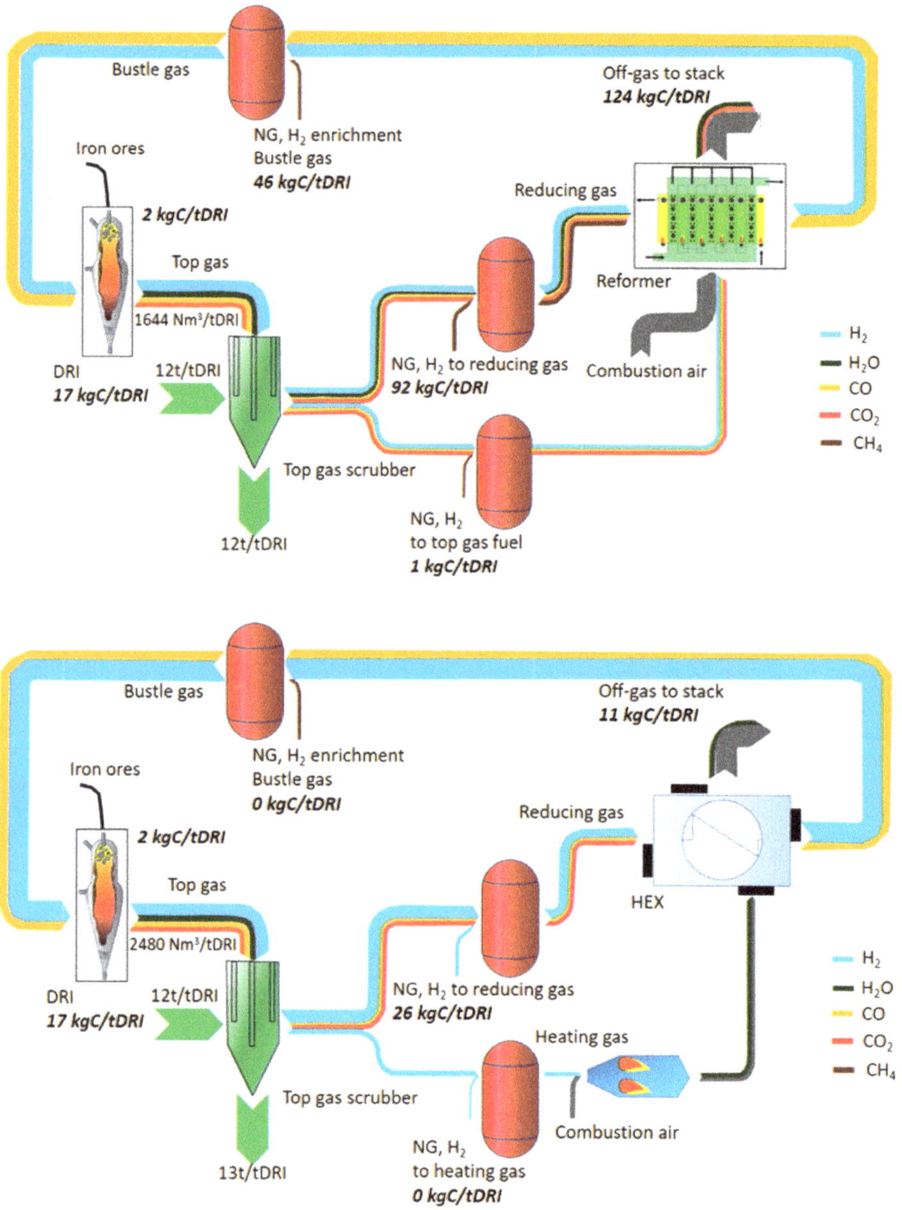

Figure 7. Schematic of the NG—(**up**) and H_2-based (**down**) Midrex plants.

3. Results and Discussion

The main unit operations of the process comprise DR-shaft, reformer, top gas scrubber, product gas compressors, and heat exchangers. The reformer allows for the treatment of the natural gas giving the energy required for the syngas production. The top gas scrubber allows one to reduce the water vapor inside the reactor in order to control the temperature during the reducing operations. As a matter of fact, temperature and pressure inside the reactor can be easily controlled. NG can be injected at different process stages: NG as input to the reforming gas, as energy input for heating the reformer or reduction gas heater as well as directly to the bustle gas before entering the shaft furnace. In this configuration, hydrogen is injected into the natural gas before the treatment in the reformer. Once hydrogen is employed instead of natural gas, the reformer is replaced by a gas heater that is a combination of a gas burner, mixer, and heat exchanger model (this is the main solution designed for the Midrex plants). The energy supplied to the heater can also be provided by NG. Both the modelled configurations are based on the same assumptions in order to have comparable results; these are the same solid input material compositions and amounts as well as the same basic assumptions for the DR shaft (e.g., reduction degree, carburization behavior, temperature distribution, and so on), top gas scrubber (temperatures and pressure levels) and gas burner (excess air). In these configurations, about 30% of the NG can be replaced by hydrogen without any process changes. Once hydrogen is added to the natural gas, carbon monoxide decreases in the reducing flow. In fact, if natural gas is employed, carbon monoxide and hydrogen are produced in the reformer; so, as the hydrogen addition is increased, less carbon monoxide is employed as reducing gas. Only a small and constant amount of NG is added for the enrichment of the bustle gas before entering the shaft. In the case of hydrogen directly injected in the reactor, it needs to be previously heated as indicated before. The second schematic belongs to a design for an input of about 95% hydrogen. In this case, natural gas is employed just to retain the desired temperature levels and the carburation of the direct reduced iron. In fact, if only hydrogen is employed, higher recirculating gas flows are necessary to retain the required process temperatures.

Natural gas that is employed during the reforming operations, the enrichment for iron carburization or the gas heating represents the main carbon source in the natural gas-based direct reduction technology. In this configuration, the main carbon output is measured in the reformer off-gas. About 124 kg C/t DRI, which is equivalent to about 453 $kgCO_2$/tDRI, is emitted in this section. In comparison, the carbon output of the DR-H_2 process is almost equally distributed between the DRI and stack emissions (released by using the top gas as combustion gas for the heater), representing 17 and 11 kg C/tDRI, respectively. The main carbon source in this case is also NG, which is required for maintaining the carbon content of the DRI (Rechberger et al., 2020).

As mentioned, the TENOVA plant can work with different concentrations of hydrogen in the feeding gas. The plant data as a function of the feeding gas composition are listed in Table 2.

Table 2. DRI (direct reduced iron) properties and plant characteristics as a function of the gas composition (MTZ is the metallization percentage).

	100% NG		55% H_2		73% H_2
Product	**Hot DRI**	**Cold DRI**	**Hot DRI**	**Cold DRI**	**Cold DRI**
DRI quality	94% MTZ, 3.5% C	94% MTZ, 3.5% C	94% MTZ, 2.5% C	94% MTZ, 1.3% C	94% MTZ, 1.0% C
NG consumption (GJ/t_{DRI})	9.96	9.58	3.97	3.9	2.27
H_2 consumption (GJ/t_{DRI})	-	-	4.86	4.85	5.99
Total energy (GJ/t_{DRI})	9.96	9.58	8.83	8.75	8.26
Electricity consumption (kWh/t_{DRI})	73	73	60	63	60
Oxygen consumption (Nm^3/t_{DRI})	58	57	-	-	-
H_2/CO ratio in the reactor	50/12	50/12	70/5	70/5	75/4

By increasing the hydrogen content in the gas mixture, the total energy required for the process decreases. So, high levels of electricity saving are recorded.

By employing this DRI reactor products as feedstock materials in the BF or in the EAF, a strong reduction in the carbon dioxide emissions can be underlined. The results compared to the traditional BF-BOF route are described in Figure 8.

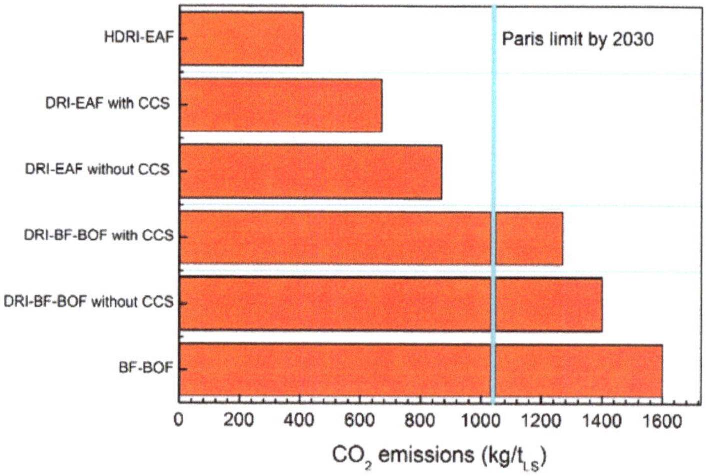

Figure 8. Carbon dioxide emissions comparison for different routes.

The HDRI (Hydrogen DRI) data refer to a gas mix with 70% H_2 and 30% NG. All the data take into account direct and indirect emissions with the assumption of 0.5 $kgCO_2/kWh$. The HDRI-EAF route has a carbon dioxide emissions level of 25% with respect to the traditional BF-BOF route. Hydrogen ironmaking is then considered the future of this very broad field [12].

When talking about the conversion of steel production from the traditional BF-BOF route toward the DRI-based one, it is fundamental to focus on the raw material use, availability and efficiency. First of all, by employing electric arc furnaces, the greenest way is the steel recycling. Irrespective of this, many recent reports show that the concentration of tramp elements in scrap is increasing and incompatible with many high-quality steel grades, so dilution with virgin iron (HBI, DRI, Hot Metal, Pig Iron) is needed. An example is given in Table 3 with special attention on copper concentration.

Table 3. EAF (electric arc furnace) data with different charge mixes for various liquid steel qualities.

			Grade 1	Grade 2	Grade 3	Grade 4	Grade 5
Cu content range (liquid steel)	ppm		<500	700–1000	1000–1500	1500–2000	2500–3000
Cu content target (liquid steel)	ppm		450	850	1250	1750	2750
Design charge mix		Cu content	Grade 1	Grade 2	Grade 3	Grade 4	Grade 5
Bushelling	%	1000 ppm	12	7	18	25	0
Obsolete scrap	%	3500 ppm	5	18	13	19	50
Obsolete scrap, heavy melting	%	4000 ppm	1	1	1	1	1
Collected scrap	%	4500 ppm	0	0	9	14	15
Internal return	%	1500 ppm	4	4	4	4	4
HBI	%	0	48	41	25	8	0
Pig iron	%	0	30	30	30	30	30
EAF output data			Grade 1	Grade 2	Grade 3	Grade 4	Grade 5
Tap to tap	min		46	46	46	46	46
Electricity consumption	kWh/t		404	398	387	377	368
Electrode consumption	kg/t		1.18	1.16	1.14	1.11	1.1
Oxygen consumption	Nm^3/t		39.2	38.8	36.9	35.1	35.6
Carbon consumption	kg/t		12	12	12	12	12
NG consumption	Nm^3/t		2.5	2.5	2.4	2.3	2.3
Lime	kg/t		33.8	32.5	28.8	28.8	28.1
Dolomite	kg/t		13.8	15.6	15.6	15	12.5

All the data belong to a new-generation electric arc furnace with the following characteristics: type, full-platform AC furnace with EBT and shell diameter of 8900 mm; tap size, 240 t; yearly productivity, 2.2 Mt; transformer rating, 240 MVA; injection system, 17,000 Nm^3/f of oxygen.

Now, recent studies show that the forecast for the availability of DR-grade pellets is very flat. The ideal chemical composition for DR-grade pellets has Fe > 67%, $SiO_2 + Al_2O_3 + TiO_2 < 3\%$ and $p < 0.03\%$. The estimated seaborne DR-grade pellet demand and exports are shown in Figure 9 [13].

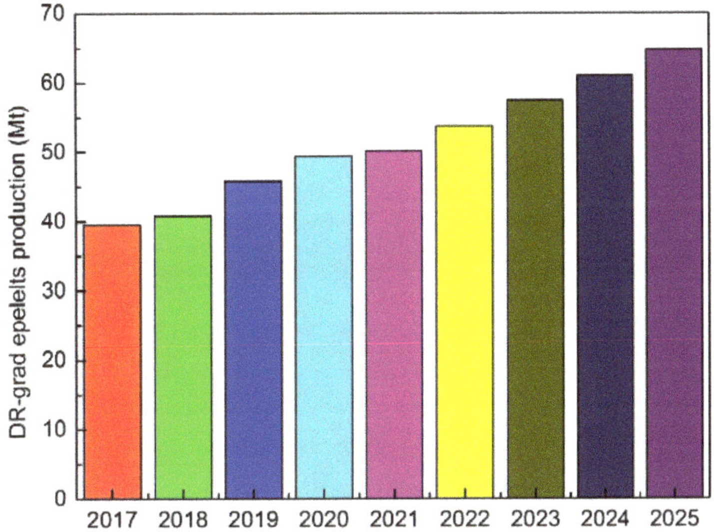

Figure 9. Direct reduced (DR)-grade pellet production forecast.

So, the availability of DR-grade pellets is limited with respect to global steel production. So, the main obstacles to the direct conversion of steel production are mainly represented by the availability of raw materials that meet the grade requirements. In fact, successful and productive operation of a DR-EAF line requires using high-grade pellets (gangue less than 5%, possibly basic). In addition, some limitations are indicated for fitting the DR-EAF route in integrated plants. By considering the power grid, 250–300 t EAFs (matching the tapping size of large converters to fit the existing CCM) fit transformers of 200–260 MVA, which has massive impacts on the power grid for flicker generation. The integration of BF and EAF is not so simple; the typical cycle times of BF and EAF do not match. EAF typically requires 45–55 min and weekly maintenance stops for refractory repair/relining. With regard to the installation, a 300 t EAF requires a building having at least 26 mt high crane rails and sufficient lifting capacity to lift the complete shell (about 550 t considering complete shell + lifting jig). Finally, by considering the plant logistic, the hot charge of DRI in EAF implies that the DR module is close by, which is typically difficult if the EAF is to be installed in an existing BOF shop to feed the existing downstream equipment.

Taking into account all the described aspects, a good solution appears to be the integration of direct reduction with large smelting furnaces. In this way, BF-grade pellets could be reduced in the DR reactor by overcoming the problem of the availability of high-quality DR-grade pellets. The reduced material is known as DRP (Direct Reduced Pellets). Their refining in the large OSBF allows one to solve the electricity and logistic limitations of the use of EAFs. Given that scientific and technical information is limited, in the following we will give a detailed description of this large smelting furnace.

It is generally observed that the smelting is easier and efficient as the slag liquidus temperature and viscosity are low. These two aspects favor the material separation and flow by lowering the smelting energy requirements. Submerged arc furnaces (SAF) processes are self-regulating, and the material gains heat up to the optimal conditions to flow. All the reducing reactions take place on the formed coke bed. The temperatures strongly vary in different zones, going from 1700 °C at the electrode tip to 500 °C in the upper part of the burden. This is very important because the temperature distribution governs the process efficiency. In this context, the electrode regulation is fundamental for both the furnace productivity and the electrode life in service.

Other factors governing the temperature distributions are the current density, the slag composition and the electrode to metal distance. As a matter of fact, as the basicity of the slag increases, its melting point increases and consequently the reaction zone temperature increases. Both temperature and basicity are fundamental for the evolution of the reduction reaction kinetics. In general, an increase in temperature and basicity leads to reduction increase and to reduction rate decrease [14].

The power input can be controlled by varying the electrode position; this allows one to vary the arc resistance and thereby control the arc current. As the temperature increases, the reduction degree increases if enough coke is provided for the CO formation. The carbon also improves the bath conductivity by increasing the process efficiency.

The main difference between OSBF and SAF is that, in the first one, the electrodes are positioned at the top of the furnace, so they are not submerged in the burden. This arc configuration is known as brush arc or open arc. This arc is produced by varying the position of the electrode tip (Figure 10).

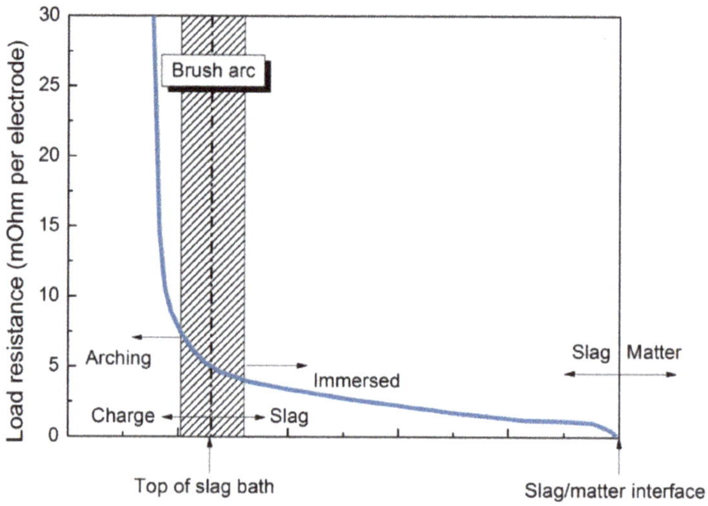

Figure 10. Operational resistance vs. electrode tip position.

This technology offers some key advantages. The mixture inside the furnace does not influence the electrical current because the electrodes are not submerged. For this reason, fines can be also easily melted. On the contrary, in SAF the burden permeability can be a limit for the process. Given that an open slag bath is created, slag and metal separate very quickly and easily. The feed mixture influences the process chemistry (that can be precisely controlled) and then the process efficiency. Given that the electrodes are not submerged, it is possible to fix the power input; so, the burden properties do not influence the process. Therefore, as a general behavior, precise and improved furnace control is allowed. The reduction kinetics can be easily governed by the DRI percentage injection.

Irrespective of this, by operating with an open arch, some limitations should be underlined. The open bath area results in high heat losses through the roof. This leads to energy dispersion and rapid refractories deterioration that can be reduced by appropriate cooling.

The material is fed to the furnace through the feed pipes. The feed piles that are produced are consumed from the bottom of the furnace as well as from the bottoms of the piles. The combination of large power input and arching on the more refractory slag allows for superheating the iron (Figure 11).

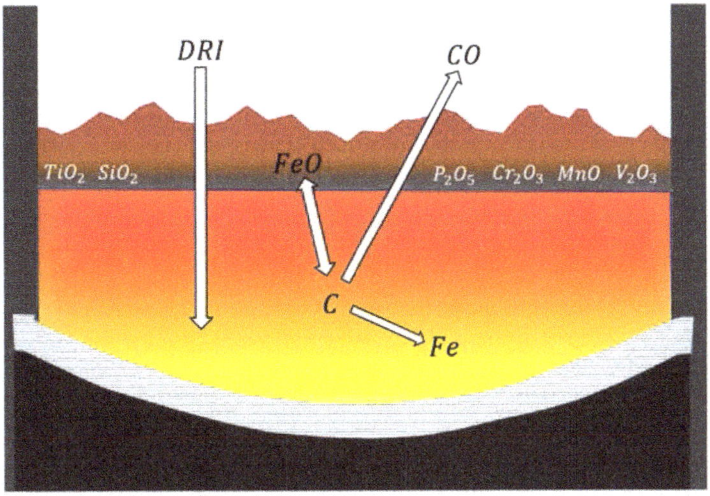

Figure 11. DRI (direct reduced iron) melting.

As the reaction temperature increases, higher carbon contents are revealed in the iron through dissolution and because of the reduction of the iron melting point. This tends to increase the iron yield, since launder and ladle sculling are minimized. The OSBF also achieves the improvement of sulfur removal thanks to the employed higher slag basicity:

$$[S] + \left(O^{2-}\right) = \left(S^{2-}\right) + [O] \tag{1}$$

The main reaction taking place in the OSBF is obviously the reduction of iron oxide:

$$FeO + C \rightarrow Fe + CO \tag{2}$$

Some iron oxide normally remains unreduced and passes to the slag. In addition, depending on the oxygen potential of the slag, the other reduction reactions involving the additional charge compounds are:

$$TiO_2 + 2C \rightarrow Ti + 2CO \tag{3}$$

$$TiO_2 + \frac{1}{2}C \rightarrow TiO_{1.5} + \frac{1}{2}CO \tag{4}$$

$$SiO_2 + 2C \rightarrow Si + 2CO \tag{5}$$

$$V_2O_3 + 3C \rightarrow 2V + 3CO \tag{6}$$

$$MnO + C \rightarrow Mn + CO \tag{7}$$

$$Cr_2O_3 + 3C \rightarrow 2Cr + 3CO \tag{8}$$

$$P_2O_5 + 5C \rightarrow 2P + 5CO \tag{9}$$

Obviously, the reduction efficiency depends on the oxygen activity and on the furnace temperature.

OSBF is a successful method to produce hot metal from DRI and provides the option to modify slag chemistry, thereby giving greater operational flexibility. OSBF large bath surface area results in a lower slag rise, which reduces the risk of slag foaming. OSBF has a much longer campaign life, reducing the logistic requirements associated with an EAF relining (crane, building, frequent downtime, etc.). Electrical consumption is similar (10% in difference), but the operating costs using Sőderberg electrodes used on OSBF compared to graphite electrodes is lower, although variable. The OSBF off-gas stream has a high CO content that can be used as fuel for the DRI process gas heater. The BF pellets to be employed in the DRI-OSBF route allows one to reduce the raw material cost by 20% (from 150 to 120 USD/t from DR-grade to BF-grade pellets). In addition, it is demonstrated that the HM produced by the new DRP-OSBF can be merged with the flow coming from the existing BFs. HM analysis and tapping temperature can be adjusted to optimize compatibility and minimize CO_2 generation.

The last results belonging to a 2.5 Mt/year DRI reactor integrated with OSBF and BOF show the CO_2 emissions reduction indicated in Figure 12.

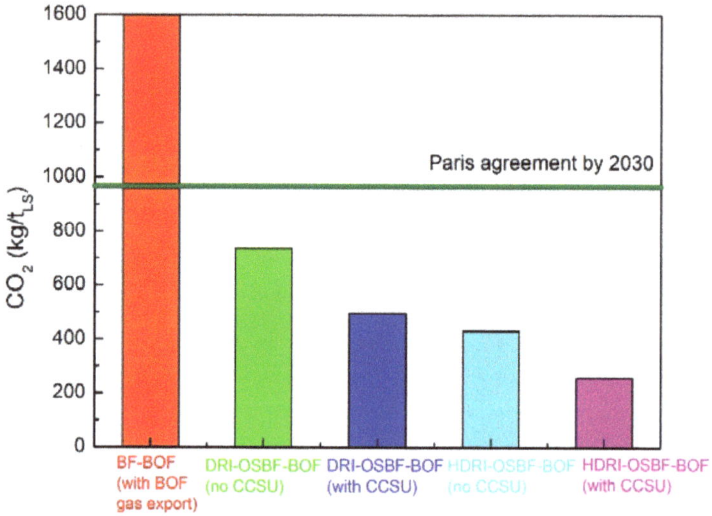

Figure 12. CO_2 emissions for different routes.

Moreover, the costs related to the operations are reduced because of the long-term steady-state operations of OSBF that require refractories replacement each 5–7 years. On the contrary, shell maintenance and repairs in EAFs are required after every 100–120 heats (every 2 weeks). A detailed analysis of the whole process shows the significance of employing the raw materials listed in Table 4.

As mentioned above, the temperature control is fundamental. Important energy losses can be recorded once new material is fed into the furnace, the power is needed to melt the solid new material and it cannot be sufficient to retain the optimal temperature in the already meld bath. This severe condition (that can lead to a reduction of 30% in the power efficiency) is known as "overcharged condition". Obviously, the charge flow must be precisely controlled because, on the contrary, insufficient charge can lead to undesired increases in the temperature with consequent damage in the refractories. This aspect must be carefully considered, and the main factors affecting the furnace reduction reactions are the fed material chemistry, the temperature profile in the furnace and the secondary melting effects. By considering the material chemistry, it is fundamental to continuously monitor the material flow and the composition; both of these factors influence the reaction evolution. The effect of the fresh ores' flow on the furnace chemistry can lead to the following three different conditions: overcharging accompanied with a drop in the energy provided to the material in the furnace, undercharging accompanied with an increase in the energy and furnace temperature, and balanced charging accompanied with an optimal melting of the material with sufficient power and heating.

Table 4. Raw materials employed in the DRP-OSBF route.

Raw Materials Analyses (% by Mass)		
	Iron Ore DRI	
[Fe]	74.44	
FeO	6.1	
SiO_2	11.75	
Al_2O_3	0.57	
MgO	1.03	
CaO	0.51	
TiO_2	0.05	
Mn	0.05	
P	0.01	
S	0.003	
[C]	4.7	
K_2O	0.1	
Na_2O	0.1	
LOI	0.47	
	Fluxes	
	Burnt lime	Dolomite
Fe_2O_3	0.9	1.8
SiO_2	3.4	1.9
Al_2O_3	1.7	1.9
MgO	1.2	35.4
CaO	92.7	59.1
H_2O	0.5	0.5
	Reductants	
	Coke	Coal
FC	80.9	54.7
Volatiles	1.5	20.3
Ash	17	25
P	0.02	0.02
S	0.6	0.6
H_2O	0.5	0.5
	Ash	
Fe_2O_3	7.3	7.9
SiO_2	50.1	33.4
Al_2O_3	35.7	19.9
MgO	1.2	0.7
CaO	5.2	36
TiO_2	0	0.9
K_2O	0	0.9
P_2O_5	0.5	0.3

The precise correlation between these conditions and the charge feed rate is shown in Figure 13 [15]. Essentially, if the power input is constant, once new material is fed into the furnace, the energy provided to the melt decreases. In overcharged conditions, the material tends to accumulate toward the electrodes by producing the bath shrinkage. In these conditions, the temperature decreases and the reduction extent is consequently delayed. The first indicator of this condition is that the percentage of titanium oxide under reduction rapidly decreases. In addition, the equilibrium shifts toward more oxidizing conditions and an increased percentage of iron oxide is revealed in the slag. In addition, very different reacting conditions are experienced in the bath. Carbon accumulates by producing the refractoriness of the bath and the increase in its resistance. So, the slag control is fundamental to the overall process. It is believed that the slag chemistry is governed by the magnesium and calcium oxides as shown in Figure 14.

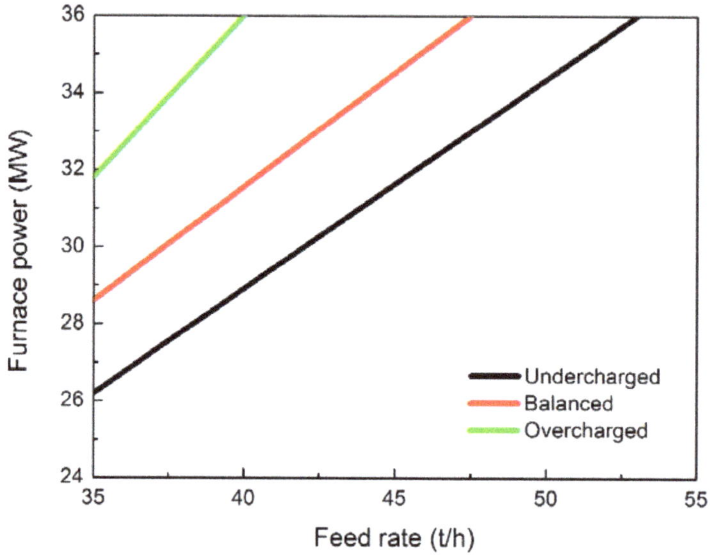

Figure 13. Relationships between the material feed rate and power for the different conditions.

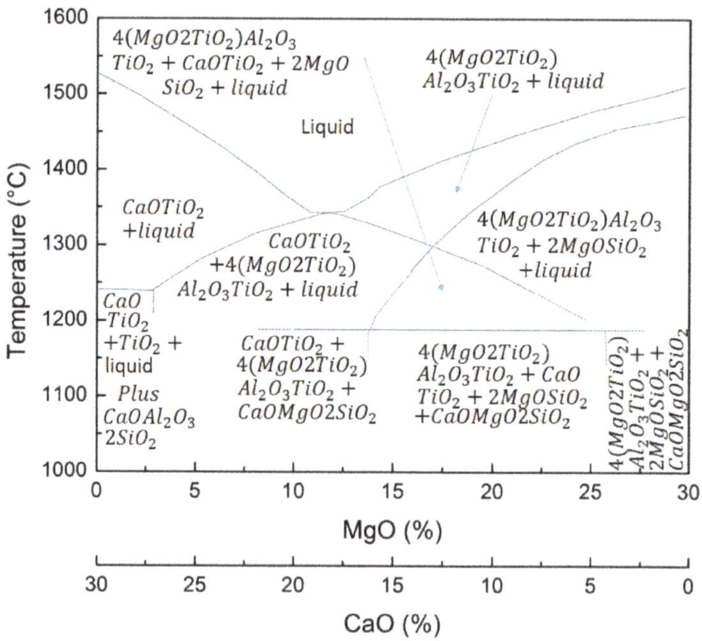

Figure 14. Effect of CaO and MgO percentages on the liquidus temperature of the slag.

The liquidus temperature is controlled by the CaO percentage irrespective of MgO/SiO_2 ratio. By varying the CaO percentage from 20 to 14, the liquidus temperature decreases from 1600 to 1350 °C. The SiO_2 percentage governs the spinel transformation and the liquidus temperature (going from 1500 to 1350 °C if the percentage varies from 18 to 24%). Taking into account the combined effect of CaO and MgO, it can be underlined that the slag liquidus temperature varies from 1520 °C at 30% CaO and 0% MgO to the minimum of 1340 °C at 18% CaO and 12% MgO to a new increase at 1505 °C at 0% CaO and 30% MgO.

Many studies indicate that the ratio Ti^{3+}/Ti^{4+} leads to the variation of the slag temperature. The relationship between the ratio and the TiO_2 content is shown in Figure 15.

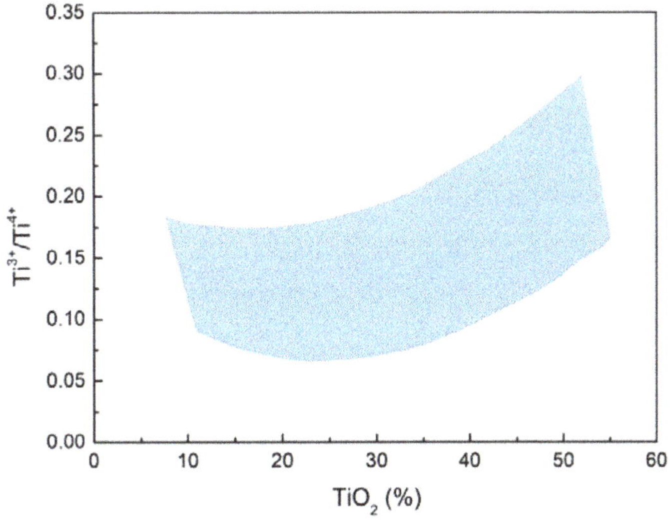

Figure 15. Effect of TiO_2 content on the oxidation state of titanium.

As can be seen from the figure below, the belt variation is due to the basicity of the slag. This is a key factor for the present analyses. Generally, small variation in the basicity of the slag leads to important differences in the liquidus temperature. The summary for selected conditions is described in Figure 16.

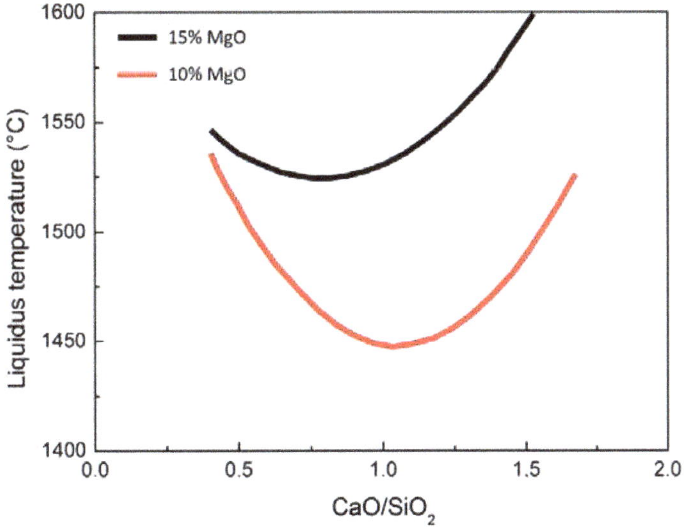

Figure 16. *Cont.*

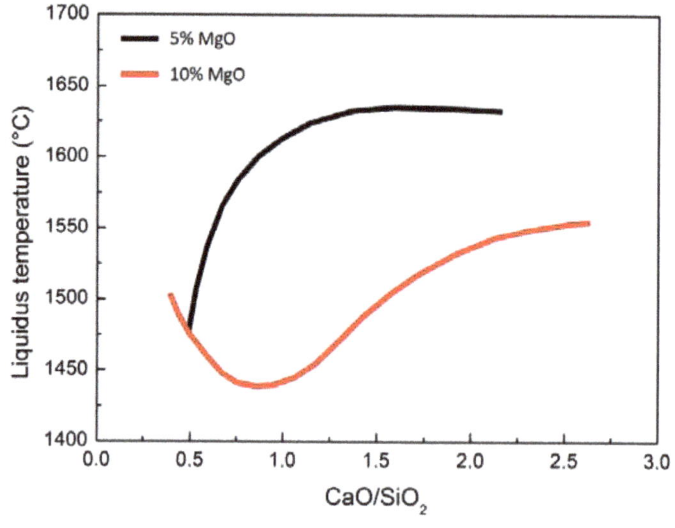

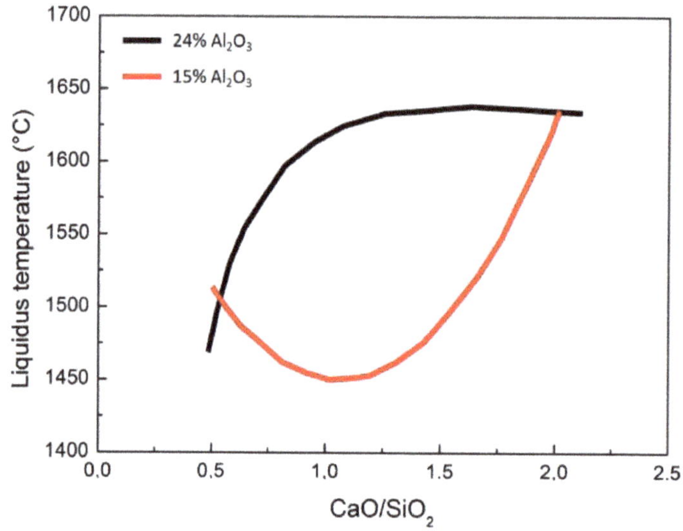

Figure 16. Effect of slag basicity with diffferent percentages of magnesium oxide (**up and central pictures**) and alumina (**down**) on the liquidus temperature.

The most important reduction state variable is the slag basicity. Now, the liquidus temperature is influenced by the contemporary effect of the slag basicity and oxidation state of titanium; it can be concluded that the liquidus temperature in the case of slags with high basicity ($CaO/SiO_2 = 1.4$) decreased with increased reduction (higher Ti^{3+}/Ti^{4+} ratios). As a consequence, in the case of low-basicity slags ($CaO/SiO_2 < 0.8$), the liquidus temperature decreases and the reduction increases. So, it is possible to modify the liquidus temperature and the Ti^{3+}/Ti^{4+} ratio by changing the basicity ratio. The furnace chemistry is normally controlled by precisely controlling the material input and the electrode power. During open slag bath furnace operations, product quality and productivity are governed by the tuning of the power/feed ratio. First of all, the raw material composition sensitively influences the overall process. In addition, the feed rate leads to strong variations in the energy consumption. Obviously, in the case of pre-reduced iron ores, the compositional control is fundamental to the optimal evolution of the metal and slag transformations.

As a general trend, the integrated BF-BOF traditional route is not so sensitive to the gangue elements percentage (SiO_2, Al_2O_3 and TiO_2) in comparison to the new developed DR-EAF route. During BF transformations, the main percentage of the gangue compounds are transferred to the slag (in the order of 150–300 kg/t) with a total basicity in the order of 1 and with very low iron content (around 1%). In this way, the silicon content in the liquid metal transferred to the BOF falls in the range 0.2–0.7%. As a general trend, as the silicon content in the BF decreases, energy demand decreases and the productivity increases [16]. Much research evidence (the most recent belonging to the blast furnace in Luleå) shows that the minimum limit of SiO_2 in BF pellets is around 2%.

Replacing sintered material with pellets has exceptional influences on the energy consumption in the BF. In this configuration, the silica content in the pellets is not so important for the BF route. On the other hand, in the case of the direct reduction route, this content is very important because the process takes place at the solid state. In this case, all the material is sent to the electric furnace, where the slag increases as the gangue results higher.

The silicon in the hot metal and silica and other gangue elements in the DRI affect the amount of slag formers that has to be added to the BOF or the EAF in order to reach the MgO saturation of the slag, and the basicity is typically around 2 or more. The MgO saturation varies with the temperature of the slag and composition as a function of the different compounds, such as FeO, CaO, SiO_2, P_2O_5, Al_2O_3, TiO_2 and VO_2. In this way, the process must be controlled through lime addition, so, as the gangue content increases, the lime percentage must be increased [17].

Pellet reduction is strongly influenced by the main properties, such as porosity, particle size and chemical composition, of the phases present in the ore. This must be maximized in order to reach high reducibility, productivity and high metallization degrees of the direct reduced iron.

One of the technical limitations is that, as the metallization degree increases, iron losses increase because of the gangue formation. The metallization obviously depends on the tailing grade of the ores that can fall in the range 10–20%.

Obviously, the slag formation in the EAF is largely influenced by the quality of the scraps employed during the process. This leads to the formation of 70 kg of slag per ton of steel in the case of using 100% of scraps in the EAF. For this reason, scraps must be well treated before entering the EAF in order to reduce this inconvenience.

Given this, a reasonable amount of slag at the end of the DR-EAF process is around 150 kg/t of steel. This is common in the case of low-carbon steel production, where high percentages of FeO can be revealed in the slag.

Again, the high percentage of slag leads to a reduced productivity as well as to increased energy and lime consumption with unbeneficial effects on the process economy. By using DRI from higher grade iron ore feed, slag/steel ratios in the order of 100–120 kg/t are possible for high DRI percentage in the charge (over 80%) with corresponding improvements on the abovementioned factors and the resulting economy.

In the new proposed solution, the DR reactor is a TENOVA HYL-type with a design base of 900,000 t/a of DRP. The OSBF has a shell diameter of 16.500 mm equipped with 1700 mm Soderberg electrodes and a wet gas scrubbing plant. The output material shows the composition listed in Table 5.

A deep analysis was conducted in order to evaluate the economic profitability of the solution. First of all, Table 6 indicates the costs of the Hot DRI or Hot DRP to be employed in the EAF and in the OSBF, respectively.

Table 5. Furnace output in the DRP-OSBF route.

Furnace Output (% by the Mass)	
Alloy	
Fe	95.73
Si	0.2
C	4
P	0.012
S	0.05
Mn	0.05
Slag	
FeO	1
SiO_2	39.37
Al_2O_3	3.12
MgO	13.1
CaO	43.18
P	0.001
S	0.01
TiO_2	0.17
Dust	
Fe	50.67
SiO_2	20.68
Al_2O_3	0.59
MgO	6.68
CaO	8.12
P	0.024
S	0.006
C	3.45

Table 6. Direct reduction costs.

Process	**(NG USD/GJ)**		**DR-EAF**		**DR-OSBF**	
Product Characteristics			**Hot DRI**		**Hot DRP**	
Metallization			94%		94%	
Carbon			2%		5%	
Temperature			600 °C		600 °C	
Specific consumption DR plant	Unit/t	Unit cost (USD/Unit)	Specific consumption (Unit/t DRI)	Cost (USD/t DRI)	Specific consumption (Unit/t DRP)	Cost (USD/t DRP)
DRI pellets	tonne	150	1.45	217.5	-	-
BF pellets	tonne	120	-	-	1.4	168
NG	Gcal	29.13	2.4	69.9	2.52	73.4
Cocking coal	kg	0.140				
Cocking cost	kg	0.017				
PCI	kg	0.120				
Electricity	kWh	0.06	108	6.48	113	6.78
Oxygen	Nm^3	0.1	66	6.6	55	5.5
Water	m^3	0.02	1.4	0.03	1.4	0.03
Fluxes/binders	USD	0.04				
Labour	USD/m-h	20	0.15	3	0.15	3
Refractories	USD	0.5				
Maintenance	USD	1	3.3	3.3	3.3	3.3
Nitrogen	Nm^3	0.05	22	1.1	22	1.1
Other consumables	USD	1	0.97	0.97	0.97	0.97
Credit CO_2	tonne	30	−0.8	−24	−0.8	−24
Steel direct cost	USD/t steel		284.9		238.1	

First of all, the costs related to the employment of BF pellets are markedly lower with respect to the DRI ones. The costs related to the NG are slightly lower in the case of DRI pellet reduction. A reduction of almost 20% in oxygen consumption is recorded in the case of DRPs. The costs of DRP are 18% lower with respect to DRI ones. In Table 7, the costs belonging to the EAF and OSBF routes are listed.

Table 7. Costs comparison between DRI-EAF and DRP-OSBF.

Process	(USD/t HM)		Hot DRI-EAF		Hot DRP-OSBF	
Specific Consumption	**Unit/t**	**Unit Cost (USD/Unit)**	**Specific Consumption (Unit/t LS)**	**Cost (USD/t LS)**	**Specific Consumption (Unit/t HM)**	**Cost (USD/t HM)**
Return scrap charge	kg	0	50	0	-	-
Imported scrap charge	kg	0.245	175	42.9	-	-
DRI charge	kg	0.285	930	265	-	-
DRP charge	kg	0.238	-	-	1150	273.8
Electricity	kWh	0.06	420	25.2	660	39.6
Graphite	kg	0.13				
Lime	kg	0.08				
Dolo-lime	kg	0.03	32	0.96	10	0.3
Magnesite	kg	0.49	5	2.45	1.4	0.69
Oxygen	Nm^3	0.1	35	3.5	-	-
Electrodes (graphite)	kg	5	1.1	5.5	-	-
Electrodes (Soderberg)	kg	0.43	-	-	3	1.28
Refractories	USD			4		1.3
Maintenance	USD			5		1.2
Direct labour	USD			4		2.7
Cost for slag disposal	USD	18	0.18	3.2	-	-
Credit granulated slag	USD	22.2	-	-	−0.17	−3.77
Total LS cost	USD/t		361.7		317.1	

First of all, no scraps should be employed in the OSBF route. The electrode consumption and the corresponding cost differences should be underlined.

Actually, the BOF route must be considered, the OSBF metal being very similar to cast iron. The costs relative to the BOF operations for the DRP-OSBF material are listed in Table 8.

Table 8. Costs for the OSBF-BOF route.

Process			OSBF-BOF	
Specific Consumption	**Unit/t**	**Unit Cost (USD/Unit)**	**Specific Consumption (Unit/t LS)**	**Cost (USD/t LS)**
Return scrap charge	kg	0	50	0
Imported scrap charge	kg	0.245	150	36.8
Hot metal-SAF	kg	0.317	912	289.2
DRP charge	kg	0.238	-	-
Lime	kg	0.08	60	5
Dolo-lime	kg	0.03	40	1.2
Oxygen	Nm^3	0.1	55	5.5
Refractories	USD			1.65
Other costs	USD			2.5
Maintenance	USD			3.3
Direct labour	USD			2
Cost for slag disposal	USD	18	0.13	2.34
Total LS cost	USD/t		349.3	

Given the low quality of raw materials, the problem of volatile trace elements is crucial for the environment issues and for the final quality of liquid steel [18]. During the steelmaking processes, many heavy metals and metalloids are emitted into the environment. Due to the high prices and low availability of high-grade ores, industries will go to employ increased quantities of low-grade ores characterized by increased impurity percentages (containing high Pb, Cd, Cr and As). These heavy metals and metalloids are highly toxic, non-degradable and highly dangerous once released into the ground and water sources. On the other hand, the volatile compounds are generally captured and treated in the scrubbers [19].

The volatilization degree as a function of temperature (*T*) and of the volatilization extent (α) can be expressed as follows:

$$\frac{d\alpha}{dt} = k(T)f(\alpha) \tag{10}$$

where *k*(*T*) is the volatilization rate constant in the unit s^{-1}, and $f(\alpha)$ is the mechanism function.

The volatilization rate constant *k*(*T*) has an Arrhenius-type dependence from the temperature:

$$k(T) = A * exp\left(-\frac{E}{RT}\right) \tag{11}$$

where *A* is the pre-exponential factor (s^{-1}), *E* is the apparent activation energy (kJ mol^{-1}), and *R* is the universal gas constant. In non-isothermal conditions, the heating rate can be considered constant:

$$\beta = \frac{dT}{dt} = constant \tag{12}$$

In this way, the volatilization degree can be expressed by:

$$\frac{d\alpha}{dt} = \frac{A}{\beta} * exp\left(-\frac{E}{RT}\right) * f(\alpha) \tag{13}$$

By integrating the differential form $f(\alpha)$, the reaction model can be obtained:

$$G(\alpha) = \int_0^{\alpha} f(\alpha)^{-1} d\alpha = \frac{A}{\beta}\int_0^{T} exp\left(-\frac{E}{RT}\right)dT \tag{14}$$

Performing transformations and approximations including the Coats–Redfern integration:

$$ln\left[\frac{G(\alpha)}{T^2}\right] = ln\left(\frac{AR}{\beta E}\right) - \frac{E}{RT} \tag{15}$$

The parameters *E*, *A* and G(α) (or $f(\alpha)$) are the kinetic triplet to be determined during the kinetic analysis.

In general, activation energies higher than 40 kJ/mol indicate that the rate-controlling step is the chemical reactions; on the other hand, for energy values lower than 20 kJ/mol, the diffusion is the rate-controlling step.

Many experimental results show that the volatilization of elements such as S, Li, Sn and Pb can be described through a diffusion model. In fact, the activation energy values of Pb and S volatilization are 26.75 and 19.22 kJ/mol, respectively, while Li and Sn show a lower value of 11.65 kJ/mol.

Now, during the overall process, the iron ore particles are cracked through many different mechanisms, such as thermal gradients inside each particle, variations of swelling rates due to the minerology of the ores, local pressures due to volatilization and diffusion processes. This macroscopic thermal cracking takes place with iron oxide reduction and with all the other chemical reactions. So, the volatile element behavior influences the overall process. They can be directly volatilized due to the temperature increase or they can be directly volatilized during the thermal cracking. The rate-controlling step for elemental volatilization could be one of these two mechanisms.

Obviously, each element acts in different way because of its intrinsic nature. As a matter of fact, elements such as As, S, Li, Sn and Pb show continuous volatilization from room temperature up to 1000 °C. As, Pb and S were volatilized in huge quantities after 500 °C; this is in agreement with the clay decomposition and iron oxide transformation of hematite–magnetite–wustite. On the other hand, Li and Sn start to volatilize at lower temperatures, in the range 200–500 °C. Any other increase in temperature does not accelerate volatilization.

4. Conclusions

A new route for steel production has been described in the present paper. An HYL TENOVA direct reduction reactor employing hydrogen as reducing gas was employed for the reduction of blast-furnace-grade pellets. The reduced material was then melted in an open slag bath furnace for further operations. The employment of hydrogen reduced the energy consumption with respect to the natural gas direct reduction route. Through the OSBF route, high carbon liquid metal is obtained with a remarkable difference with respect to the direct reduction of iron ores. Then, the reduced blast furnace pellets can be processed in the BOF by largely reducing the material costs and the energy consumption as well as reducing the overall greenhouse emissions of the steelmaking process. This route allows for steel production at competitive costs compared to the DRI-EAF route with very similar total dangerous emission. It is believed to be an important solution, especially in those situations where the raw material quality can be a problem for the DRI route.

Author Contributions: Conceptualization, P.C., P.S. and P.D.; methodology, P.C, A.P., A.S.; validation, P.C., P.S. and P.D.; formal analysis, P.C., P.S. and P.D.; investigation, A.P., A.S.; resources, P.C., P.S. and P.D.; data curation, P.C., P.S. and P.D.; writing—original draft preparation, P.C.; writing—review and editing, P.C.; visualization, P.C., P.S. and P.D.; supervision, P.C. All authors have read and agreed to the published version of the manuscript.

Funding: This research received no external funding.

Institutional Review Board Statement: Not applicable.

Informed Consent Statement: Not applicable.

Data Availability Statement: Data will be available by contacting the corresponding author.

Conflicts of Interest: The authors declare no conflict of interest.

Abbreviations

BF	blast furnace
BOF	basic oxygen furnace
CCM	continuous casting machine
CCS	carbon capture and storage
DRI	direct reduced iron
DRP	direct reduced pellets
EBT	eccentric bottom tapping
EAF	electric arc furnaces
HM	hot metal
HDRI	hydrogen DRI
HBI	hot briquetted iron
MTZ	metallization
NG	natural gas
OSBF	open slag bath furnace
PCI	pulverized coal injection
PG	pressurized gas
SAF	submerged arc furnace

References

1. Hamadeh, H.; Mirgaux, O.; Patisson, F. Detailed Modeling of the Direct Reduction of Iron Ore in a Shaft Furnace. *Metals* **2018**, *11*, 1865. [CrossRef] [PubMed]
2. Cavaliere, P. *Clean Ironmaking and Steelmaking Processes-Efficient Technologies for Greenhouse Emissions Abatement*; Springer Nature: Cham, Switzerland, 2019; pp. 1–596. [CrossRef]
3. Jiang, X.; Wang, L.; Shen, F.M. Shaft furnace direct reduction technology—Mid-rex and Energiron. *Adv. Mater. Res.* **2013**, *805–806*, 654–659. [CrossRef]
4. Schenck, J.; Lüngen, H.B. Review of application of DRI processes in EC countries. *Chernye Met.* **2017**, *2*, 25–31.
5. Khodabandeh, E.; Ghaderi, M.; Afzalabadi, A.; Rouboa, A.; Salarifard, A. Parametric study of heat transfer in an electric arc furnace and cooling system. *Appl. Therm. Ener.* **2017**, *123*, 1190–1200. [CrossRef]
6. Duarte, P.; Martinez, J. Improving performances and decreasing CO_2 emissions in blast furnaces installations with high-carbon DRI/HBI. In Proceedings of the AISTech-Iron and Steel Technology Conference Proceedings, Warrendale, PA, USA, 26–27 June 2017; pp. 815–824.
7. Chevrier, V. Midrex H_2 TM: Ultimate low-CO_2 ironmaking and its place in the new hydrogen economy. In Proceedings of the AISTech-Iron and Steel Technology Conference; 2018; pp. 725–729. Available online: https://www.midrex.com/tech-article/midrex-h2-ultimate-low-co2-ironmaking-and-its-place-in-the-new-hydrogen-economy/ (accessed on 3 January 2022).
8. Zare Ghadi, A.; Valipour, M.S.; Biglari, M. CFD simulation of two-phase gas-particle flow in the Midrex shaft furnace: The effect of twin gas injection system on the performance of the reactor. *Int. J. Hydrogen Energy* **2017**, *42*, 103–118. [CrossRef]
9. Hölling, M.; Weng, M.; Gellert, S. Analysis of sponge iron production with H_2. *Chernye Met.* **2018**, *3*, 6–11.
10. Prammer, J. The actual problems of current decarbonization. *Chernye Met.* **2019**, *1*, 55–59.
11. Ranzani da Costa, A.; Wagner, D.; Patisson, F. Modelling a new, low CO_2 emissions, hydrogen steelmaking process. *J. Clean. Prod.* **2013**, *46*, 27–35. [CrossRef]
12. Patisson, F.; Mirgaux, O. Hydrogen Ironmaking: How It Works. *Metals* **2020**, *10*, 922. [CrossRef]
13. Poveromo, J. Raw Materials and Iromnaking Global Consulting, «Outlook for DR Grade Pellet Supply for DRI Shaft Furnace Processes. In Proceedings of the STI Webinar, Online, 21 October 2020.
14. Dishwar, R.K.; Sinah, O.P. Effect of basicity on the activation energy during reduc-tion of highly fluxed iron ore pellets. *Fuel* **2021**, *296*, 120640. [CrossRef]
15. Steinberg, W.S.; Pistorius, P.C. Control of open slag bath furnaces at Highveld Steel and Vanadium Ltd: Development of operator guidance tables. *Ironmak Steelmak.* **2009**, *36*, 500–504. [CrossRef]
16. Cavaliere, P.; Perrone, A. Optimization of blast furnace productivity coupled with CO2 emissions reduction. *Steel Res. Int.* **2014**, *85*, 89–98. [CrossRef]
17. Gyllenram, R.; Arzpeyma, N.; Wei, W.; Jönsson, P.G. Driving investments in ore benefi-ciation and scrap upgrading to meet an increased demand from the direct reduction-EAF route. *Min. Econ.* **2021**. [CrossRef]
18. Kan, T.; Strezov, V.; Evans, T.; Zhou, X.; Theiss, F.; Frost, R. Volatilisation of trace el-ements during reduction of iron ore by hydrogen: Statistical analysis, kinetic study and environmental assessment. *J. Clean. Prod.* **2020**, *271*, 122524. [CrossRef]
19. Cavaliere, P. *Ironmaking and Steelmaking Processes: Greenhouse Emissions, Control, and Reduction*; Springer Nature: Cham, Switzerland, 2016; pp. 1–466. [CrossRef]

metals

Article

Effect of Mold Width on the Flow Field in a Slab Continuous-Casting Mold with High-Temperature Velocity Measurement and Numerical Simulation

Jian-Qiu Liu [1], Jian Yang [1,*], Chao Ma [1], Yi Guo [1], Wen-Yuan He [2], Chang-Liang Zhao [2], Ren-Bo Jiang [3] and Yin-Tao Guo [3]

[1] State Key Laboratory of Advanced Special Steel, School of Material Science and Engineering, Shanghai University, Shanghai 200444, China; liujianqiu@shu.edu.cn (J.-Q.L.); mc1457769184@shu.edu.cn (C.M.); guoyi@shu.edu.cn (Y.G.)
[2] Steelmaking Department, Shougang Jingtang United Iron and Steel Co., Ltd., Tangshan 063200, China; hewy2012@sgjtsteel.com (W.-Y.H.); zhaocl5270@sgjtsteel.com (C.-L.Z.)
[3] Steelmaking Department, Tangshan Stainless Steel Co., Ltd., Tangshan 063000, China; jiangrenbo@hbisco.com (R.-B.J.); gyt3133@163.com (Y.-T.G.)
* Correspondence: yang_jian@t.shu.edu.cn; Tel.: +86-21-6613-6580

Abstract: In this paper, the effects of the width of the mold on the surface velocity, flow field pattern, turbulent kinetic energy distribution, and surface-level fluctuation in the mold were studied with measurement of the flow velocity near the surface of the mold at high temperature with the rod deflection method and numerical calculation with the standard *k*-ε model coupled with the discrete-phase model (DPM) model for automobile exposed panel production. Under the conditions of low fixed steel throughput of 2.2 ton/min, a nozzle immersion depth of 140 mm, and an argon gas flow rate of 4 L/min, as the width of the mold increases from 880 mm to 1050 mm and 1300 mm, the flow velocity near the surface of the mold decreases. The flow direction changes from the positive velocity with the mold widths of 880 mm and 1050 mm to the unstable velocity with the mold width of 1300 mm. The calculated results are in good agreement with the measured results. The turbulent kinetic energy near the submerged entry nozzle (SEN) gradually increases, and the risk of slag entrainment increases. Under the conditions of high fixed steel throughput of 3.5 ton/min, the SEN immersion depth of 160 mm, and the argon gas flow rate of 10 L/min, as the width of the mold increases from 1600 mm to 1800 mm and 2000 mm, the velocity near the mold surface decreases. The flow velocity at 1/4 of the surface of the mold is positive with the mold width of 1600 mm, while the velocities are negative with the widths of 1800 mm and 2000 mm. The calculated results are basically consistent with the measured results. The high turbulent kinetic energy area near the nozzle expands to a narrow wall, and the risk of slag entrainment is significantly increased. In both cases of low and high fixed steel throughput, the change rules of the flow field in the mold with the width are basically the same. The argon gas flow rate and the immersion depth of SEN should be adjusted reasonably to optimize the flow field in the mold with different widths under the same fixed steel throughput in the practical production.

Keywords: mold width; flow field in mold; high-temperature measurement; numerical simulation; surface velocity

Citation: Liu, J.-Q.; Yang, J.; Ma, C.; Guo, Y.; He, W.-Y.; Zhao, C.-L.; Jiang, R.-B.; Guo, Y.-T. Effect of Mold Width on the Flow Field in a Slab Continuous-Casting Mold with High-Temperature Velocity Measurement and Numerical Simulation. *Metals* **2021**, *11*, 1943. https://doi.org/10.3390/met11121943

Academic Editors: Pasquale Cavaliere and Alexander McLean

Received: 3 October 2021
Accepted: 28 November 2021
Published: 1 December 2021

1. Introduction

In the production process of automobile exposed panels, improper control of the flow field in the slab continuous-casting mold tends to bring about mold flux entrainment, large inclusions, and bubble-type inclusions in the continuous-casting slab, which may result in linear defects on the hot-dip-galvanized automobile exposed panel, giving rise to serious surface quality problems [1]. The continuous-casting process parameters, such as casting speed, argon gas flow rate, immersion depth, and nozzle structure of the submerged entry nozzle (SEN), can be usually adjusted with the results of mathematical simulation and water modeling to optimize the mold flow field for improving the surface quality of automobile exposed panels.

In recent years, the size and distribution of bubbles in liquid steel and the influence of bubble size on gas–liquid two-phase flow has been widely studied. Li [2] used the 1:4 water model to investigate the two-phase flow characteristics and the bubble size distribution. A mathematical model based on the Euler–Euler approach was developed to analyze the bubble aggregation and breakage in the bubbly flow. Liu [3] used the population balance equations combined with the Eulerian–Eulerian two-phase model to predict the polydispersed bubbly flow inside the slab continuous-casting mold, and the effects of the water flow rate and the gas flow rate on the average bubble size were studied. Cho [4] used the 1:3 water model with the aid of a high-speed video camera and analytical models to investigate the gas pressure, initial bubble size, bubble descending velocity, bubble residence time, and bubble size distribution. Meanwhile, mathematical modeling has been developed to accurately simulate the mechanism of various defects, such as inclusions and the formation of oscillation marks [5–10]. Wu [7,8] considered the evolution of the solid shell, including a fully solidified strand and a partially solidified dendritic mushy zone, which strongly interacts with the turbulent flow and in the meantime is subject to continuous deformation due to the funnel-type mold. Liu [9] developed a Euler–Euler model to predict the argon-steel-slag three-phase flow and the formation of the exposed slag eye in a slab continuous-casting mold.

With regard to the influence of the casting speed, the argon gas flow rate and the immersion depth of the SEN on the mold flow field, Zhang [11] used numerical simulation to study the fluctuation of molten steel on the mold surface at different casting speeds. When the casting speed is low, the change in casting speed has little effect on the fluctuation of the steel-slag liquid level. However, when the casting speed is high, a small change in casting speed will aggravate the liquid-level fluctuation. In addition, when the casting speed is changed uniformly, the surface fluctuation of the liquid level is small. Salas [12] and Zhang [13] researched the effect of SEN immersion depth on the flow field pattern, free-surface fluctuation, and slag layer exposure in the mold by combining water modeling and numerical simulation. With increasing immersion depth, the surface velocity of molten steel, the fluctuation range of the free surface, and the exposed area of the slag layer decrease, but the number of bubbles and the fluctuation of liquid surface near the SEN increase.

In view of the influence of the SEN structure on the flow field of the mold, Cho [14] found that when the nozzle inclination angle is 15° downward, the flow field in the mold is a typical single-roll flow (SRF), the liquid level fluctuates sharply, and the risk of mold flux entrapment increases. When the nozzle inclination angle is 30° downward, the flow field in the mold is a double-roll flow (DRF). Lee [15] reported that for the nozzle downward inclination in the range of 0°–15°, the flow velocity in the upper circulation area is unstable and the liquid level of the mold fluctuates sharply. When the downward inclination angle is greater than 20°, the flow velocity on the surface of the mold decreases and the fluctuation range of the mold liquid level decreases. When argon gas is injected, owing to the buoyancy of the gas bubbles, the unstable range of the nozzle inclination is extended to 20°. Ismael [16,17] indicated that under the same conditions, compared with a square nozzle, the turbulent jet formed by a round nozzle increases, and the risk of mold flux entrainment increases from the results of the large eddy simulation (LES) model and water modeling.

Generally, since the width of the slab mold used for different widths of automobile exposed panels is also different, it is necessary to optimize the mold flow fields of different widths to improve the surface quality of automobile exposed panels with various widths. In view of the characteristics of the flow fields in the mold with different widths, in our recent works [18–20], the flow fields of slab molds with narrow, medium, and wide widths have been studied through rod deflection high-temperature measurement combined with numerical simulation. For the narrow width [18], increasing casting speed increases the subsurface velocity of molten steel and shifts down the position of the vortex core in the downward circulation zone. For the medium width [19], when the casting speed is increased from 1.0 to 1.3, 1.5, and 2.0 m/min, the flow pattern in the mold changes from SRF to unstable flow (UF) and then to DRF. In the case of wide width [20], when the argon gas flow rate is increased from 6 L/min to 10 L/min and 14 L/min, the flow pattern changes from DRF to SRF, and the risk of slag entrainment increases. With the casting speed increased from 1.0 m/min to 1.1 m/min, the flow field pattern changes from SRF to strong DRF.

With regard to the influence of the width on the flow field of the mold, the research methods can be divided into two kinds. One is fixing the casting speed. Ma [21] found that the larger is the mold width, the greater is the impact pressure of the molten steel stream on the narrow wall, and the greater is the turbulent kinetic energy on the mold liquid surface. Zhang [22] concluded that under the conditions of the same casting speed and other parameters, the liquid-level fluctuation in the narrow-width mold is severer than that of the wide-width mold and the impact pressure of the liquid steel stream on the narrow wall is also larger. Tang [23] reported that with increasing mold width from 1400 mm to 1600 mm and 2150 mm under the same casting speed, the surface velocity gradually decreases and the point where the molten steel stream hits the narrow wall gradually moves downward. The second research method is fixing the steel throughput. Deng [24] found that when the steel throughput is 3.6 ton/min, increasing the mold width can reduce the average wave height of the liquid level and decrease the surface flow velocity to reduce the incidence of mold flux shearing entrainment.

The flow velocity of molten steel on the surface of the mold has an important influence on the surface quality of the automobile exposed panel. When the surface velocity of the steel–slag interface is greater than the critical velocity of slag entrainment, shearing instability (Kelvin–Helmholtz instability [25]) will occur on the steel–slag interface, resulting in shearing entrainment. At the same time, too low surface flow velocity will cause uneven slag consumption due to excessively low molten steel surface temperature and poor slagging. Therefore, it is necessary to know the flow velocity on the surface of the mold. However, it is difficult to measure the surface velocity of molten steel at high temperature. In early research works, the methods for measuring the velocity of molten steel on the surface of the mold [26–28] included flow control sensors, Karman eddy current probes, and sub-meniscus velocity control sensors. Subsequently, the nail-dipping method [25,29] was used to measure the surface velocity of molten steel. Szekely [30] developed a mechanical force reaction probe method to measure the velocity in a turbulent, electromagnetically driven recirculating low-melting-alloy system. A 19-mm-diameter non-ferromagnetic stainless steel disc was fastened to the end of a spring-loaded rod. The end of the rod was connected to a Hewlett-Packard model 7DC-3000 Linear Voltage Differential Transformer (Hewlett-Packard Company, Palo Alto, CA, USA). The pressure on the disc was converted to the (stagnation) velocity. To measure the near-surface velocity of the liquid metal, a drag-form strain gauge system [31] was also developed at MEFOS (Swedish National Metallurgical Research Institute).

In recent years, we and others [32] developed the rod deflection method. When the velocity-measuring rod is inserted into molten steel in the mold to reach a stable deflection under gravity, buoyancy, and the impact force of molten steel, the deflection angle is read and the surface velocity of molten steel can be calculated. The method can accurately and conveniently measure the flow velocity near the surface of molten steel at high temperature.

The flow in a continuous-casting mold is highly turbulent and extremely complex. The geometry and a number of process parameters exert a considerable influence on the flow behavior. Since the surface quality of the automobile exposed panel is highly dependent on the flow field in the mold, an in-depth understanding of these relationships is important for optimization and effective control of the process. For this reason, a large number of studies have already been carried out on this subject in the past. However, under fixed steel throughput, no papers reported the effect of mold width on the mold flow field assisted with the quantitative measurement of mold surface velocity at high temperature, which is beneficial not only for improving the industrial operation of continuous casting but also obtaining knowledge for the scientific community.

In this paper, the effect of mold width on the flow field in the mold is studied at fixed steel throughputs of 2.2 ton/min and 3.5 ton/min. The rod deflection method is used to measure velocity near the mold surface for mold widths of 880 mm, 1050 mm, 1300 mm, 1600 mm, 1800 mm, and 2000 mm. The numerical calculation is conducted by use of the standard k-ε model coupled with the discrete-phase model (dpm) model, and the calculation and measurement results are compared and analyzed. The influence of mold width on the surface velocity, flow field pattern, turbulent kinetic energy, and surface fluctuation of molten steel in the mold is clarified, which will provide technical guidance for improving the surface quality of automobile exposed panels with different widths.

2. High-Temperature Velocity Measurement Method

The high-temperature velocity measurement results were obtained by the rod deflection method in the steel plant of Shougang Jingtang Iron & Steel Co., Ltd. (Tangshan, China) and Tangshan Stainless Steel Co., Ltd. in China. As shown in Figure 1, the speed-measuring device of the rod deflection method is composed of a balance block, a deflection bearing, a deflection angle indicator, and a stainless steel detecting rod. The function of the balance block is to make the center of gravity of the entire speed-measuring device fall near the center of the deflection axis, which enables the subtle changes in the velocity of the molten steel to be sensitively detected by the change in the deflection angle of the detecting rod. When measuring the flow velocity near the surface of the mold, the stainless steel detecting rod is inserted into the place near 1/4 of the wide surface of the mold, and the flowing molten steel deflects the detecting rod at a certain angle. By use of the measured deflection angle combined with Equation (1), the surface velocity of the mold can be calculated [32].

$$U_0 = \sqrt{\frac{2(GL_1 \tan\theta - FL_2 \tan\theta)}{L_2 C_D \rho A}} \tag{1}$$

where U_0 is the velocity near the mold surface (m/s). G is the gravity of the detecting rod (N). L_1 is the gravity arm (m). F is the buoyancy (N). L_2 is the buoyancy arm (m). C_D is the resistance coefficient of the flow. ρ is the density of molten steel (kg/m^3). A is the projected area of the immersion part of the detecting rod in the direction vertical to the flow of molten steel (m^2). The detailed description of the rod deflection method and the error discussion are available in our previous paper [32].

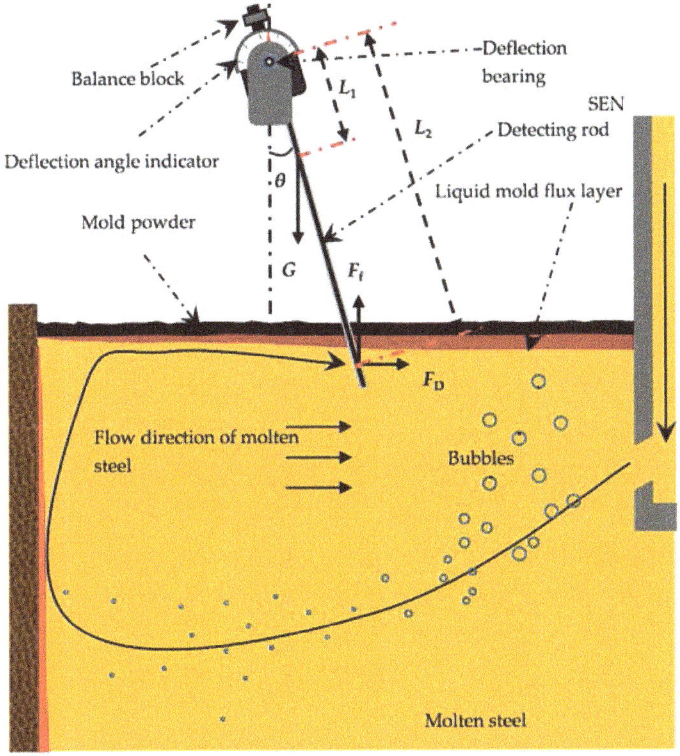

Figure 1. Schematic diagram of measuring velocity near the mold surface by the rod deflection method.

3. Mathematical Model

3.1. Numerical Simulation

In this paper, the standard k-ε model and the DPM model were used to simulate the turbulent flow in the submerged entry nozzle and continuous-casting mold. To simplify the complex solution process, the following assumptions were made. Molten steel is an incompressible fluid. Argon gas bubbles are all rigid spheres with uniform diameter and are not affected by pressure and temperature, ignoring the merging and rupture of bubbles. The discrete phase did not occupy the volume in the calculation domain and was regarded as a mass point. The influence of the solidified shell on the flow field was ignored. The influence of mold flux on liquid-level fluctuation and flow field was ignored.

3.2. Governing Equation

In the calculation of the two-phase flow of molten steel and argon gas bubbles in the mold, the continuous phase of molten steel was calculated using the Euler method, the discrete phase of argon gas bubbles was calculated using the Lagrangian method, and the discrete and continuous phases were coupled through the source term in the governing equation. There was bidirectional coupling between the continuous phase and the discrete phase. Namely, there was an interaction between the bubbles and the continuous phase, and the movement of the bubbles was affected by the continuous phase. Conversely, the bubbles also had an impact on the movement of the continuous phase.

The movement of the continuous phase is described by Equation (2) of the conservation of mass and Equation (3) of the conservation of momentum.

$$\frac{\partial}{\partial t}(\alpha_1\rho_1) + \nabla \cdot (\alpha_1\rho_1\vec{v}_1) = 0 \quad (2)$$

$$\frac{\partial}{\partial t}\left(\alpha_1\rho_1\vec{v}_1\right)+\vec{v}_1\cdot\nabla(\alpha_1\rho_1\vec{v}_1)=-\nabla p+\nabla\cdot[\alpha_1(\mu_1+\mu_t)]\nabla\vec{v}_1+\vec{F}_k \tag{3}$$

where $\vec{v}_1$ is the flow velocity (m/s). ρ_1 is the density of molten steel (kg/m^3). p is the pressure (N/m^2). μ_1 is the liquid viscosity (Pa·s). μ_t is the turbulent viscosity (Pa·s). $\mu_1+\mu_t$ is the effective viscosity (Pa·s). g is the gravitational acceleration (m/s^2). F_k is the force of argon gas bubbles acting on molten steel (N/m^3). α_1 is the liquid-phase volume fraction, which is expressed as

$$\alpha_1=1-\frac{\sum_i V_{d,i}}{V_{cell}} \tag{4}$$

where $V_{d,i}$ is the discrete-phase volume (m^3). V_{cell} is the volume of the grid cell (m^3).

The standard k-ε model was used to simulate turbulent flow, where the turbulent viscosity μ_t of the liquid could be expressed as a function of turbulent kinetic energy of k and turbulent energy dissipation rate of ε.

$$\mu_t=C_\mu\rho_1\frac{k^2}{\varepsilon} \tag{5}$$

The influence of molecular viscosity was ignored in the standard k-ε model, and at the same time, a model of transmission equation for k and ε was established as follows:

$$\alpha_1\rho_1\left(\frac{\partial k}{\partial t}+\vec{u_1}\cdot\nabla k\right)=-\nabla\left(\alpha_1\frac{\mu_t}{\sigma_k}\nabla k\right)+\alpha_1 G_k-\alpha_1\rho_1\varepsilon \tag{6}$$

$$\alpha_1\rho_1\left(\frac{\partial\varepsilon}{\partial t}+\vec{u_1}\cdot\nabla\varepsilon\right)=-\nabla\left(\alpha_1\frac{\mu_t}{\sigma_\varepsilon}\nabla\varepsilon\right)+\alpha_1 C_1\frac{\varepsilon}{k}G_k-\alpha_1 C_2\rho_1\frac{\varepsilon^2}{k} \tag{7}$$

where the constants in the model are C_μ = 0.09, σ_k = 1.00, σ_ε = 1.30, C_1 = 1.44, and C_2 = 1.92 [13].

G_k is the amount of turbulent kinetic energy generated, which is expressed as

$$G_k=\mu_t\left(\frac{\partial u_{i,j}}{\partial x_j}+\frac{\partial u_{i,j}}{\partial x_i}\right)\frac{\partial u_{i,j}}{\partial x_j} \tag{8}$$

In this paper, the discrete-phase model was used to calculate the influence of argon bubbles on the flow field in the mold. The discrete-phase model is a multiphase flow model that tracks the dispersed phase in the Lagrangian coordinate system. Newton's second law is used to calculate the movement of bubbles in the mold [33].

$$m_d\frac{d\vec{v}_d}{dt}=\vec{F}_d+\vec{F}_p+\vec{F}_b+\vec{F}_{vm}+\vec{F}_g+\vec{F}_L \tag{9}$$

where m_d is the bubble mass (kg). $\vec{v}_d$ is the bubble velocity (m/s). $\vec{F}_d$ is the drag force (N). $\vec{F}_p$ is the pressure (N). $\vec{F}_b$ is the buoyancy force (N). $\vec{F}_{vm}$ is the virtual mass force (N). $\vec{F}_g$ is gravity (N). $\vec{F}_L$ is the lift force (N).

$\vec{F}_d$ represents drag force and is defined as

$$\vec{F}_d=C_d\frac{\rho_1\left|\vec{v}-\vec{v}_d\right|\left(\vec{v}-\vec{v}_d\right)}{2}\frac{\pi d_d^2}{4} \tag{10}$$

where the drag coefficient C_d is defined as $C_d=\frac{24}{Re_d}$ when $Re_d<2$, $C_d=\frac{24}{Re_d}(1+0.15Re_d^{0.687})$ when $2<Re_d<500$, and $C_d\approx 0.44$ when $Re>500$, where Re_d is defined as $Re_d=\frac{\rho_f d_d\left|\vec{v}-\vec{v}_d\right|}{\mu_l}$.

$\vec{F}_p$ is the pressure gradient force and defined as

$$\vec{F}_p=\frac{1}{6}\pi d_d^3\rho_1\frac{d\vec{v}}{dt} \tag{11}$$

$\vec{F}_b$ is buoyancy, which can be defined as

$$\vec{F}_{\rm b} = -\frac{1}{6}\pi d_{\rm d}^3 \rho_{\rm l} \cdot \vec{g} \tag{12}$$

The virtual mass force of $\vec{F}_{\rm vm}$ is related to the acceleration of the liquid around the bubble, which can be defined as

$$\vec{F}_{\rm vm} = \frac{1}{6}\pi d_{\rm d}^3 C_{\rm vm} \rho_{\rm f} \frac{d}{dt}\left(\vec{v} - \vec{v}_{\rm d}\right) \tag{13}$$

Among them, for spherical objects moving in the liquid phase, $C_{\rm vm} = 0.5$.

$\vec{F}_{\rm g}$ is the gravity force (N) and is defined as

$$\vec{F}_{\rm g} = \frac{1}{6}\pi d_{\rm d}^3 \rho_{\rm d} \cdot \vec{g} \tag{14}$$

Owing to the horizontal velocity gradient, when a bubble rises in the liquid, a lateral lift force of $\vec{F}_{\rm L}$ (N) acts on it, which is defined as

$$\vec{F}_{\rm L} = -\frac{1}{6}\pi d_{\rm d}^3 C_{\rm L} \rho_{\rm l} \left(\vec{v}_{\rm d} - \vec{v}\right)\left(\nabla \cdot \vec{v}\right) \tag{15}$$

$C_{\rm L}$ is the lift coefficient, $\lim\limits_{x \to \infty} C_L = 0.5$.

The random walk model was used to consider the influence of turbulent velocity fluctuation on bubble motion. The trajectory of the bubble is determined as

$$\vec{x}_{d,i} = \int \vec{v}_{d,i} dt \tag{16}$$

3.3. Calculation Area and Boundary Conditions

Figure 2a is a schematic diagram of the calculation domain and grid division, including the velocity inlet, the pressure outlet, the immersion nozzle, and the entire calculation domain in the mold. The calculation domain was divided into 540,000 cells, and the mesh sizes at the immersion nozzle were refined in particular. The core part of the fluid flow was calculated more accurately to improve the reliability of the calculation results, as shown in Figure 2b. The maximum grid size of the immersion nozzle was 16 mm, and the maximum grid size of the remaining parts was 32 mm. The specific experiment and simulation parameters of each part are listed in Table 1. The chemical compositions of the steel grade involved in the numerical simulation are shown in Table 2.

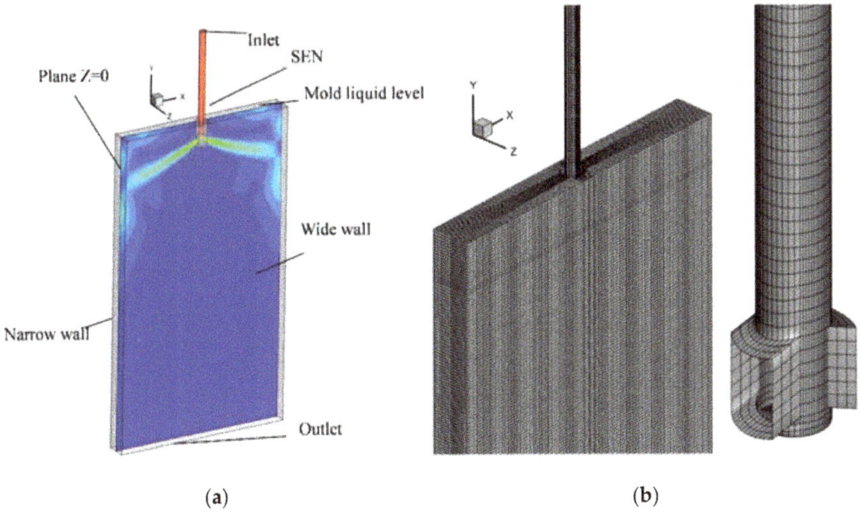

Figure 2. Schematic diagram of (**a**) computational domain and (**b**) mesh division and local refinement.

Table 1. Experiment and simulation parameters.

Parameters	Values
Mold width (mm)	880, 1050, 1300, 1600, 1800, 2000
Mold thickness (mm)	200, 237
Steel throughput (ton/min)	2.2, 3.5
Casting speed ($m \cdot min^{-1}$)	Calculated from the steel throughput
Immersion depth of nozzle (mm)	140, 160
Density of molten steel ($kg \cdot m^{-3}$)	7020
Viscosity of molten steel ($kg \cdot m^{-1} \cdot s^{-1}$)	0.0055
Argon gas flow rate ($L \cdot min^{-1}$)	4, 10
Nozzle port size (mm × mm)	65 × 80, 70 × 90
Nozzle port inclination angle (°)	15, 20
Bottom shape of nozzle	Concave bottom
Argon bubble diameter (mm)	1
Argon density ($kg \cdot m^{-3}$)	0.56

Table 2. Chemical composition of the steel grade (mass%).

C	Si	Mn	P	S	Al	Nb	Ti
0.002	0.005	0.66	0.039	0.01	0.042	0.006	0.0052

For the continuous phase of molten steel, the inlet boundary condition of the immersion nozzle was to set the inlet velocity as a fixed value based on the mass balance of inlet and outlet. The inlet velocity was obtained by Equation (17), and the outlet boundary condition at the bottom of the calculation domain was set as the pressure outlet. The boundary conditions of the free surface of the mold top surface were set as stable sliding surfaces, and the boundary conditions of other wall surfaces were set as stable non-slip surfaces.

$$Q = V_{inlet} \cdot A_{inlet} = V_{casting} \cdot A_{outlet} \quad (17)$$

For discrete-phase argon bubbles, the argon bubble was set to be a rigid sphere with a diameter of 1 mm. The bubbles were all set to escape on the free surface, at the immersion nozzle inlet and mold outlet as their boundary conditions, which means that the discrete phase of bubbles will escape out of the computational domain when they contact these areas. The boundary condition of the wall surface of the immersion nozzle was set as reflection. Under this condition, the bubbles will bounce off the wall with the coefficient of restitution specified by the user. In this paper, the coefficient of recovery was set to be 0.3. The boundary conditions of all other walls were set to capture. When bubbles touch these walls, they will be adsorbed on the walls.

4. Results and Discussion

In this paper, the rod deflection method was used to measure the flow velocity of molten steel near the surface of the mold with different widths under the condition of the fixed steel throughput, and the numerical simulation method was used to calculate the flow field in the mold. The influence of the mold width on the flow velocity of molten steel near the surface of the mold, flow field pattern, and turbulent kinetic energy distribution in the mold was investigated.

4.1. Influence of Width on Mold Flow Field under Low Fixed Steel Throughput

Under the conditions that the steel throughput was 2.2 ton/min, the argon gas flow rate was 4 L/min, the immersion depth was 140 mm, the submerged entry nozzle port shape was rectangular, and the nozzle inclination angle was 15°. Figure 3 gives the flow velocities near the mold surface with different mold widths under low steel throughput. Figure 3a shows the numerical simulation results of the molten steel flow rate near the mold surface, and Figure 3b shows a comparison between the measurement result and the numerical simulation result at 1/4 of the width of the mold. If the velocity value was greater than 0 m/s, the molten steel near the mold surface flowed from the narrow wall to the SEN and the flow direction in the mold was the positive velocity. If the velocity value was less than 0 m/s, the molten steel near the mold surface flowed from the SEN to the narrow wall and the flow direction in the mold was the negative velocity [21]. When the mold width was 880 mm, the velocity peak value of molten steel near the mold surface appeared near 1/4 of the width of the mold. When the mold width was 1050 mm, the velocity peak value deviated from 1/4 of the width and approached the narrow wall. When the mold width was 1300 mm, the molten steel velocities value near the surface of the mold were mostly below 0 m/s. Figure 3b shows a comparison between the measured and calculated values of the molten steel flow velocities at 1/4 of the width of the mold. Namely, as the mold width increased, the molten steel flow velocity decreased. When the widths of the mold were 880 mm and 1050 mm, the velocity values were the positive velocity; when the width was 1300 mm, the velocity value was the unstable velocity. The calculated results are in good agreement with the measured results.

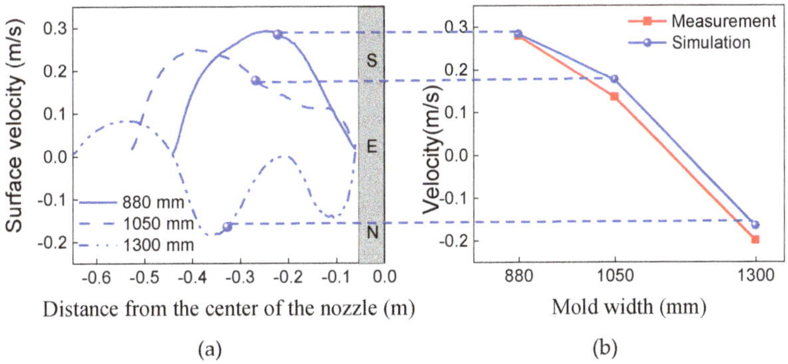

Figure 3. Flow velocities near the mold surface with different mold widths under the low steel throughput (**a**) distribution in the width direction and (**b**) comparison between measured and calculated values.

Figure 4 shows the velocity contour on the center longitudinal section of the mold with different mold widths under low steel throughput and under the same other conditions as described above. As the width of the mold increased, the velocity of the molten steel in the region of the upper circulation flow in the mold decreased significantly, and the velocity of the molten steel rising up near the nozzle increased obviously. This is because an increase in the mold width leads to an increase in the distance between the side port and the narrow wall of the mold. Therefore, the flow kinetic energy decreases, which directly leads to weakening of the main stream of the upper circulating flow.

Figure 5 shows the streamlines on the center longitudinal section of the mold with different mold widths under low steel throughput and under the same other conditions as described above. It clearly shows that when the widths of the mold were 880 mm and 1050 mm, the flow direction in the mold was the positive velocity. However, when the width of the mold was 1300 mm, the stream A of the molten steel near the SEN wall had a tendency to flow upward and directly impact the top free surface of the mold. Stream B of the molten steel flowed toward the narrow wall. After impinging the narrow wall, it split into an upward flow stream and a downward flow stream. The upward flow stream collided with stream A on the free surface of the mold, which greatly increased the risk of the entrainment of mold flux.

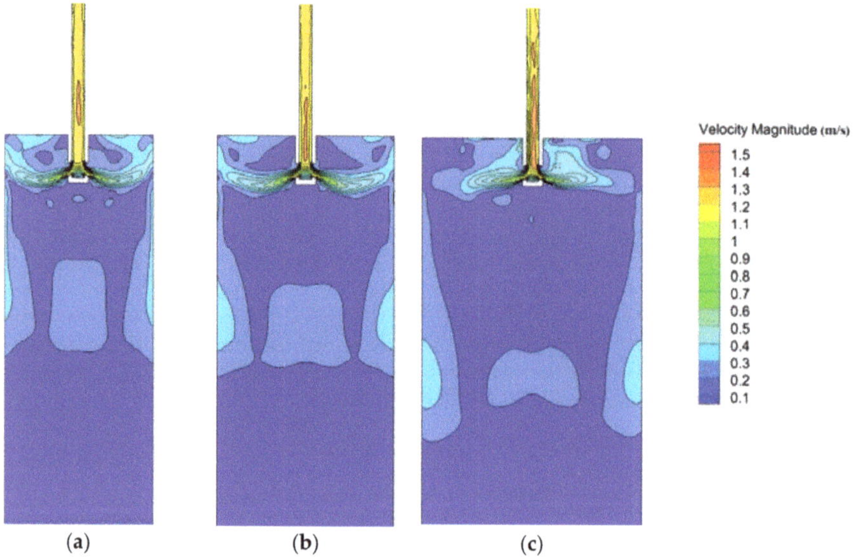

Figure 4. Velocity contour on the center longitudinal section of the mold with different mold widths under low steel throughput: (**a**) 880 mm, (**b**) 1050 mm, and (**c**) 1300 mm.

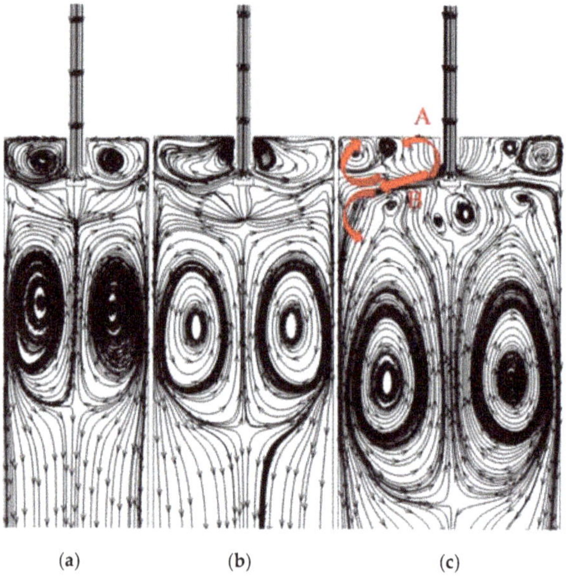

Figure 5. Streamlines on the center longitudinal section of the mold with different mold widths under low steel throughput: (**a**) 880 mm, (**b**) 1050 mm, and (**c**) 1300 mm.

Figure 6 shows the contour of turbulent kinetic energy on the free surface of the mold with different widths under low steel throughput and under the same other conditions as described above. When the width of the mold was 880 mm, the turbulent kinetic energy on the free surface was distributed uniformly, and the area with higher turbulent kinetic energy was mainly concentrated at the position of 1/4 of the width of the mold. When the width of the mold was 1050 mm, the turbulent kinetic energy near the nozzle on the free surface of the mold was significantly higher than that in the other regions. When the mold width further increased to 1300 mm, the high turbulent kinetic energy region occupied the entire area from the SEN wall to over 1/4 of the mold width, and the risk of slag entrainment increased obviously.

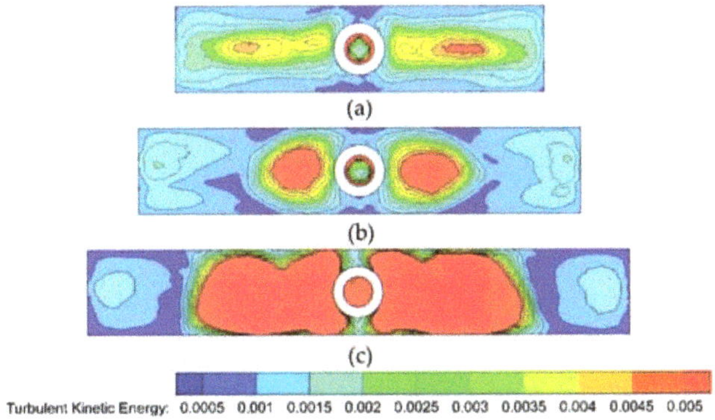

Figure 6. Contour of turbulent kinetic energy on the free surface of the mold with different widths under low steel throughput: (**a**) 880 mm, (**b**) 1050 mm, and (**c**) 1300 mm.

The degree of fluctuation of the free surface of molten steel is an important index demonstrating the stability of continuous casting, which is crucial to the surface quality of the automobile exposed panel. The height of the free surface, H, is calculated by simple potential energy balance [34]:

$$H = \frac{P_i - P_{\text{mean}}}{\left(\rho_{\text{steel}} - \rho_{\text{slag}}\right) \times g} \tag{18}$$

where P_i is the pressure of a certain position on the free surface (N/m^2). P_{mean} is the average value of the pressure on the entire free surface (N/m^2). ρ_{steel} is the density of molten steel (kg/m^3). ρ_{slag} is the density of liquid slag (kg/m^3). g is the acceleration of gravity (m/s^2).

Figure 7 is the top surface-level profile with different widths under low steel throughput and under the same other conditions as described above. When the width of the mold was 880 mm, the surface fluctuation of the molten steel in the mold was relatively smooth. When the width of the mold was 1050 mm, the fluctuation of the top surface level at the position of 1/4 of the width of the mold increased. When the width of the mold was 1300 mm, the top surface-level fluctuation deteriorated with the position of the peak top surface level changing from 1/4 of the width to the SEN vicinity of the mold, which increased the risk of slag entrainment.

4.2. *Influence of Width on Mold Flow Field under High Fixed Steel Throughput*

Under the conditions that the steel throughput was 3.5 ton/min, the argon gas flow rate was 10 L/min, the immersion depth was 160 mm, the submerged entry nozzle port shape was rectangular, and the nozzle inclination angle was 20°. Figure 8 shows the surface velocity near the mold surface with different mold widths under high fixed steel throughput and a comparison between the measured and calculated values of molten steel flow velocity near the surface of the mold with different widths. Figure 8a shows the numerical simulation results of the molten steel flow rate near the mold surface. As shown in Figure 8a, when the width of the mold was 1600 mm, the surface velocity was greater than 0 m/s, and the velocity distribution in the width direction of the mold had two peaks. The maximum value of the surface velocity appeared between 1/4 of the width of the mold and the narrow wall, and another peak value appeared near the SEN. When the mold width was 1800 mm, the velocity value of the molten steel at the distance between 0.5 m and 0.9 m from the SEN center was greater than 0 m/s and smaller than 0.2 m/s, indicating that the flow direction of molten steel is from the narrow wall to the SEN. However, the velocity value of molten steel between 0 and 0.5 m from the SEN center was smaller than 0, indicating that the flow direction of the molten steel is from the SEN to the narrow wall. When the width of the mold was 2000 mm, the surface velocity was negative at all points in the width direction, indicating that the direction of the molten steel near the free surface of the mold is all from the SEN to the narrow wall of the mold.

Figure 8b shows a comparison between the measurement result and the numerical simulation result at 1/4 of the width of the mold. As shown in Figure 8b, the simulated results of the molten steel velocity at the 1/4 position of the mold width under the same steel throughout decreased with increasing mold width. However, the measurement results show that the surface flow velocity decreases first and then slightly increases as the width increases. When the width of the mold was 1600 mm, the calculated and measured results of the flow velocity at 1/4 of the surface of the mold were both positive. When the widths were 1800 mm and 2000 mm, both of the measured and calculated results had negative values, which increases the risk of slag entrainment. The calculated results are basically consistent with the measured results.

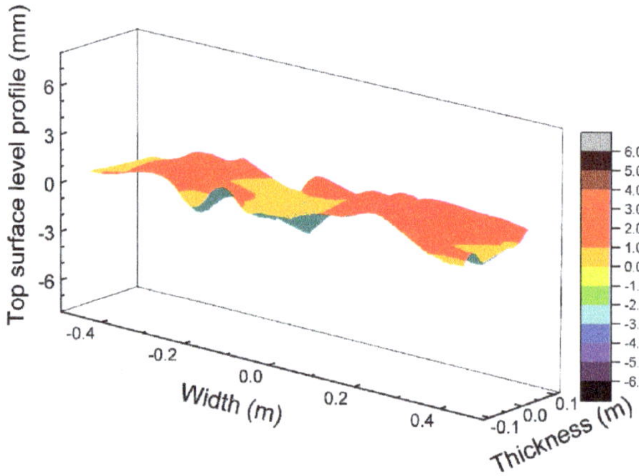

(a)

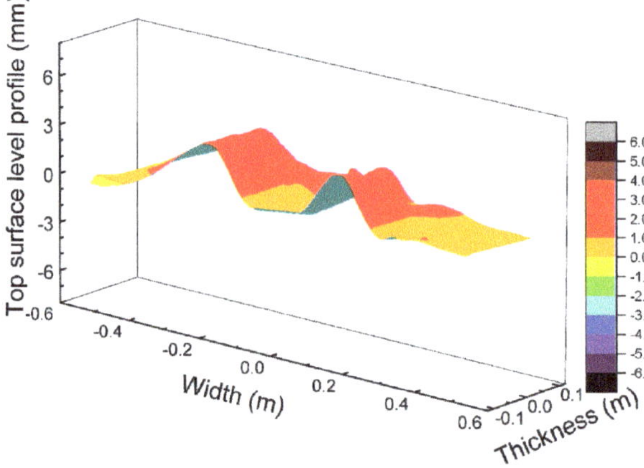

(b)

Figure 7. *Cont.*

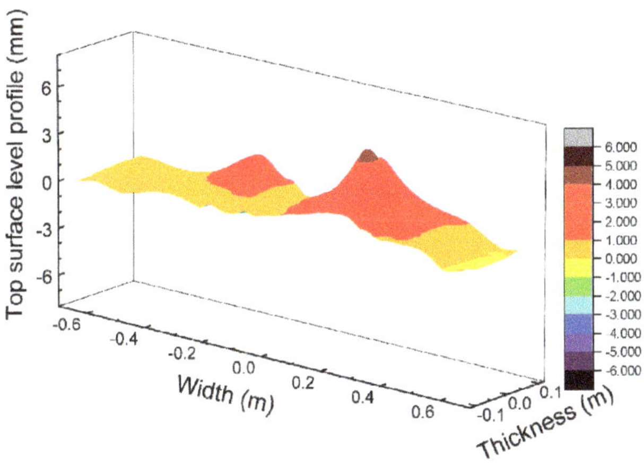

(c)

Figure 7. Top surface-level profile with different widths under low steel throughput: (**a**) 880 mm, (**b**) 1050 mm, and (**c**) 1300 mm.

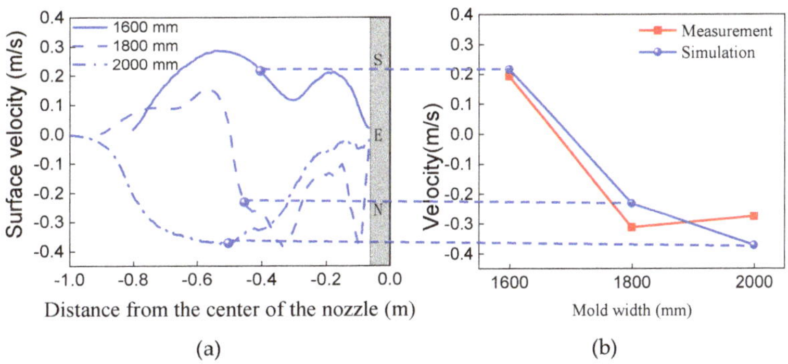

Figure 8. Surface velocity near the mold surface with different mold widths under the high fixed steel throughput (**a**) distribution in the width direction and (**b**) comparison between measured and calculated values.

Figure 9 shows the streamlines on the center longitudinal section of the mold with different widths under high fixed steel throughput and under the same other conditions as described above. As the width of the mold increased, the flow field also changed significantly. When the width of the mold was 1600 mm, the flow direction in the mold was a positive velocity. The molten steel flowed out of the side ports of the SEN to the narrow wall. After impacting the narrow wall, it was divided into upper and lower circulating flows to form the positive velocity. Under the action of the rising argon gas bubbles at a gas flow rate of 10 L/min, part of molten steel near the SEN had a tendency to directly flow upward to the top free surface of the mold, but it was suppressed by the molten steel stream flowing up from the narrow wall and along the top free face in the direction to the SEN, as shown in Figure 9a. This also explains why there are two peaks in the distribution curve of the surface velocity in the width direction of the 1600 mm mold in Figure 8a. When the width of the mold was 1800 mm, the flow direction in the mold was the unstable velocity, and the unstable velocity was a flow direction between the positive velocity and the negative velocity [24,27]. The molten steel flowed out from the side ports of the SEN, a part of it flowed to and then impacted the narrow wall to form the upward stream A and the downward stream B. The surface flow stream C toward the narrow wall met stream A at a distance of 0.5 m from the SEN center to form the unstable velocity. When the width of the mold was 2000 mm, the flow direction in the mold was a typical negative velocity. After the molten steel flowed out from the side ports of the immersion nozzle, it rose upward to the top surface, flowed along the top surface to the narrow wall, and then flowed downward along the narrow wall to form the negative velocity. According to previous research results, the risk of slag entrainment will increase when the flow direction in the mold is the negative velocity and the unstable velocity to increase the incidence ratio of linear defects on the automobile exposed panel [35–37].

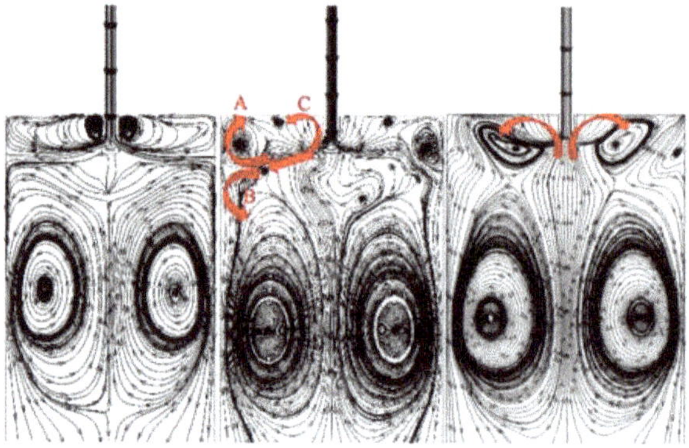

(**a**) (**b**) (**c**)

Figure 9. Streamlines on the center longitudinal section of the mold with different widths under high fixed steel throughput: (**a**) 1600 mm, (**b**) 1800 mm, and (**c**) 2000 mm.

Figure 10 shows the turbulent kinetic energy contour on the top surface of the mold with different widths under high fixed steel throughput and under the same other condition as described above. When the width of the mold was 1600 mm, the turbulent kinetic energy near the SEN and the narrow wall was higher than those of other areas, but it was generally at a reasonable level. When the mold widths were 1800 mm and 2000 mm, the high turbulent kinetic energy region near the SEN gradually expanded to the narrow wall. Since the distance of the molten steel flowing from the side ports of the SEN to the narrow wall of the mold was increased, the molten steel stream became weak. Under the action of the rising argon bubbles, the molten steel was easily lifted up to the top surface of the mold. As a result, a high turbulent kinetic energy region appeared in the area near the SEN and expanded to the narrow wall with increasing the mold width, increasing the risk of slag entrainment.

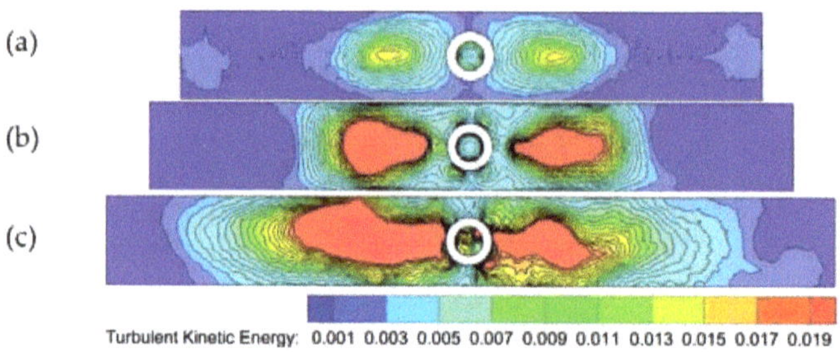

Figure 10. Turbulent kinetic energy contour on the top surface of the mold with different widths under high fixed steel throughput: (**a**) 1600 mm, (**b**) 1800 mm, and (**c**) 2000 mm.

Figure 11 shows the top surface-level profile of the mold with different widths under high fixed steel throughput and under the same other conditions as described above. In addition, under high steel throughput, as the width of the mold increased, the area with the high-level fluctuation gradually shifted from 1/4 of the width of the mold to the vicinity of the SEN. Especially, when the cross-sectional width was 2000 mm, the peak value of the fluctuation near the SEN reached about 7 mm, which easily led to slag entrainment and thus caused surface defects in the automobile exposed panel.

From the above description, it is found that regardless of whether it is low fixed steel throughput or high fixed steel throughput, the change rule of the flow field in the mold with different widths is basically the same. Therefore, in actual production, the argon gas flow rate and the immersion depth of the SEN should be adjusted reasonably to optimize the flow field in the mold with different widths under the same fixed steel throughput.

In fact, the flow field in the slab mold is quite complex, which is affected by a lot of parameters of continuous casting. In our previous papers, the effects of casting speed, argon gas flow rate, immersion depth, and structure of the SEN on the flow field of the slab mold with narrow [15], medium [16], and large [17] widths have been investigated. The coupling of different influencing variables is also important, which we will study in the future.

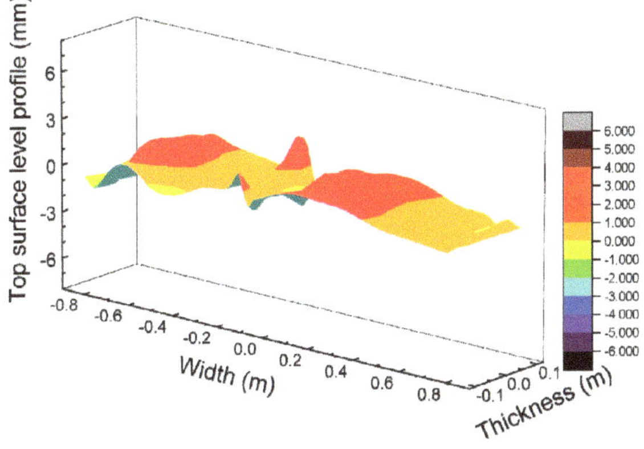

(**a**)

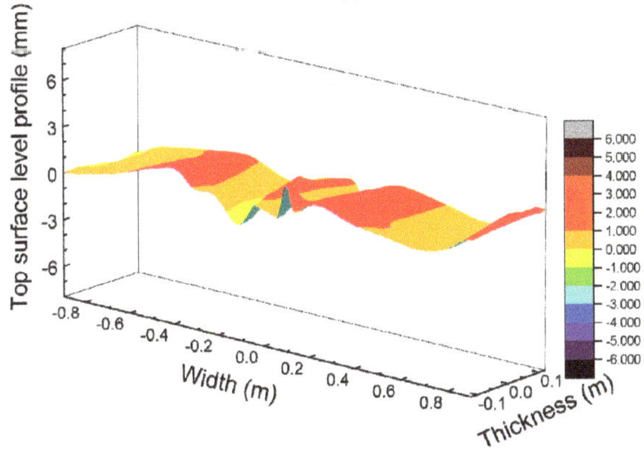

(**b**)

Figure 11. *Cont.*

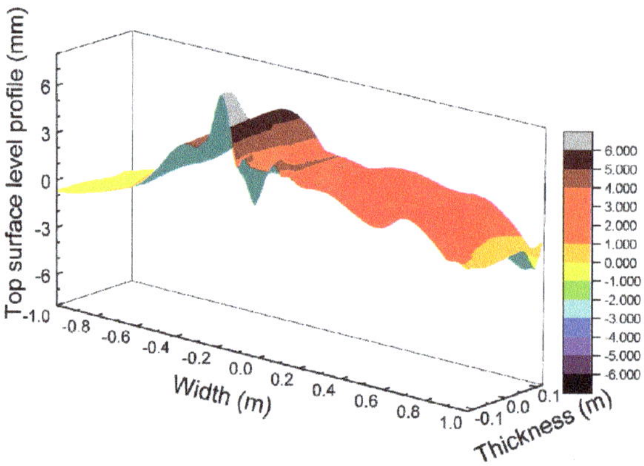

(c)

Figure 11. Top surface-level profile of the mold with different widths under high fixed steel throughput: (**a**) 1600 mm, (**b**) 1800 mm, and (**c**) 2000 mm.

5. Conclusions

In this paper, the rod deflection method was used to measure the flow velocity near the surface of the mold at high temperature, and the standard k-ε model coupled with the discrete-phase model (DPM) was used for numerical calculation. The effects of the width of the mold on the surface velocity, flow field pattern, turbulent kinetic energy distribution, and surface-level fluctuation in the mold were studied. The main conclusions are as follows:

(1) Under the conditions of low fixed steel throughput of 2.2 ton/min, a nozzle immersion depth of 140 mm, and an argon gas flow rate of 4 L/min, as the width of the mold increases from 880 mm to 1050 mm and 1300 mm, the flow velocity near the surface of the mold decreases. When the widths of the mold are 880 mm and 1050 mm, the surface velocity is positive, forming the positive velocity. When the width is 1300 mm, the surface velocity is negative, forming the negative velocity. The calculated results are in good agreement with the measured results. The turbulent kinetic energy near the submerged entry nozzle (SEN) gradually increases, and the risk of slag entrainment increases.

(2) Under the conditions that the high fixed steel throughput is 3.5 ton/min, the immersion depth of the SEN is 160 mm, and the argon gas flow rate is 10 L/min, as the width of the mold increases from 1600 mm to 1800 mm and 2000 mm, the velocity near the mold surface decreases. When the width of the mold is 1600 mm, the flow velocity at 1/4 of the surface of the mold is positive. When the widths are 1800 mm and 2000 mm, both surface velocities are negative. The calculated results are basically consistent with the measured results. The high turbulent kinetic energy area near the nozzle shows a trend of expanding to a narrow wall, and the risk of slag entrainment is significantly increased.

(3) Regardless of whether it is a low fixed steel throughput or a high fixed steel throughput, the change rules of the flow field in the mold with different widths are basically the same. Therefore, in actual production, the argon gas flow rate and the immersion depth of the SEN should be adjusted reasonably to optimize the flow field in the mold with different widths under the same fixed steel throughput.

Author Contributions: Conceptualization, J.-Q.L. and J.Y.; methodology, J.Y.; software, J.-Q.L.; validation, J.-Q.L.; formal analysis, J.-Q.L.; investigation, J.-Q.L.; resources, C.-L.Z., W.-Y.H., R.-B.J., and Y.-T.G.; data curation, C.M. and Y.G.; writing—original draft preparation, J.-Q.L.; writing—review and editing, J.Y.; visualization, J.-Q.L.; supervision, J.Y.; project administration, J.Y.; funding acquisition, J.Y. All authors have read and agreed to the published version of the manuscript.

Funding: The authors gratefully acknowledge the financial support from the Natural Science Foundation of China (No. U1960202).

Data Availability Statement: Data supporting reported results can be found in this paper.

Acknowledgments: The authors also gratefully acknowledge the financial support from Shougang Jingtang United Steel Co., Ltd. and Tangshan Stainless Steel Co., Ltd.

Conflicts of Interest: The authors declare no conflict of interest.

References

1. Pan, X.-Q.; Yang, J.; Park, J.; Ono, H. Distribution characteristics of inclusions along with the surface sliver defect on the exposed panel of automobile: A quantitative electrolysis method. *Int. J. Miner. Met. Mater.* **2020**, *27*, 1489–1498. [CrossRef]
2. Li, L.-M.; Liu, Z.-Q.; Li, B.-K. Modelling of bubble aggregation, breakage and transport in slab continuous casting mold. *J. Iron Steel Res. Int.* **2015**, *22*, 30–35. [CrossRef]
3. Liu, Z.; Qi, F.; Li, B.; Cheung, S.C. Modeling of bubble behaviors and size distribution in a slab continuous casting mold. *Int. J. Multiph. Flow* **2016**, *79*, 190–201. [CrossRef]
4. Cho, M.S.; Thomas, B.G.; Kim, H.S. Bubble Behavior and Size Distributions in Stopper-Rod Nozzle and Mold during Continuous Casting of Steel Slabs. *ISIJ Int.* **2018**, *58*, 1443–1452. [CrossRef]
5. Lee, P.D.; Ramirez-Lopez, P.E.; Mills, K.C.; Santillana, B. Review: The "butterfly effect" in continuous casting. *Ironmak. Steelmak.* **2012**, *39*, 244–253. [CrossRef]
6. Mills, K.C.; Ramirez-Lopez, P.; Lee, P.; Santillana, B.; Thomas, B.; Morales, R. Looking into continuous casting mould. *Ironmak. Steelmak.* **2014**, *41*, 242–249. [CrossRef]
7. Wu, M.; Vakhrushev, A.; Ludwig, A.; Kharicha, A. Influence of forced convection on solidification and remelting in the developing mushy zone. *IOP Conf. Ser. Mater. Sci. Eng.* **2016**, *117*, 012045. [CrossRef]
8. Vakhrushev, A.; Wu, M.; Ludwig, A.; Tang, Y.; Hackl, G.; Nitzl, G. Numerical Investigation of Shell Formation in Thin Slab Casting of Funnel-Type Mold. *Met. Mater. Trans. A* **2014**, *45*, 1024–1037. [CrossRef]
9. Liu, Z.; Li, B.; Vakhrushev, A.; Wu, M.; Ludwig, A. Physical and Numerical Modeling of Exposed Slag Eye in Continuous Casting Mold using Euler–Euler Approach. *Steel Res. Int.* **2018**, *90*, 1800117. [CrossRef]
10. Liu, Z.; Vakhrushev, A.; Wu, M.; Kharicha, A.; Ludwig, A.; Li, B. Scale-Adaptive Simulation of Transient Two-Phase Flow in Continuous-Casting Mold. *Met. Mater. Trans. A* **2018**, *50*, 543–554. [CrossRef]
11. Zhang, Q.-Y.; Wang, X.-H. Numerical Simulation of Influence of Casting Speed Variation on Surface Fluctuation of Molten Steel in Mold. *J. Iron Steel Res. Int.* **2010**, *17*, 15–19. [CrossRef]
12. Saldaña-Salas, F.; Torres-Alonso, E.; Ramos-Banderas, J.; Solorio-Díaz, G.; Hernández-Bocanegra, C. Analysis of the Depth of Immersion of the Submerged Entry Nozzle on the Oscillations of the Meniscus in a Continuous Casting Mold. *Metals* **2019**, *9*, 596. [CrossRef]
13. Zhang, H.; Fang, Q.; Xiao, T.; Ni, H.; Liu, C. Optimization of the Flow in a Slab Mold with Argon Blowing by Divergent Bifurcated SEN. *ISIJ Int.* **2019**, *59*, 86–92. [CrossRef]
14. Cho, S.-M.; Thomas, B.G.; Kim, S.-H. Effect of Nozzle Port Angle on Transient Flow and Surface Slag Behavior during Continuous Steel-Slab Casting. *Met. Mater. Trans. A* **2018**, *50*, 52–76. [CrossRef]
15. Lee, W.-H.; Yi, K.-W. Relationship between Fluid Flow Stability and Submerged Entry Nozzle Port Angle in a Conventional Slab Continuous-Casting Mold. *Met. Mater. Int.* **2020**, *27*, 4168–4181. [CrossRef]
16. Calderón-Ramos, I.; Morales, R.D. The Role of Submerged Entry Nozzle Port Shape on Fluid Flow Turbulence in a Slab Mold. *Met. Mater. Trans. A* **2015**, *46*, 1314–1325. [CrossRef]
17. Calderón-Ramos, I.; Morales, R.D.; Salazar-Campoy, M. Modeling Flow Turbulence in a Continuous Casting Slab Mold Comparing the use of Two Bifurcated Nozzles with Square and Circular Ports. *Steel Res. Int.* **2015**, *86*, 1610–1621. [CrossRef]
18. Zhang, T.; Yang, J.; Xu, G.J.; Liu, H.J.; Zhou, J.J.; Qin, W. Effects of operating parameters on the flow field in slab continuous casting molds with narrow widths. *Int. J. Miner. Metall. Mater.* **2021**, *28*, 238–248. [CrossRef]
19. Jiang, P.; Yang, J.; Zhang, T.; Xu, G.; Liu, H.; Zhou, J.; Qin, W. Optimization of Flow Field in Slab Continuous Casting Mold with Medium Width Using High Temperature Measurement and Numerical Simulation for Automobile Exposed Panel Production. *Metals* **2019**, *10*, 9. [CrossRef]
20. Ma, C.; He, W.-Y.; Qiao, H.-S.; Zhao, C.-L.; Liu, Y.-B.; Yang, J. Flow Field in Slab Continuous Casting Mold with Large Width Optimized with High Temperature Quantitative Measurement and Numerical Calculation. *Metals* **2021**, *11*, 261. [CrossRef]
21. Ma, F.J.; Wen, G.H.; Li, G. Numerical Simulation of Liquid Steel Flow in Slab Continuous Casting Mould. *Iron Steel Res.* **2000**, *36*, 10–14.

22. Zhang, L.; Yang, S.; Cai, K.; Li, J.; Wan, X.; Thomas, B.G. Investigation of Fluid Flow and Steel Cleanliness in the Continuous Casting Strand. *Metall. Mater. Trans. B* **2007**, *38*, 63–83. [CrossRef]
23. Tang, D.C.; Feng, J.; Li, Y.L.; Zhu, L.X. Numerical Simulation of the Influence of Width Variation on the Flow Field of Slab Continuous Casting Mould. *Iron Steel* **2011**, *46*, 37–39.
24. Deng, X.X.; Ji, C.X.; An, Z.Q.; Zhu, G.S.; Suo, J.L.; Huang, F.X. Influence of width on liquid level characteristics of slab continuous casting mold. *Iron Steel* **2017**, *52*, 48–54.
25. Liu, R.; Thomas, B.; Sengupta, J.; Chung, S.D.; Trinh, M. Measurements of Molten Steel Surface Velocity and Effect of Stopper-rod Movement on Transient Multiphase Fluid Flow in Continuous Casting. *ISIJ Int.* **2014**, *54*, 2314–2323. [CrossRef]
26. Lee, G.-G.; Thomas, B.; Kim, S.-H. Effect of refractory properties on initial bubble formation in continuous-casting nozzles. *Met. Mater. Int.* **2010**, *16*, 501–506. [CrossRef]
27. Iguchi, M.; Terauchi, Y. Karman Vortex Probe for the Detection of Molten Metal Surface Flow in Low Velocity Range. *ISIJ Int.* **2002**, *42*, 939–943. [CrossRef]
28. Domgin, J.-F.; Gardin, P.; Galpin, J.-M.; Dez, A. Effect of process parameters variation on CC mould hydrodynamics and inclusions behaviour. *Metall. Res. Technol.* **2005**, *102*, 703–710. [CrossRef]
29. Cho, S.-M.; Kim, S.-H.; Thomas, B. Transient Fluid Flow during Steady Continuous Casting of Steel Slabs: Part I. Measurements and Modeling of Two-phase Flow. *ISIJ Int.* **2014**, *54*, 845–854. [CrossRef]
30. Szekely, J.; Chang, C.W.; Ryan, R.E. The measurement and prediction of the melt velocities in a turbulent, electro-magnetically driven recirculating low melting alloy system. *Metall. Mater. Trans. B* **1977**, *8*, 333–338. [CrossRef]
31. El-Kaddah, N.; Szekely, J.; Carlsson, G. Fluid flow and mass transfer in an inductively stirred four-ton melt of molten steel: A comparison of measurements and predictions. *Met. Mater. Trans. A* **1984**, *15*, 633–640. [CrossRef]
32. Zhang, T.; Yang, J.; Jiang, P. Measurement of Molten Steel Velocity near the Surface and Modeling for Transient Fluid Flow in the Continuous Casting Mold. *Metals* **2019**, *9*, 36. [CrossRef]
33. Zhang, T.; Luo, Z.; Liu, C.; Zhou, H.; Zou, Z. A mathematical model considering the interaction of bubbles in continuous casting mold of steel. *Powder Technol.* **2014**, *273*, 154–164. [CrossRef]
34. Zhang, L.F.; Wang, Y.F.; Zuo, X.J. Flow transport and inclusion motion in steel continuous-casting mold under sub-merged entry nozzle clogging condition. *Metall. Mater. Trans. B* **2008**, *39*, 534–550. [CrossRef]
35. Kunstreich, S.; Dauby, P.H. Effect of liquid steel flow pattern on slab quality and the need for dynamic electromagnetic control in the mould. *Ironmak. Steelmak.* **2005**, *32*, 80–86. [CrossRef]
36. Dauby, P.H. Continuous Casting: Make Better Steel and More of It! *Rev. Métallurgie Int. J. Metall.* **2012**, *109*, 113–136. [CrossRef]
37. Ramos, I.C.; Morales, R.D.; Garcia-Hernandez, S.; Ceballos-Huerta, A. Effects of Immersion Depth on Flow Turbulence of Liquid Steel in a Slab Mold Using a Nozzle with Upward Angle Rectangular Ports. *ISIJ Int.* **2014**, *54*, 1797–1806. [CrossRef]

metals

MDPI

Article

Influence of Bio-Coal Properties on Carbonization and Bio-Coke Reactivity

Asmaa A. El-Tawil [1,*], Bo Björkman [1], Maria Lundgren [2], Astrid Robles [2] and Lena Sundqvist Ökvist [1,2]

1 MiMeR, Luleå University of Technology, 971 87 Luleå, Sweden; Bo.Bjorkman@ltu.se (B.B.); lena.sundqvist-oqvist@ltu.se (L.S.Ö.)
2 Swerim AB, Box 812, 971 25 Luleå, Sweden; Maria.Lundgren@swerim.se (M.L.); Astrid.Robles@swerim.se (A.R.)
* Correspondence: asmaa.el-tawil@ltu.se; Tel.: +46-920-493131

Abstract: Coke corresponds to 2/3–3/4 of the reducing agents in BF, and by the partial replacement of coking coals with 5–10% of bio-coal, the fossil CO_2 emissions from the BF can be lowered by ~4–8%. Coking coal blends with 5% and 10% additions of bio-coals (pre-treated biomass) of different origins and pre-treatment degrees were carbonized at laboratory scale and with a 5% bio-coal addition at technical scale, aiming to understand the impact on the bio-coal properties (ash amount and composition, volatile matter content) and the addition of bio-coke reactivity. A thermogravimetric analyzer (TGA) connected to a quadrupole mass spectroscope monitored the residual mass and off-gases during carbonization. To explore the effect of bio-coal addition on plasticity, optical dilatometer tests were conducted for coking coal blends with 5% and 10% bio-coal addition. The plasticity was lowered with increasing bio-coal addition, but pyrolyzed biomass had a less negative effect on the plasticity compared to torrefied biomasses with a high content of oxygen. The temperature for starting the gasification of coke was in general lowered to a greater extent for bio-cokes produced from coking coal blends containing bio-coals with higher contents of catalyzing oxides. There was no significant difference in the properties of laboratory and technical scale produced coke, in terms of reactivity as measured by TGA. Bio-coke produced with 5% of high temperature torrefied pelletized biomass showed a similar coke strength as reference coke after reaction.

Keywords: bio-coals; carbonization; gasification; reactivity; dilatation; fluidity

Citation: El-Tawil, A.A.; Björkman, B.; Lundgren, M.; Robles, A.; Sundqvist Ökvist, L. Influence of Bio-Coal Properties on Carbonization and Bio-Coke Reactivity. *Metals* **2021**, *11*, 1752. https://doi.org/10.3390/met11111752

Academic Editor: Pasquale Cavaliere

Received: 28 September 2021
Accepted: 26 October 2021
Published: 31 October 2021

1. Introduction

The iron-ore-based blast furnace (BF) process is still the most dominant method for producing metallic iron units for steelmaking [1]. In the BF, the high-quality metallurgical coke that is used is considered crucial for the process. Besides working as a reducing agent for iron ore, coke also serves as the structural support for the burden in the furnace and provides passages for the upward movement of reducing gases [1]. The total consumption of coke is about 300 kg/t hot metal [1], depending on the amount of auxiliary reducing agents used (coke, coal, oil natural gas, etc.) [2]. The iron and steel industry aims to decrease the use of fossil carbon to minimize CO_2 emissions. According to the World Steel Association, the iron and steel industry accounts for approximately 7–9% of total world CO_2 emissions [3].

The use of pre-treated biomass (bio-coal) as a part of the raw material blend for cokemaking could be one possible way of reducing the fossil CO_2 emissions linked to the BF process. The advantage of replacing part of the fossil coal with bio-coal is that the biomass regeneration time is comparatively short [4] and the effects on global warming can be reduced as the carbon cycle is closed [5]. It has been reported that with the partial replacement of coking coals with 5–10% of bio-coal, the fossil CO_2 emissions at the BF can be lowered by ~4–8% [6]. Using bio-coke (bio-coal-containing coke) in the BF has the potential to lower the thermal reserve zone temperature (TRZT) of the BF, as the gasification reaction

(reaction of carbon in coke with CO_2) can proceed at a lower temperature in the more reactive bio-coke. A lowering in the TRZT shifts the gas composition in the equilibrium of FeO/Fe towards higher CO_2 and lower CO contents. This leads to higher utilization of the input carbon for the reduction of the iron ore and lowers the required carbon input [7]. A higher reactivity will, however, also reduce the coke strength after reaction, which is not desirable, as it may cause disintegration and have an impact on the permeability.

During cokemaking the coking coals go through chemical and physical changes including softening, swelling, shrinking, evolution of volatile matter (VM), and, finally, re-solidification to obtain coke [8]. The thermoplastic properties of the coals are important for the quality of the final coke [9]. In bio-coke research, several types of biomass-based materials have been studied for evaluation of their effect on coal fluidity. This material includes different raw biomasses, torrefied biomass, and pyrolyzed biomass. The addition of any type of biomass decreases the fluidity of the coal blend [10–14]. Ueki et al. [12] reported that the addition of raw woody biomass during carbonization results in the formation of voids between coal particles during carbonization, due to the release of biomass VM, and this results in brittle bio-coke. The coal particles were seen to be connected to woody biomass when they had a lower volatile content [12]. Diez et al. [11] reported that the thermal decomposition of raw biomasses (sawdust) occurs at low temperature ~150–400 °C and the majority of VM are released before the plastic stage of coking coal occurs at around 400–500 °C; this means that there is a lower amount of VM and remaining biomass that will interact with the coal during the fluidity development. On the other hand, Guerrero et al. [10] found that the devolatilization stages of coal and charcoal overlap partially and charcoal emits VM during carbonization that can block the fluidity, by establishing cross-linked O-C bonds. Fraga et al. [13] and Solar et al. [14] found that the addition of charcoal to coking coal blend caused a reduction in thermoplastic properties, and more so with increasing amounts. During the coal plastic stage charcoal acts as an inert material, which does not soften and melt, and as an active material that binds the components from the plasticized coal [11]. It was stated that chemically active additives, such as tar, petroleum residues, and oils, may act as fluidity enhancers or inhibitors, depending on their ability to donate or accept transferable hydrogen, respectively [9]. Generally, the blend of coal is required to have a maximum Gieseler fluidity and maximum dilatation greater than 400 ddpm and 55%, respectively, to produce high-quality metallurgical coke [15,16].

The quality of coke affects the performance of the BF; thus, it must be strictly controlled. Mechanical strength, coke reactivity index (CRI), and coke strength after reaction (CSR) are the most important parameters used to define coke quality, which are measured according to ISO standards 1889-4 and 556, respectively. It has been reported that there is a strong correlation between CRI and CSR. When the CRI is low, the CSR is maintained at a high level [17].

Ash components such as oxides of alkali and alkaline earth metal and iron are known to act as catalysts in the gasification reaction [18–20]. Nomura et al. [7,21] found that coke containing a high content of Ca in non-coking coal has a high reactivity and that Fe and Ca act as catalysts during gasification. Grigore et al. [20] reported that the coke reactivity increased with the increasing total amount of mineral phases containing Ca. Babich et al. [22] found that coke with a higher porosity has a higher reactivity, which was explain by the fact that microporosity also affects the reactivity, because it enables the access of CO_2 molecules to the inner part of the coke.

The effect of bio-coal particle size on the quality of bio-coke, in terms of CRI and CSR values, is reported in the literature [23–27]. The addition of 0.3–5% [14], 2–5% [23,24], 5% [25], 2–10% [26], 8% [27], and 5–25% [28] charcoal or (4.5–9%) lignite [29] to the coking coal blend was studied. It was found that the CRI of bio-coke is higher in comparison to reference coke, and vice versa for CSR. The increase in the CRI was more pronounced with fine charcoal, due to the presence of calcium oxide, which acts as a catalyst that promotes the reaction of carbon with CO_2, being more dispersed in the coke pieces compared to when using coarse charcoal [14,23,24,26]. It was shown that bio-coke has a higher reactivity when charcoal with a size <0.07 mm was added, compared to when coarse charcoal (2–9 mm) was used, with 5% charcoal addition [23,24], and further increased with 8% [27] and 10% charcoal additions [26]. The cold strength of bio-coke was strongly affected by the addition of coarse charcoal, while the coke strength was preserved when adding fine charcoal [23–27]. The main drawbacks related to bio-cokes are the lower mechanical strength and strength after reaction compared to cokes produced from coking coals only. The low mechanical strength of bio-coke contributes to difficulties during handling and charging, as well as higher dust formation in the upper part of the BF. A low strength after reaction may result in the generation of fines within the BF, especially in the lower part.

The carbonization behavior of coking coal blends containing raw or pre-treated biomass was reported in [12,14,23,24,26,28,30–33]. Matsumura et al. [30] studied the carbonization of raw biomass with one type of coking coal. Montiano et al. [32,33] studied the carbonization of industrial coal blends, containing more than six different coals, with the addition of two types of raw woody biomass: chestnut sawdust and pine sawdust. Solar et al. [14] studied the carbonization of an industrial coking coal blend with the addition of 0.3–5% of one type of charcoal. Ueki et al. [12], Florenonti et al. [31], and Yustanti et al. [28] studied the carbonization of coking coals with bio-coals produced from woody raw biomass or agricultural waste pre-treated at different temperatures and, therefore, with different VM and ash contents. Others [23,24,26] have studied the carbonization of charcoal with different types of coals. No studies were found in which the aim was to understand the effect on carbonization and coke reactivity from bio-coal with different origins (different ash composition and cell structure) and different pre-treatments (pyrolysis degree, compaction through pelletizing) when adding up to 10% of the bio-coal in a coking coal blend consisting of three typical coking coals in the ratio used at industrial coking plants for the production of high-quality coke. Such a study would give information about the relative impact of applying bio-coal additions at an industrial coking plant.

In this study the effect on carbonization from the addition of different types of pre-treated biomass to coking coal blends consisting of three coking coals, as typically used in European industrial plants, was investigated. The bio-coals used were of different pyrolysis degrees and origin, therefore, their properties differed, e.g., in terms of structure, VM, and ash composition, and this may influence their individual effects during coking and the properties of the produced bio-coke. The aim was to understand the impact of different bio-coal types on the carbonization and quality of bio-coke, depending on their properties, i.e., cell structure and reactivity, at added amounts of 5% or 10%.

2. Materials and Methods

2.1. Materials

Three different types of coking coals, i.e., with low, medium, and high content of VM, were used in the coking coal blend, with or without the addition of bio-coal. Four types of bio-coals were selected based on their differences in VM and ash contents due to different pre-treatment temperatures and the origin of materials. The proximate and ultimate analysis, as well as the ash composition, for the bio-coals and the coking coals was analyzed according to the standard methods of ALS Scandinavia AB [34] and Swedish steel producer SSAB Europe in Luleå [35], respectively.

Carbonaceous materials are shown in Table 1; namely, torrefied biomass (TSD and TFR) from BioEndev AB (Holmsund, Sweden) [36] and Sveaskog AB (Stockholm, Sweden) [37], respectively, high-temperature torrefied biomass (HTT) from BioEndev AB and pyrolyzed biomass (CC) from Vindelkol AB (Vindeln, Sweden) [38], and three types of coking coals provided by SSAB Europe in Luleå, Sweden. HTT was densified through pelletizing before torrefaction.

Table 1. Selected carbonaceous materials with abbreviations and their approximate pre-treatment temperatures.

Carbonaceous Materials	Biomass Type	Temperature (°C)	Abbreviation
Torrefied forest residue	Wood chips of tops and branches of pine/spruce	270	TFR
Torrefied sawdust	Pine/Spruce	290	TSD
High temperature torrefied	Pelletized saw dust of pine/spruce	400	HTT
Charcoal	Mixture of pieces of pine, birch, alder, aspen	550	CC
High Volatile coal	-	-	HV
Medium volatile coal	-	-	MV
Low volatile coal	-	-	LV

The proximate and ultimate analyses are presented in Table 2. The pre-treatment temperature of biomass affects their properties, e.g., VM, oxygen, and fixed carbon (C_{fix}). The bio-coals prepared at low temperature (TFR and TSD) had a high content of VM and a low content of C_{fix}; the opposite was the case for biomass pre-treated at high temperature (HTT and CC). The ash compositions are stated in Table 3. TFR has higher contents of catalytic components in terms of K_2O, CaO, Fe_2O_3, and Na_2O in comparison to TSD. The content of the catalytic components K_2O, Fe_2O_3, and Na_2O is higher in HTT than in CC.

Table 2. Proximate and ultimate analysis for carbonaceous materials (dry base).

Abbreviation	Proximate Analysis (wt%)			Ultimate Analysis (wt%)				
	*C_{fix}	VM	Ash	C_{tot}	H	N	S	O
TFR	25.1	72.7	2.20	58.0	5.30	0.48	0.029	34.0
TSD	26.1	73.5	0.40	57.5	5.50	<0.1	<0.01	36.5
HTT	69.5	29.2	1.30	79.0	4.00	0.11	0.008	15.5
CC	80.7	18.6	0.70	87.0	3.40	0.25	<0.004	8.30
HV	61.5	32.3	6.10	81.3	5.15	1.57	0.85	4.96
MV	67.4	24.0	9.0	81.4	4.46	1.86	0.50	3.10
LV	70.0	19.4	10.6	79.7	4.27	1.79	0.63	3.16

C_{fix} fixed carbon; VM volatile matter, C_{tot} total carbon; H hydrogen; N nitrogen; S sulfur; O oxygen.
* C_{fix} = 100% − (%Ash + %VM).

Table 3. Oxide content in the carbonaceous materials (wt%, dry basis).

Abbreviation	Al_2O_3	CaO	SiO_2	Fe_2O_3	K_2O	MgO	MnO	Na_2O
TFR	0.006	0.71	0.062	0.009	0.23	0.10	0.0053	0.018
TSD	0.005	0.16	0.028	0.007	0.073	0.023	0.017	-
HTT	0.020	0.31	0.26	0.086	0.14	0.058	0.039	0.02
CC	0.006	0.32	0.028	0.009	<0.002	0.11	0.044	<0.009
HV	1.79	0.074	3.28	0.507	0.010	0.049	-	0.0015
MV	2.85	0.11	4.68	0.482	0.007	0.067	-	0.0033
LV	3.11	0.12	6.25	0.409	0.012	0.069	-	0.0070

The true density of the bio-coals was measured in a Micrometric Accupyc 1340 pycnometer (Micromeritics Instrument Corporation, South San Francisco, CA, USA). The sample was placed in a cell with a volume of ~37.67 cm^3 and the pores were filled with helium gas. The prepared sample had a particle size of 1–2 mm and was dried before measurements. The true densities were 1.34, 1.41, 1.44, and 1.51 g/cm^3 for TFR, TSD, HTT, and CC, respectively, and the trend was for increasing density with higher pre-treatment temperature.

The received data for the three different types of high-quality coking coals generated from Gieseler and dilatometer tests showed a max. dilatation and max. fluidity of 278%, 107%, and 69% and 30,000 ddpm, 1092 ddpm, and 182 ddpm, for HV, MH, and LV coals, respectively. The Gieseler softening and re-solidification temperature for these coking coals was within the temperature range of ~386–500 °C.

2.2. Methodology

2.2.1. Thermoplastic Properties of Coking Coals Evaluated Using an Optical Dilatometer

The heating of a basic coal blend, BB, with and without a 5% and 10% addition of bio-coals was performed in an optical dilatometer from Leitz (Ernst Leitz Gmbh, Wertzlar, Germany), with an automatic image analysis system from Hesse Instruments (Hesse, Osterode am Harz, Germany). To secure the correct temperature readings, measurements were also conducted on zinc metal wire, of purity 99.996%, and a thin piece of a bismuth-tin alloy, with known melting points of 419.3 °C and 137 °C, respectively. The measured melting points were 409 °C and 137 °C, respectively.

The change in sample height and area with temperature was recorded and evaluated for similar parameters, as described in the standard ISO 23873. The main difference in comparison to the standard method is the lack of an applied load on the sample and the significantly smaller sample size. Due to the sample size and inhomogeneity of coals and bio-coals, variations in measurements can be expected. The variation test was repeated three times for two different coking coal blends with bio-coal, the standard deviation and standard error for the max dilatation were 0.98 and 0.56, respectively, for the first sample, and for the second sample 2.00 and 1.15, respectively. The samples in these tests were 3 mm in diameter and 3 mm in height, compared to 60 mm in length and 8 mm in diameter in the standard method. The raw materials were finely pulverized in a mortar and mixed with water to achieve smooth sample surfaces. The sample was heated at a heating rate of 3 °C/min to a final temperature of 550 °C. Based on the area measurement of the sample, the swelling index, SI (%), was calculated according to Equation (1).

$$\text{SI (\%)} = \frac{\text{Change in area}}{\text{Original area}} \times 100 \tag{1}$$

2.2.2. Preparation of Coke

The recipes with relative additions of coking coals and bio-coal in the coal blend given in percentage weight are presented in Table 4. As seen, the ratios of bio-coals added were 5% or 10%. The blends with different bio-coal types added and the reference blend are presented in Table 5. The carbonaceous materials used were sieved and added to the coal blend in the proportions 19–21 wt% of +2.8 mm fraction, 30–34 wt% of a fraction −0.5 mm, and the rest of the material was in between, +2.8–0.5 mm. The coking coals and bio-coal were mixed carefully, and water was added to reach ~7.5% moisture. The mix was transferred into a graphite crucible with an inner diameter of 70 mm and an inner height of 210 mm, and the crucible was then placed in a Tamman furnace (Berlin, Germany), as shown in Figure 1. Two-thirds of the crucible were packed with moist carbonaceous materials to achieve a bulk density of ~800 kg/m^3. Furthermore, to record the temperature of carbonaceous materials during the coking, a thermocouple (k-type) in a protection tube of alumina was placed in the middle of the sample bed. A thermocouple (s-type) was used between the graphite crucible and the graphite tube to monitor the heating of the Tamman furnace.

Table 4. The mixing ratio of coal blend with bio-coal (wt%).

Coal Blend	Recipe (wt%)			
	HV	MV	LV	Bio-Coal
Basic coal blend (BB)	28.0	32.0	40.0	-
BB + 5% bio-coal	26.6	30.4	38.0	5.0
BB + 10% bio-coal	25.2	28.8	36.0	10.0

Table 5. Added bio-coals and abbreviations of coking coal blends for carbonization in the Tamman furnace and the produced cokes.

Bio-Coals	Coking Coal Blend	Coke
-	HV, LV and MV	BB
TFR	BB + 5%TFR	TFR5
	BB + 10%TFR	TFR10
TSD	BB + 5%TSD	TSD5
	BB + 10%TSD	TSD10
HTT	BB + 5% HTT	HTT5
	BB + 10% HTT	HTT10
CC	BB + 5% CC	CC5
	BB + 10% CC	CC10

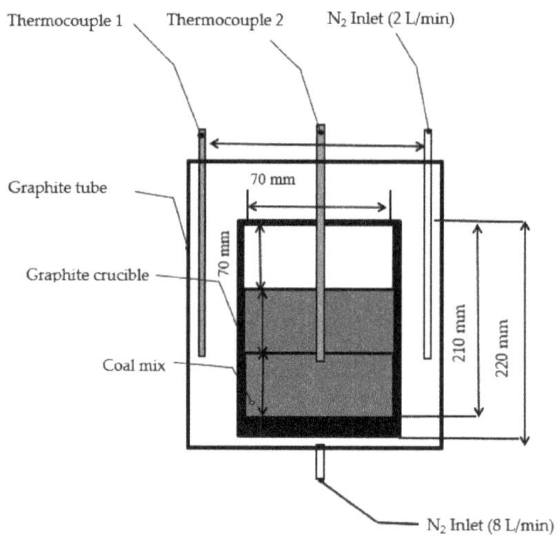

Figure 1. Schematic setup for cokemaking in the Tamman furnace.

The coal blends were heated under inert conditions using nitrogen gas (purity 99.996% and a flow rate of 10 L/min), according to the heating profile stated in Figure 2. The temperature increase was initially slow, and the samples were kept for ~1.5 h at 100 °C, then the sample was heated up to 1050 °C, and the sample was maintained for 1.5 h at 1050 °C, and finally the furnace was turned off and the coke was allowed to cool in N_2 gas atmosphere. The produced coke was discharged from the crucible, crushed, and screened into a fraction of 1–2 mm and kept in a desiccator until characterization was carried out.

The coke yield was calculated according to Equation (2) and is presented in Table 8.

$$\text{Coke yield } (\%) = \frac{m_1}{m_0(1 - \text{moisture } \% \times 0.01)} \quad (2)$$

m_1 is the remaining mass of coking coal blend after carbonization and m_0 is the amount of coking coal blend before carbonization.

Selected coal blends, BB, TSD5, and HTT5, were also carbonized in a 10-kg-retort at DMT GmbH & Co. KG in Essen, Germany [39]. The carbonaceous materials used were sieved to achieve the proportions 18–20 wt% of +3.15 mm a fraction, and 30–34 wt% of a fraction −0.5 mm, and the rest was between 3.15 and 0.5 mm. The small-scale coking test retort that was charged with approximately ~11 kg of coking coal blend had a height of 50 mm and a diameter of 180 mm [40,41]. The set points for moisture content and bulk density (wet) were ~9% and 800 kg/m^3, respectively. The duration of carbonization at DMT was approximately 4 h, and the final coke temperature was ~1020 °C. The coke was analyzed for CRI, CSR (ISO 1889-4), and mechanical strength, using the Micum test, according to ISO556, but modified for smaller sample sizes according to [40]. CRI corresponds to the weight loss due to the carbon solution loss reaction of carbon in coke under a CO_2 atmosphere at 1100 °C for 2 h. Due to the gasification reaction consuming carbon, the outer layer of coke particles becomes porous and the mechanical strength against abrasion drops. To measure this effect, the reactivity test was followed by determination of CSR in a tumbler test, which corresponds to the percentage of particles that remain larger than 10 mm after 600 rotation [1]. The M40 and M10 deduced from the Micum test describe the resistance to physical degradation and were calculated from the percentage material remaining of +40 and −10 mm, respectively, after 100 revolutions. In Europe, the Micum (M40/M10) test for cold mechanical strength is commonly used and the acceptable ranges are M40 > 80% and M10 < 7% [17].

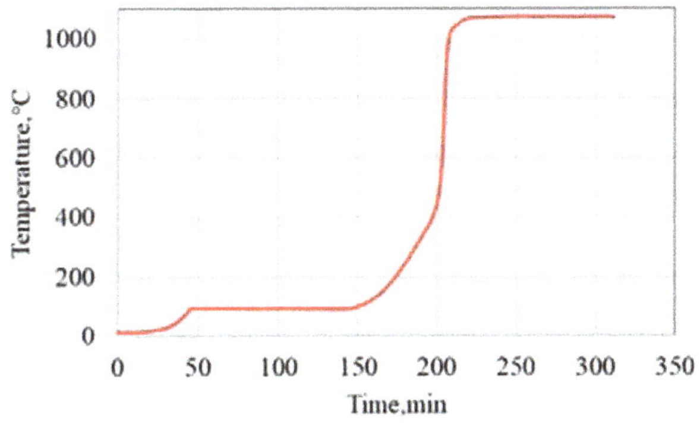

Figure 2. Typical temperature profile during carbonization in the Tamman furnace.

The textures of bio-coal, coke, and bio-cokes were investigated using a light optical microscope (LOM, Nikon ECLIPSE E600 POL, Tokyo, Japan). The samples were mounted in epoxy resin and the surfaces were polished before the studies. The chemical composition of coke prepared in the laboratory and technical scale was determined by SSAB Luleå using Thermo an ARL 9900 X-ray fluorescence (XRF) instrument (Zagreb, Croatia) with a rhodium tube at 50 kV and 50 mA.

2.2.3. Thermogravimetric Analysis

In this study two thermogravimetric analyzers were used. A TGA, Netzsch STA 409 instrument (sensitivity ±1 µg, Columbus, OH, USA) attached to a Quadruple Mass Spectroscope (QMS, Netzsch, Selb, Germany) was used to monitor the mass loss and off-gas analysis, respectively, during the pyrolysis of carbonaceous materials. The TGA/QMS used in this study is described in detail in reference [42]. A TGA equipment Netzsch STA 409, with graphite furnace [8], was used when studying the reactivity of the coke and bio-coke.

Carbonization was done using TGA/QMS equipment. A sample of ~150 mg was placed in an alumina crucible and heated according to the heating profile presented in

Figure 3. The carbonization was performed in an inert atmosphere of N_2 gas (purity 99.996%) having a flow rate of 200 mL/min.

Reactivity tests were performed using a CO_2 atmosphere at a rate of 300 mL/min for the coke and bio-coke samples that were carbonized in the Tamman furnace and at DMT. Particles of 1–2 mm size were used. Approximately 40–50 mg of the sample was placed in an alumina crucible with low edges, to avoid accumulation of formed CO in the sample and heated in TGA equipment and heated at 20 °C/min from room temperature up to 600°C, and at 3 °C/min between 600 and 1100 °C.

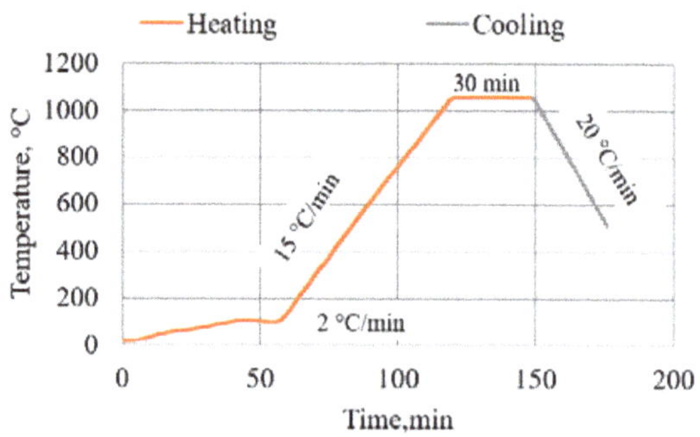

Figure 3. Heating profile used for the carbonization of carbonaceous materials in the TGA tests.

The accuracy of the balance and temperature for the graphite furnace was checked using calcium oxalate and gold in an argon atmosphere up to 1000 °C and 1100 °C, respectively. The mass loss for calcium oxalate was 12.3%, 19.1%, and 30.3% up to 300 °C, 600 °C and 1000 °C, respectively, which can be compared with the reference data of 12.3%, 18.9%, and 30.2%, with a tolerance of ±0.2%. The known melting point for gold is 1064 °C, and the measured one was 1069 °C. The influence of the possible inhomogeneity in the coke sample was checked in a TGA, Netzsch STA 409 Spectroscope (QMS, Netzsch, Selb, Germany), and three reactivity tests was performed for one sample, while the TGA result showed consistent results without significant variations; the standard deviation for mass loss was ±1%.

3. Results

3.1. Bio-Coals and Coal Properties

3.1.1. Structure of Bio-Coals

Figure 4 shows LOM images of the four different types of bio-coals used in the coking coal blend. It is noted that the structure of the TFR is characterized by a spherical and compact structure, while samples of TSD showed a duct structure with elongated pores, as seen in Figure 4a and 4b, respectively. From the areas studied with the microscope, it was indicated that the TSD sample contained cells of larger size, 26–48 µm, compared to TFR, which had cells of sizes around 9–24 µm. The structure of CC had a regular distribution of pores in a size between 40 and 79 µm, and it had a duct structure, as seen in Figure 4d. The texture of the HTT sample contained more collapsed pores, with a pore size between 6 and 11 µm, as seen in Figure 4c.

3.1.2. Carbonization Behavior of Single Coals or Bio-Coals

Table 6 shows the mass loss for carbonaceous materials at different temperatures. TSD and TFR had the most mass loss in the temperature range 200–400 °C, while the CC and HTT had a main mass loss above 500 °C. It is seen that TSD had a higher mass loss than TFR at 200–400 °C. At temperatures above 400 °C, TFR had a slightly higher mass loss than TSD. In general, the mass loss of bio-coals is higher than for coals. The maximum mass loss for coals occurs in the temperature range 400–500 °C.

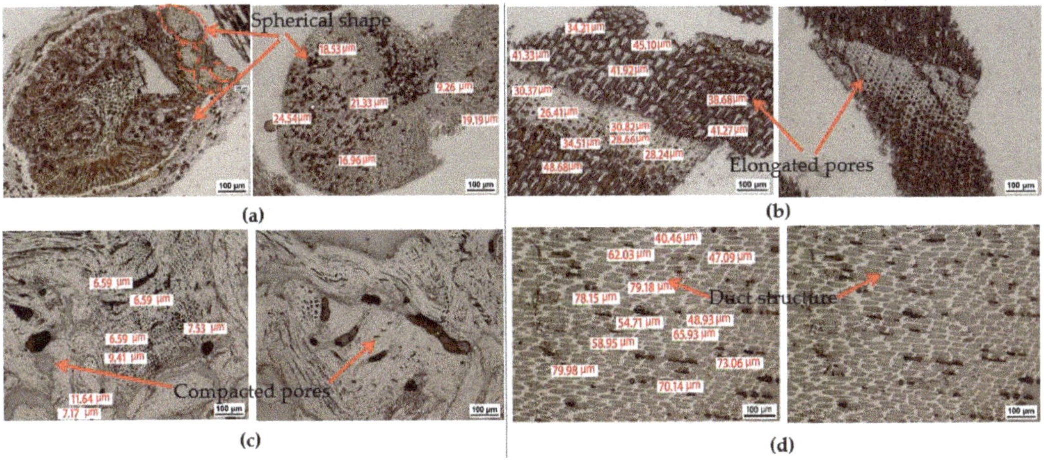

Figure 4. Microstructure images of bio-coals (**a**) TFR, (**b**) TSD, (**c**) HTT, and (**d**) CC.

Table 6. Mass loss (wt%) for coals and bio-coals during carbonization.

Abbreviation	200–400 °C	400–500 °C	500–800 °C	800–1050 °C
TFR	42.7	10.8	6.69	1.10
TSD	56.3	7.36	5.46	0.80
HTT	4.30	5.66	14.9	1.20
CC	3.63	3.93	13.3	1.22
HV	2.27	14.5	7.87	1.40
MV	1.79	11.1	8.83	1.17
LV	1.11	7.14	8.54	1.25

During the thermal decomposition of bio-coals and coals the gases H_2, CO, CO_2, H_2O, and ionized hydrocarbons such as CH_4^+, $C_2H_5^+$ were detected. However, the lengths of the carbon chains in the released hydrocarbons were probably initially longer, before thermal decomposition and excitation in the QMS. The highest ion currents recorded by QMS were for m/z of 2, 18, 15, 29, and 44; judged to represent H_2, H_2O, CH_4^+, $C_2H_5^+$, and CO_2, respectively. The recorded ion currents for H_2, H_2O, CO_2, and CH_4^+ are shown in Figure 5. Hydrocarbon such as CH_4^+ was detected at 290 °C for torrefied biomass (TFR and TSD) and at 455 °C for pyrolyzed biomass (HTT and CC), see Figure 6a. Overlapping between the peaks for ionized hydrocarbons CH_4^+ was seen within the temperature range 430–630 °C for coals and all types of bio-coals. The release of H_2 started at 310 °C for TFR and TSD, while it started at 500 °C for pyrolyzed bio-coals and coals, see Figure 5b. CO_2 was detected only for TFR and TSD within a temperature range of 119–770 °C, with a slightly higher intensity peak at 350 °C for TSD, see Figure 5c. H_2O was detected from 115 °C until 1050 °C for all types of bio-coals, but TFR and TSD had a higher intensity peak at 340 °C, see Figure 5d.

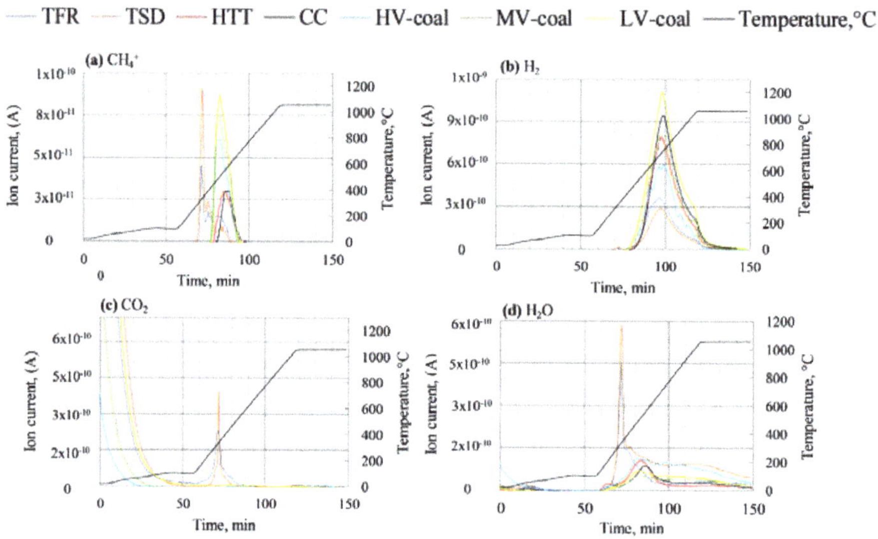

Figure 5. Off gas analysis of (**a**) CH_4^+; (**b**) H_2; (**c**) CO_2; and (**d**) H_2O for different types of bio-coals and coals at carbonization conditions in N_2 atmosphere.

3.2. Carbonization of Selected Coking Coal Blends in TGA

The TGA results for the carbonization of BB and coal blend with a 10% addition of TFR and TSD are presented in Figure 6. The total mass loss achieved for TFR10 and TSD10 was higher than for BB. The devolatilization of TFR10 and TSD10 started at ~290 °C and 340 °C, respectively, compared to at ~400 °C for the BB. DTG analysis shows that the TFR10 and TSD10 had two devolatilization steps, while the main devolatilization for BB occurred in one step, above 400 °C.

Hydrocarbon, such as $C_2H_5^+$, was detected at 300 °C for TFR10 and TSD10, while it was detected at 400 °C for BB, as seen in Figure 7a. CH_4^+ was detected in the temperature range of ~400–600 °C, and the release of H_2 started at ~430 °C for all coking coal blends, see Figure 7b and 7c, respectively. The release of CO_2 was negligible in all coking coal blends and is therefore not presented.

Plastic Properties of Coking Coal Blends

The results from the optical dilatometer tests for BB and blends with 5 or 10% addition of bio-coals are shown in Table 7. The results show that the max. dilatation of CC5 and HTT5 was quite similar to BB but dropped for the other coking coal blends. The drop in max. dilatation was more pronounced for coking coal blends containing 10% of bio-coals, especially when using the torrefied bio-coals TSD and TFR. The swelling index, SI, decreased with the addition of bio-coals to the BB, and the decrease was more significant for the addition of TFR, TSD, and HTT in comparison to CC. The coking coal blends produced from the mixtures BB, TSD5, and HTT5 were also analyzed at DMT using DIN 51739. The max. dilatation was found to be 48, −10, and 41 for BB, TSD5, and HTT5, which was a similar trend as in the modified tests in the optical dilatometer.

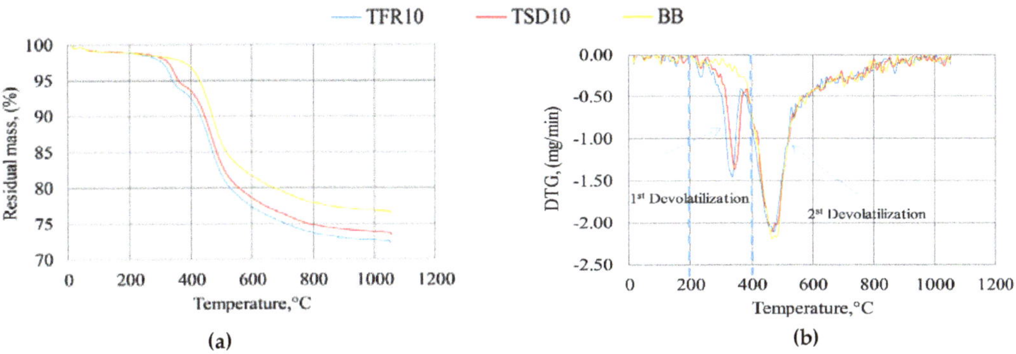

Figure 6. (**a**) TGA (**b**) DTG for TFR10, TSD10, and BB at carbonization conditions in N_2 atmosphere.

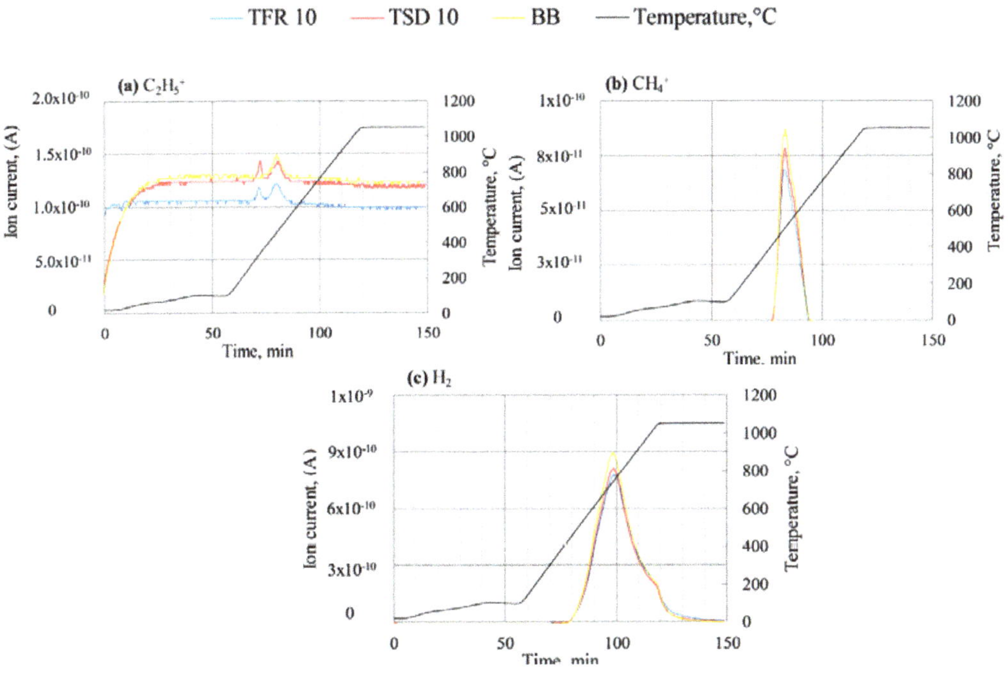

Figure 7. Off gases detected by QMS when testing coking coal blends at carbonization conditions in the TGA, (**a**) $C_2H_5^+$, (**b**) CH_4^+, and (**c**) H_2; for TFR10, TSD10, and BB.

Table 7. Thermoplastic parameters for the basic coal blend, BB, and coal blends with the addition of bio-coals, deduced from optical dilatometer tests.

Coal Blend	Max. Dilatation		SI
	°C	%	%
BB	436	15.8	68
TFR5	425	10.4	22
TSD5	430	8.29	19
HTT5	423	15.2	20
CC5	421	16.3	31
TFR10	346	−0.38	-
TSD10	406	−1.50	2
HTT10	417	8.03	2
CC 10	428	9.77	18

SI: Swelling index.

3.3. Properties of Coke

3.3.1. Laboratory Cokemaking

Table 8 shows the coke yields for the produced coke and bio-coke. As could be expected, the coke yield was lower for bio-coke-containing torrefied bio-coals, TFR, and TSD, in comparison to bio-coke containing the pyrolyzed ones, HTT and CC.

Table 8. Coke yield (%) of coke and bio-coke.

BB	TFR5	TFR10	TSD5	TSD10	HTT5	HTT10	CC5	CC10
76.5	70.4	74.0	75.1	73.6	78.5	77.0	77.7	76.5

3.3.2. Structure of Bio-Cokes and Chemical Composition

The structure of the produced coke at laboratory and technical scales was examined in LOM. Figure 8 shows the TSD5 sample, which indicates that bio-coals possibly retained their structure after carbonization. The elongated duct cell shape seen for the TSD was also found in the coke, as seen in Figures 8a and 8b, respectively.

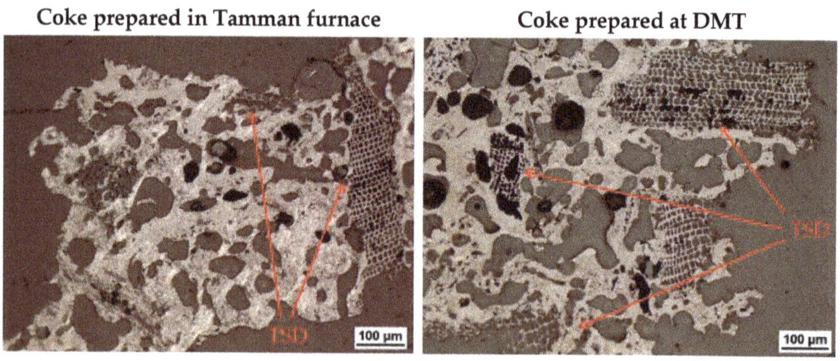

(a) (b)

Figure 8. LOM images showing the microstructure for coke and bio-coke samples for (**a**) TSD5 and (**b**) TSD5-DMT.

Analyses using XRF of bio-coke ash showed that there was no significant difference in the ash amount or composition between the produced bio-cokes when 5 or 10% bio-coal was added, as the coals in the basic blend were mainly influencing the average ash composition. However, local variations in ash composition due to the presence of bio-coal ash can still be important.

3.3.3. Reactivity of Coke Prepared at Lab and Technical Scale

Table 9 shows the mass loss of the sample during the reactivity test of coke produced from BB and bio-coke containing 5% and 10% bio-coal. The TGA analysis showed that bio-coke containing 5% and 10% bio-coal is more reactive than the BB. HTT5 is an exception, with similar gasification as BB. TFR5 had a higher mass loss than other bio-cokes, with 5% bio-coal addition. Bio-coke TSD5 and CC5 had quite similar mass losses. HTT10 had a higher mass loss than the other bio-cokes with 10% bio-coal addition, already occurring at lower temperatures. The gasification of bio-coke containing 5% bio-coal started at lower temperatures in comparison to BB and the start of reaction temperature further decreased with 10% bio-coal in the coking coal blend.

Table 9. Temperature for the start of gasification and accumulated mass loss at different temperatures, as measured in TGA.

Types	Samples	Start of Gasification (°C)	Accumulated Mass Loss (%), Up to Each Temperature			
			950 °C	1000 °C	1050 °C	1100 °C
Technical scale coke produced in retort at DMT	BB	953	0.00	0.96	4.36	11.5
	TSD5	943	0.86	3.13	7.69	15.9
	HTT5	908	1.80	4.25	7.94	14.5
Laboratory coke produced in a Tamman furnace	BB	964	0.23	1.63	4.94	11.2
	TFR5	895	2.72	5.30	10.8	19.8
	TSD5	912	1.42	3.34	7.65	15.4
	HTT5	905	1.49	3.30	6.36	11.9
	CC5	919	1.50	4.31	9.65	17.9
	TFR10	869	3.45	6.51	12.1	20.2
	TSD10	888	2.00	4.48	9.90	18.9
	HTT10	884	4.49	8.68	13.2	20.7
	CC10	875	2.44	6.07	12.0	20.9

The quality of coke carbonized at technical scale was analyzed for CRI, CSR, and Micum test, see Table 10. The results showed a higher mass loss for TSD5 compared to HTT5. The same trend was found for the reactivity measured by TGA for lab and technical scale produced coke.

Table 10. Quality of coke prepared at technical scale.

Abbreviations	CRI/CSR Test		Drum Test/Micum	
	CRI	CSR	M40	M10
BB	28.6	57.6	80.7	7.0
TSD5	36.5	46.6	76.4	7.7
HTT5	31.8	57.8	78.5	7.8

The drum test results for bio-cokes prepared at DMT showed that the M40 for bio-coke with 5% bio-coal addition of HTT was quite similar to BB, and this indicates that HTT5 had a high strength. The true density was measured for bio-coke samples carbonized at laboratory and technical scales, and the results were quite equal, 1.89 and 1.82 g/cm^3, respectively.

4. Discussion

4.1. Influence of Bio-Coal Carbonization on the Thermoplastic Properties of Coking Coal

The maximum dilatation for coking coal blends decreased with 5% bio-coal addition of torrefied bio-coals (TFR, TSD), whereas it was not significantly changed for CC and HTT. The maximum dilatation decreased further with a 10% bio-coal addition, especially for TFR and TSD. It has been stated in the literature that the release of H_2 [9,43,44] can influence fluidity. However, the release of H_2 from bio-coals occurred in a similar temperature range as for coking coals, and can, thus, not explain the impact on fluidity from bio-coal addition. From the results, it is noted that the release of hydrocarbons (for example CH_4^+ detected in

QMS) from HTT and CC occurred in the plasticity temperature range for the coals, possibly contributing to the better development of plasticity in these blends.

In comparison to HTT or CC, a significantly larger effect on max dilatation was seen when adding TSD and TFR, especially at 10% addition, which could have been due to the higher O_2 content in TFR and TSD, 34% and 36.5%, respectively, compared to 15.5% and 8.3% for HTT and CC. Mochizuki et al. and Tsubouchi et al. [45,46] found that heteroatoms such as oxygen have a deleterious effect on the plastic properties of coal. From these results, the VM released from bio-coals including a high amount of oxygen is likely to inhibit the thermoplastic properties of the coal, which was discussed in [13,14]; however, the use of charcoal with a low oxygen content mainly has effects as an inert material. It was reported that the addition of inert materials to coking coal may cause a reduction in thermoplastic properties, due to the effect from the high surface area of bio-coals acting as adsorbents of the primary decomposition products of coking coal, which binds the plasticizing part of the coal, favoring the inhibition of fluidity development [47].

According to the literature [48], the presence of a higher content of catalytic components, e.g., K_2O and CaO, in TFR, as mentioned in Table 3, could also have an impact on lowering the dilatation in TFR5 and TFR10. However, as the maximum dilatation is always lower for TSD, the oxygen content seems to be of greater importance. The low O_2 content in CC and HTT could be the reason for CC5 and HTT5 having similar maximum dilatations to BB. The CC5 results agree with Ng et al. [23], who found that the presence of a 5% charcoal addition to the blend did not affect the development of a plastic layer of coking coal.

4.2. *Influence of Bio-Coal Addition on Coke Quality*

There are many parameters, i.e., cell structure of bio-coal, amount of reactive carbon originating from bio-coal present in coke, and contribution of catalytic oxides in the ash, that simultaneously affect bio-coke reactivity. Therefore, it is difficult to make a precise conclusion about which parameter is most important.

The TGA results show that the start of the carbon gasification reaction was lowered by a higher bio-coal addition, and this could have been due to an increase in the content of reactive bio-coal carbon that was gasified at a lower temperature in comparison to carbon originating from coking coals. The lowering of the start of gasification temperature for bio-coke, in comparison to a reference coke, was reported in [18,25,27], and according to Suopajärvi et al. [25] it was lowered by 30–50 °C for bio-coke containing 5% charcoal. In this study, the temperature was lowered by 45 °C for a corresponding addition of CC. As expected, in bio-cokes containing bio-coal with a higher content of catalytic components in the ash, the start of gasification occurred at lower temperature and the reactivity was higher, as seen for TFR5 and TFR10, in comparison to TSD5 and TSD10.

The cell structure, i.e., open or closed, may also affect bio-coke reactivity, as it has an impact on the gas diffusion. The compact cell structure of HTT could be the reason for HTT5 having almost the same behavior as BB; but with a higher content of carbon originating from HTT in HTT10, the reactivity is higher than for BB coke, which reflects the complex set of parameters influencing coke reactivity. The CRI and CSR test results for the cokes produced at technical scale show that the difference between HTT5 and BB is minor, and this indicates that it could be possible to maintain the properties of industrial coke with a 5% HTT addition. The CRI test results for cokes produced at technical scale showed that the CRI values increased in the order BB < HTT5 < TSD5, and vice versa for CSR. The higher CRI for TSD5 could have been due to the difference discussed above; high VM and oxygen content may contribute to the higher porosity of produced bio-coke, which enhances the diffusion of CO_2. The low mechanical strength of TSD5 produced at technical scale could have been due to the higher oxygen content in TSD, and this will have a deleterious effect on the plastic properties of coal. The CRI results for the technical scale coke followed the same trend as the reactivity tests conducted in TGA. Hayashizaki et al. [49] reported that the coke strength largely correlated to the dilatation. When the coal dilatation is insufficient, the voids between coal particles will not fill up and coal particles will not be tightly bonded together, which decreases the coke strength. This could have also influenced the diffusion of gas within the coke piece.

5. Conclusions

This study aimed to understand the impact of different bio-coal types from different origins and pyrolysis degrees, with added amounts of 5% or 10% to an industrially used coal blend, on carbonization and quality of bio-coke structure and reactivity. The following conclusions can be drawn:

- Pyrolyzed bio-coals, such as charcoal and high temperature torrefied bio-coal having a devolatilization behavior such as coking coals, did not have a significant effect on the fluidity up to a 5% addition; while they had a moderate effect on lowering the fluidity at 10% addition.
- Torrefied bio-coals, such as forest residue and sawdust, with a high content of volatile matter releases hydrocarbons, such as CH_4^+, at lower temperature than coking coals and have a high O_2 content. These properties of bio-coals lead to deleterious effects on the plastic properties of coking coal blends, especially at 10% addition. However, for bio-coals with a low volatile and oxygen content, the plastic properties, as shown in the optical dilatometer test, are less influenced.
- All addition of bio-coal in cokemaking lowers significantly the starting temperature for gasification, in comparison to reference coke; with differences in the content of catalyzing compounds having a smaller effect.
- Coke produced at laboratory and technical scales from coking coal blends with 5% addition of high-temperature torrefied pelletized sawdust or 5% torrefied sawdust show a similar trend in reactivity as measured using thermogravimetric analysis.
- Technical scale produced types of bio-coke showing low reactivity in the thermogravimetric analysis also showed a high quality in standard tests for reactivity, strength after reaction, and mechanical strength, when technical scale produced bio-coke was tested. This indicates that the coking coal blend with a 5% high temperature torrefied bio-coal could be suitable for industrial use.

Author Contributions: Conceptualization, B.B., L.S.Ö. and M.L.; methodology, A.A.E.-T. and A.R.; formal analysis and investigation, A.A.E.-T.; writing—original draft preparation, A.A.E.-T.; writing—review and editing, L.S.Ö., B.B., M.L. and A.A.E.-T.; supervision, L.S.Ö. and B.B.; project administration, L.S.Ö., B.B. and M.L. All authors have read and agreed to the published version of the manuscript.

Funding: This research was funded by Swedish Research Council for Sustainable Development (FORMAS). The work was carried within the project "Bio-coal as raw material in coke for lower CO_2-emission in metal production". Grant number FR-2018/0010.

Data Availability Statement: The data presented in this study are available from the corresponding author, upon reasonable request.

Acknowledgments: Financial support from Swedish Research Council for Sustainable Development (FORMAS) is gratefully acknowledged. In addition, the additional funding provided by Center of Advanced Mining and Metallurgy (CAMM) at Luleå university of technology is acknowledged. For additional support, the following companies are acknowledged; Bioendev AB, Sveaskog AB and SSAB.

Conflicts of Interest: The authors declare no conflict of interest.

References

1. Geerdes, M.; Chaigneau, R.; Kurunov, I. *Modern Blast Furnace Ironmaking: An Introduction*; IOS Press: Amsterdam, The Netherlands, 2015; ISBN 1614994994.
2. Suopajärvi, H.; Fabritius, T. Towards more sustainable ironmaking—an analysis of energy wood availability in finland and the economics of charcoal production. *Sustainability* **2013**, *5*, 1188–1207. [CrossRef]

3. WorldSteel Association. Available online: https://www.worldsteel.org/publications/bookshop/product-details~{}Steel-s-contribution-to-a-low-carbon-future--update-version-coming-shortly-~{}PRODUCT~{}contribution-low-carbon-pp~{}.html (accessed on 18 September 2020).
4. Gupta, R.C. Woodchar as a sustainable reductant for ironmaking in the 21st century. *Miner. Process. Extr. Metall. Rev.* **2003**, *24*, 203–231. [CrossRef]
5. Özçimen, D.; Ersoy-Meriçboyu, A. Characterization of biochar and bio-oil samples obtained from carbonization of various biomass materials. *Renew. Energy* **2010**, *35*, 1319–1324. [CrossRef]
6. Sundqvist Ökvist, L.; Lundgren, M. Experiences of Bio-Coal Applications in the Blast Furnace Process—Opportunities and Limitations. *Minerals* **2021**, *11*, 863. [CrossRef]
7. Nomura, S.; Ayukawa, H.; Kitaguchi, H.; Tahara, T.; Matsuzaki, S.; Naito, M.; Koizumi, S.; Ogata, Y.; Nakayama, T.; Abe, T. Improvement in blast furnace reaction efficiency through the use of highly reactive calcium rich coke. *ISIJ Int.* **2005**, *45*, 316–324. [CrossRef]
8. Lundgren, M. Development of Coke Properties during the Descent in the Blast Furnace. Ph.D. Thesis, Luleå Tekniska Universitet, Luleå, Sweden, 2013.
9. Fernández, A.M.; Barriocanal, C.; Díez, M.A.; Alvarez, R. Influence of additives of various origins on thermoplastic properties of coal. *Fuel* **2009**, *88*, 2365–2372. [CrossRef]
10. Guerrero, A.; Diez, M.A.; Borrego, A.G. Influence of charcoal fines on the thermoplastic properties of coking coals and the optical properties of the semicoke. *Int. J. Coal Geol.* **2015**, *147*, 105–114. [CrossRef]
11. Diez, M.A.; Alvarez, R.; Fernández, M. Biomass derived products as modifiers of the rheological properties of coking coals. *Fuel* **2012**, *96*, 306–313. [CrossRef]
12. Ueki, Y.; Nunome, Y.; Yoshiie, R.; Naruse, I.; Nishibata, Y.; Aizawa, S. Effect of Woody Biomass Addition on Coke Properties. *ISIJ Int.* **2014**, *54*, 2454–2460. [CrossRef]
13. Fraga, M.; Flores, B.; Osório, E.; Vilela, A. Evaluation of the thermoplastic behavior of charcoal, coal tar and coking coal blends. *J. Mater. Res. Technol.* **2020**, *9*, 3406–3410. [CrossRef]
14. Solar, J.; Caballero, B.M.; Barriocanal, C.; Lopez-Urionabarrenechea, A.; Acha, E. Impact of the Addition of Pyrolysed Forestry Waste to the Coking Process on the Resulting Green Biocoke. *Metals* **2021**, *11*, 613. [CrossRef]
15. Kumar, P.P.; Barman, S.C.; Singh, S.; Ranjan, M. Influence of coal fluidity on coal blend and coke quality. *Ironmak. Steelmak.* **2008**, *35*, 416–420. [CrossRef]
16. Coking Coal. Available online: https://www.ispatguru.com/coking-coals/ (accessed on 21 June 2021).
17. Díez, M.A.; Alvarez, R.; Barriocanal, C. Coal for metallurgical coke production: Predictions of coke quality and future requirements for cokemaking. *Int. J. Coal Geol.* **2002**, *50*, 389–412. [CrossRef]
18. Ng, K.W.; MacPhee, J.A.; Giroux, L.; Todoschuk, T. Reactivity of bio-coke with CO_2. *Fuel Process. Technol.* **2011**, *92*, 801–804. [CrossRef]
19. Gupta, A.; Thengane, S.K.; Mahajani, S. CO_2 gasification of char from lignocellulosic garden waste: Experimental and kinetic study. *Bioresour. Technol.* **2018**, *263*, 180–191. [CrossRef] [PubMed]
20. Grigore, M.; Sakurovs, R.; French, D.; Sahajwalla, V. Influence of mineral matter on coke reactivity with carbon dioxide. *ISIJ Int.* **2006**, *46*, 503–512. [CrossRef]
21. Nomura, S.; Kitaguchi, H.; Yamaguchi, K.; Naito, M. The characteristics of catalyst-coated highly reactive coke. *ISIJ Int.* **2007**, *47*, 245–253. [CrossRef]
22. Babich, A.; Senk, D.; Gudenau, H.W. Effect of coke reactivity and nut coke on blast furnace operation. *Ironmak. Steelmak.* **2009**, *36*, 222–229. [CrossRef]
23. Ng, K.W.; Giroux, L.; MacPhee, T.; Todoschuk, T. Incorporation of charcoal in coking coal blend-A study of the effects on carbonization conditions and coke quality. In Proceedings of the AISTech 2012: Proceedings of the Iron & Steel Technology Conference, Atlanta, GA, USA, 7–10 May 2012; pp. 225–236.
24. Ng, K.W.; Giroux, L.; MacPhee, T.; Todoschuk, T. Biofuel Ironmaking Strategy from a Canadian Perspective: Short-Term Potential and Long-Term Outlook. In Proceedings of the 1st International Conference on Energy Efficiency and CO2 Reduction in the Steel Industry, Düsseldorf, Germany, 27 June–1 July 2011.
25. Suopajärvi, H.; Dahl, E.; Kemppainen, A.; Gornostayev, S.; Koskela, A.; Fabritius, T. Effect of charcoal and Kraft-lignin addition on coke compression strength and reactivity. *Energies* **2017**, *10*, 1850. [CrossRef]
26. MacPhee, J.A.; Gransden, J.F.; Giroux, L.; Price, J.T. Possible CO_2 mitigation via addition of charcoal to coking coal blends. *Fuel Process. Technol.* **2009**, *90*, 16–20. [CrossRef]
27. Flores, B.D.; Flores, I.V.; Guerrero, A.; Orellana, D.R.; Pohlmann, J.G.; Diez, M.A.; Borrego, A.G.; Osório, E.; Vilela, A.C.F. Effect of charcoal blending with a vitrinite rich coking coal on coke reactivity. *Fuel Process. Technol.* **2017**, *155*, 97–105. [CrossRef]
28. Yustanti, E.; Wardhono, E.Y.; Mursito, A.T.; Alhamidi, A. Types and Composition of Biomass in Biocoke Synthesis with the Coal Blending Method. *Energies* **2021**, *14*, 6570. [CrossRef]
29. Rejdak, M.; Bigda, R.; Wojtaszek, M. Use of alternative raw materials in coke-making: New insights in the use of lignites for blast furnace coke production. *Energies* **2020**, *13*, 2832. [CrossRef]
30. Matsumura, T.; Ichida, M.; Nagasaka, T.; Kato, K. Carbonization Behaviour of Woody Biomass and Resulting Metallurgical Coke Properties. *ISIJ Int.* **2008**, *48*, 572–577. [CrossRef]

31. Florenonti-Madiedo, L.; Casal, D.; Díaz-Faes, E.; Barriocanal, C. Effect of sawdust addition on coking pressure produced by two low vol bituminous coals. *J. Anal. Appl. Pyrolysis* **2017**, *127*, 369–376. [CrossRef]
32. Montiano, M.G.; Díaz-Faes, E.; Barriocanal, C.; Alvarez, R. Influence of biomass on metallurgical coke quality. *Fuel* **2014**, *116*, 175–182. [CrossRef]
33. Montiano, M.G.; Díaz-Faes, E.; Barriocanal, C. Effect of briquette composition and size on the quality of the resulting coke. *Fuel Process. Technol.* **2016**, *148*, 155–162. [CrossRef]
34. ALS Scandinavia AB. Available online: https://www.alsglobal.se/en (accessed on 15 September 2020).
35. SSAB. Available online: https://www.ssab.se/ssab-koncern/om-ssab/produktionsorter-i-sverige/lulea (accessed on 15 September 2020).
36. BioEndev AB. Available online: http://www.bioendev.se/ (accessed on 15 September 2020).
37. Sveaskog AB. Available online: https://www.sveaskog.se (accessed on 15 September 2020).
38. Vindelkol AB. Available online: https://vindelkol.se/ (accessed on 15 September 2020).
39. DMT. Available online: https://www.dmt-group.com/ (accessed on 21 September 2020).
40. Gajic, D. Influences on Bulk Density and Its Effect on Coke Quality and Internal Gas Pressure. In Proceedings of the AISTech—Iron and Steel Technology Conference Proceedings, Pittsburgh, PA, USA, 6–9 May 2013; pp. 307–319.
41. Gajic, D.; Kaiser, M. New findings from the DMT small-scale coking test retort regarding coke quality and coke oven wall safety. In Proceedings of the AISTech 2012: Proceedings of the Iron & Steel Technology Conference, Atlanta, GA, USA, 7–10 May 2012; pp. 237–246.
42. El-Tawil, A. Bio-Coal as an Alternative Reducing Agent in the Blast Furnace. Ph.D. Thesis, Luleå University of Technology, Luleå, Sweden, 2020.
43. Nomura, M.; Kidena, K.; Hiro, M.; Murata, S. Mechanistic study on the plastic phenomena of coal. *Energy Fuels* **2000**, *14*, 904–909. [CrossRef]
44. Grint, A.; Mehani, S.; Trewhella, M.; Crook, M.J. Role and composition of the mobile phase in coal. *Fuel* **1985**, *64*, 1355–1361. [CrossRef]
45. Mochizuki, Y.; Naganuma, R.; Tsubouchi, N. Influence of inherently present oxygen-functional groups on coal fluidity and coke strength. *Energy Fuels* **2018**, *32*, 1657–1664. [CrossRef]
46. Tsubouchi, N.; Mochizuki, Y.; Naganuma, R.; Kamiya, K.; Nishio, M.; Ono, Y.; Uebo, K. Influence of inherent oxygen species on the fluidity of coal during carbonization. *Energy Fuels* **2016**, *30*, 2095–2101. [CrossRef]
47. Loison, R.; Foch, P.; Boyer, A. *Coke: Quality and Production*; Elsevier: Amsterdam, The Netherlands, 2014; ISBN 1483165272.
48. Khan, M.R.; Jenkins, R.G. Influence of added calcium compounds on swelling, plastic, and pyrolysis behaviour of coal devolatilized at elevated pressures. *Fuel* **1986**, *65*, 1203–1208. [CrossRef]
49. Hayashizaki, H.; Hayashi, Y.; Kubota, Y.; Uebo, K.; Nomura, S. *Development of Coal Blending Technology for Improvement of Coke Quality*; Technical Report; Nippon Steel: Tokyo, Japan, 2020.

metals

Article

A Comparison of Laboratory Coal Testing with the Blast Furnace Process and Coal Injection

Julian Steer [1,*], Mark Greenslade [2] and Richard Marsh [1]

[1] Cardiff School of Engineering, Cardiff University, Queen's Buildings, The Parade, Cardiff CF24 3AA, UK; MarshR@cardiff.ac.uk
[2] Tata Steel UK, Port Talbot SA13 2NG, UK; mark.greenslade@tatasteeleurope.com
* Correspondence: SteerJ1@cardiff.ac.uk; Tel.: +44-29-20-870599

Abstract: The injection of coal through tuyeres into a blast furnace is widely adopted throughout the industry to reduce the amount of coke used and to improve the efficiency of the iron making process. Coals are selected depending on their availability, cost, and the physical and chemical properties determined by tests, such as the volatile matter content, fixed carbon, and ash content. This paper describes research comparing the laboratory measured properties of injection coals that were used over a two-month production period compared to the process variables and measurements of the blast furnace during that study period. In addition to the standard tests, a drop tube furnace (DTF) was used to compare the burnout of coals and the char properties against the production data using a range of statistical techniques. Linear regression modelling indicated that the coal type was the most important predictor of the coal rate but that the properties measured using laboratory tests of those coals were a minor feature in the model. However, comparisons of the Spearman's correlations between different variables indicated that the reverse Boudouard reactivity of the chars, prepared in the DTF from the coals, did appear to be related to some extent to the coal and coke rates on production. It appears that the constant process adjustments made by the process control systems on the furnace make it difficult to identify strong correlations with the laboratory data and that the frequency of coal sampling and the coal blend variability are likely to contribute to this difficulty.

Keywords: coal injection; blast furnace; drop tube furnace; statistical correlation; production; ironmaking

Citation: Steer, J.; Greenslade, M.; Marsh, R. A Comparison of Laboratory Coal Testing with the Blast Furnace Process and Coal Injection. *Metals* **2021**, *11*, 1476. https://doi.org/10.3390/met11091476

Academic Editors: Timo Fabritius and Pasquale Cavaliere

Received: 19 July 2021
Accepted: 15 September 2021
Published: 17 September 2021
Corrected: 10 January 2022

1. Introduction

Coal injection into a blast furnace is a very well-established technique used as a means to improve the efficiency of the iron-making process and to reduce the amount of coke charged to the furnace, which in turn reduces particulate and environmental emissions [1–3]. Coal is milled to either a pulverised or granulated particle size and carried pneumatically through a lance into the tuyeres [4]. These water-cooled nozzles direct the hot blast of air into the furnace, forming a balloon-like void known as the raceway, dependent on the blast pressure, material consumption, and injectants [5]. Although coke is necessary to support the raw material burden and provide a porous network for gas to ascend the furnace, the furnace operators try to maximise the injection of coal [6].

Many furnaces achieve coal injection rates of 200 kg/tHM (kg/tonne of hot metal) [7]; however, the scale and nature of this process can result in variations in characteristics such as temperature and pressure [8]. The impact of this can be localised hot/cold spots, which can cause damage to the furnace lining or result in variations in the rate of raw material consumption. Alternatively, it can result in erratic burden descent, causing sudden changes resulting in "hanging or slipping" of the burden [9]. Variations in these properties describe the blast furnace stability and are very important for consistent, predictable, and efficient iron production that does not cause furnace damage and prolongs its lifespan.

How much coke can be replaced by coal depends on the role it has as a structural support for the raw material burden and depending on the coke replacement ratio [7,10]. The latter is calculated using an equation based on the carbon, hydrogen, ash, and moisture content necessary to achieve a replacement of coke with an equivalent carbon content of 87.5% [4]. In addition, the volatile matter content, which is a measure of the coal thermal pyrolysis products, is known to have an important effect on the blast furnace stability by influencing the gas/char combustion and consumption of oxygen in the raceway region [11,12]. The choice of coals injected in the furnace depend on their availability, cost, and properties and are often used as blends, variations of which can impact the stability [13].

The proximate analyses (volatile matter content, ash, moisture, and fixed carbon) are used as important information for selecting coal [14,15]; however, it is not the only way to assess the suitability of injectant coals [11]. The raceway region of the furnace is likely to show much variation in dimensions and is characterised by short residence times [16] and rapidly diminishing availability of oxygen for combustion [12]. For this reason, partially burnt coal chars ascend the shaft, and their properties and reactivity play a role in the suitability of injectants. To assess this, coal samples have also been run through a drop tube furnace to measure the coal burnout and to prepare and collect partially burnt chars for further reactivity characterisation.

This paper compares the properties of coal injectant samples measured in the laboratory with blast furnace process information from runs over a study period of production at TATA Steel Port Talbot. The aim of this work was to compare any correlations between laboratory testing and production variables, using the SPSS® statistical package, by examining and modelling the information to determine relationships that could be applied in the future to help maximise coal injection rates.

2. Materials and Methods

Two separate study periods were chosen for the coal sampling and testing. The first period consisted of 24 days sampling/measurement, and the second consisted of 33 days sampling/measurement. Samples were taken directly from the production coal injection line twice daily and were tested in the laboratory for comparison with the process information (PI) of the blast furnace during the corresponding time period.

The laboratory tests for a specific coal or coal blend would correspond to the composite sample taken during a specific 12 h production period for comparison with the production variability over that period.

This study used data and analyses from three sources to understand the laboratory testing of injection coals as a potential predictor for blast furnace performance. A statistical package was used to analyse the data and investigate the relationships between these:

- Proximate and thermo gravimetric analysis (TGA) measured the ash content, volatile matter content, fixed carbon, and reverse Boudouard gasification.
- A Drop tube furnace (DTF) (manufactured by Severn Thermal Solutions, Dursley, Gloucester, UK) was used to measure the combustion burnout of production samples of coals or coal blends in the DTF and to produce partially burnt chars for TGA analysis.Blast furnace process information (PI data) was used to compare data from the blast furnace on inputs, such as coal addition rates and blast volume, and on process measurements, such as blast pressures and production rates, etc.

2.1. Proximate Analysis and Thermo Gravimetric Analysis

For the study period, production granulated coal samples were taken daily from a sample port on the coal injection line during the day and night shift. The samples were dried at 105 °C using BS11722:2013 until a constant weight and the volatile matter content was measured using standard BS15148:2005. Ash contents were carried out using the standard method BS 1171:2010.

Figure 1 shows the range of measured volatile matter contents (11–24%) for the coal samples collected over the duration of the trials, which covers low to medium volatile matter injection coals/blends. Coal D shows a particularly wide variation between 13–24%. This suggests the possibility of sample contamination, delivery contamination on the stockyard, or a particularly wide inconsistent delivery of the coal. In this case, comparisons for this coal are considered less reliable.

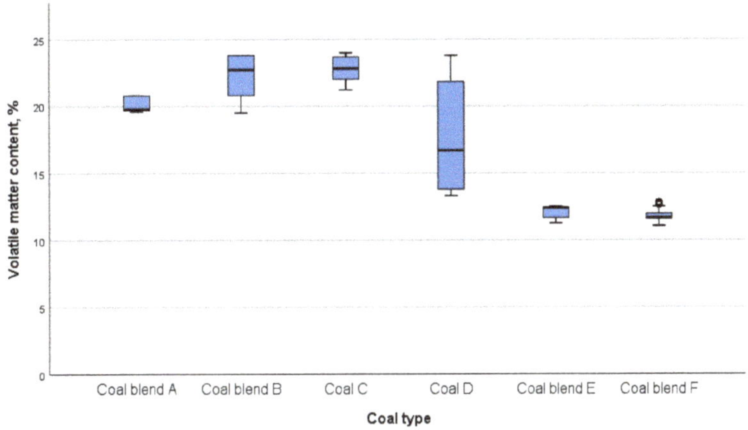

Figure 1. The range of the measured volatile matter contents for oven-dried coals and coal blends used during the study periods (○ represents outliers).

The ash contents shown in Figure 2 range from 6.8–10.9%, with the coal blend B measuring the greatest variation of 7.5–11%.

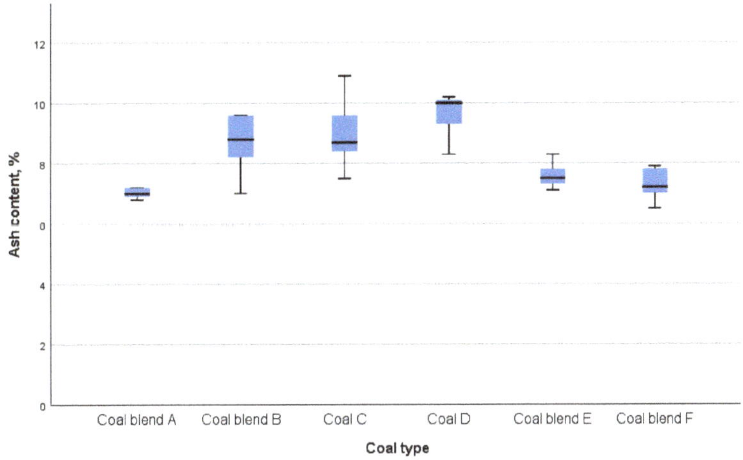

Figure 2. The ranges of the measured ash contents for the oven-dried coals and coal blends used during the study periods.

The variation of the fixed carbon content, shown in Figure 3, was much less than the other proximate analyses except for the coal D, which was expected since the volatile matter range was so wide.

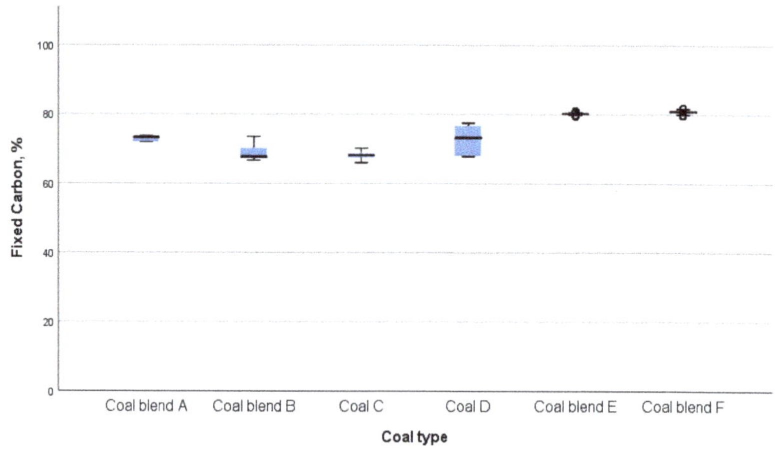

Figure 3. The ranges of the measured ash contents for the oven-dried coals and coal blends used during the study periods (○ represents outliers).

Partially burnt chars that exit the raceway of the blast furnace will ascend the furnace where thermal decomposition can continue and where gasification can occur due to the reverse Boudouard reaction when carbon dioxide (CO_2) reacts with carbon, gasifying it to carbon monoxide (CO). A Mettler–Toledo TGA/DSC 3+ was used to monitor the weight loss by first heating to 900 °C in nitrogen and holding for 7 min to devolatilise the sample then switching to a CO_2 flow rate of 100 mL/min until complete conversion was obtained. Gasification reactivity is defined as $t_{0.5}$, the time taken in minutes to achieve 50% conversion of the sample. The conversion is calculated using the following equation, where w_{intial} is the initial mass, w_t is the measured mass at any given time, and w_{final} is the final mass. Comparison of the conversion in this way allows the "like for like" comparison of chars with different ash contents.

$$Conversion\ (\%) = 100 \times \frac{w_{intial} - w_t}{w_{initial} - w_{final}}$$

Thus, more reactive chars have shorter $t_{0.5}$ gasification times compared to less reactive chars, which have longer $t_{0.5}$ gasification times.

2.2. Drop Tube Furnace

The drop tube furnace (DTF) shown in Figure 4 is a vertical tube furnace used to characterise the devolatilisation and burnout of coal samples at 1100 °C in air at a residence time of 100 ms and to prepare chars for analysis using the TGA. The high heating rate and short residence times of the DTF has characteristics similar to those when coal is injected into the blast air of the blast furnace raceway [17,18]. Samples were fed into the top at feed rates of 30 g/h, entrained in a laminar air flow at 20 L/min, and collected at the bottom by means of a cyclone collector. The ash tracer method was used to calculate the burnout of the coals [19].

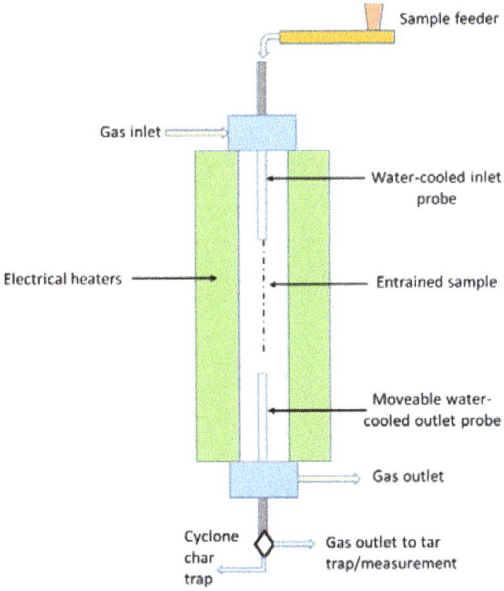

Figure 4. Schematic of drop tube furnace.

2.3. *Blast Furnace Process Information*

Process information (PI) data were obtained directly from the blast furnace process control system; this included real-time measurements taken directly from the furnace and control variables (Table 1). Because of the practicalities of organizing a large-scale trial, it was not possible to run under absolute steady state parameters or to choose predefined production conditions, so variation occurred during what was a normal processing period. Additionally, for the purposes of this work, only coal addition rates above 100 kg/tHM (kg/tonne of hot metal) were used, as levels below this were associated with reduced production regimes and problematic blockages.

Table 1. Process variables and measurements.

PI Variable	Description
Date	Both date and time (average of 5 min responses)
Dust	Blast furnace slurry mass (tonnes)
Volatiles	Coal volatile matter content (%)
Ash	Ash matter content (%)
Sulphur	Sulphur content of coal (%)
Less125mcn	Coal particle-size distribution less than 125 µm (%)
Burnout	Coal sample burnout in a drop tube furnace (%)
Gasification	Gasification reactivity time to achieve 50% conversion ($t_{0.5}$, min)
Silicon	Silicon content in metal (%)
O_2enrichment	Oxygen enrichment of hot air blast (%)
O_2 flow	Oxygen flow rate (Nm^3/h)
Prodrate	Iron production rate, tonnes of hot metal per hour (tHM/hr)
Coal	Coal or coal blend type
Cokerate	Addition rate of coke (kg/tHM)
Coalrate	Addition rate of coal (kg/tHM)
TotRedt	Total addition rate of coke and coal (kg/tHM)
BlastPressure	Hot air blast pressure (bar)

Table 1. *Cont.*

PI Variable	Description
TopPressure	Pressure at the top of the furnace (bar)
DeltaPressure	Difference in pressure between top and bottom of furnace (bar)
Flametemp	Calculated adiabatic flame temperature (°C)
HeatGain	Heat balance at the furnace walls
CO_2	CO_2 in top gas exiting the furnace (%, vol)
CO	CO in top gas exiting the furnace (%, vol)
H_2	H_2 in top gas exiting the furnace (%, vol)
N_2	N_2 in top gas exiting the furnace (%, vol)
TotalVol	Total gas volume rate (Nm^3/h)
MaxDiff	Pressure difference between top/bottom of blast furnace (bar)
TopPressSetpoint	Top pressure setpoint (bar)
Distribution	A number assigned to the pattern of burden addition
TopTemp	Temperature of gas exiting the top of the furnace (°C)
Blast	Volume rate of hot blast (Nm^3/h)
Carbon	Carbon in metal (%)

2.4. Statistical Analysis

The IBM software package SPSS® version 26 was used to interrogate the results using three statistical approaches. Firstly, Spearman's correlations to compare the relationship of laboratory test results with blast furnace variables and measurements; secondly, with box plots to examine the variation of results for the different coals and coal blends used during the production period; and thirdly, using multiple regression to form a model that describes the relationship between variables.

The SPSS™ multiple regression model is an example of a linear model used to predict a target based on linear relationships between the target and one or more predictors based on continuous variables.

3. Results and Discussion

3.1. Spearman's Correlation

Correlation coefficients are a useful statistical technique to visualise and quantify linear relationships between two variables. In this case, the laboratory measurements were related to changes in the blast furnace process variables.

Pearson's coefficients are used to measure how closely the relationship between the variables follows a straight line when plotted. However, in some cases, variables do show a relationship, but it is not a linear one, and this is where Spearman's correlation coefficients are useful. These are calculated on ranks instead of data values which allows the comparisons of relationships that are non-linear.

The data shown in Table 2 is divided into moderate and high correlations. The highest correlations are associated between the blast furnace process variables. As expected, the production rate of iron from the furnace shows a high correlation with process variables, such as the blast volume, the total reductant rate, and the oxygen flow rate (see Figure 5 scatter plot with a Spearman's correlation of 0.819), as these are variables that are essential for the process to function. In this case, it is not possible to achieve high production rates without higher blast volumes, which also then correspond with increased pressures. This reinforces the importance of these variables to control the process, often preprogramed in automatically controlled systems, as a response to production parameters and illustrating how fluctuations in coal properties are masked behind these process variables.

Some of the moderate correlations also correspond to the blast furnace process information (PI) variables, such as oxygen enrichment with oxygen flow rate (oxyflow), and the oxyflow with blast pressure, top pressure, total volume, and nitrogen. However, in this case, high oxyflow is often accompanied with oxygen enrichment to facilitate higher production rates but is not always available to the blast furnace operator. Likewise, higher oxyflow means higher gas flow and corresponds to increases in the pressure; however, other non-measured variables, such as fine particulates in the burden, can also contribute to increased measured blast furnace pressure. In terms of the relationship between the laboratory analysis of the coal and chars with the blast furnace variables, moderate correlations were recorded for the char gasification reactivity with the coal rate (0.600 **) and coke rate (−0.685 **).

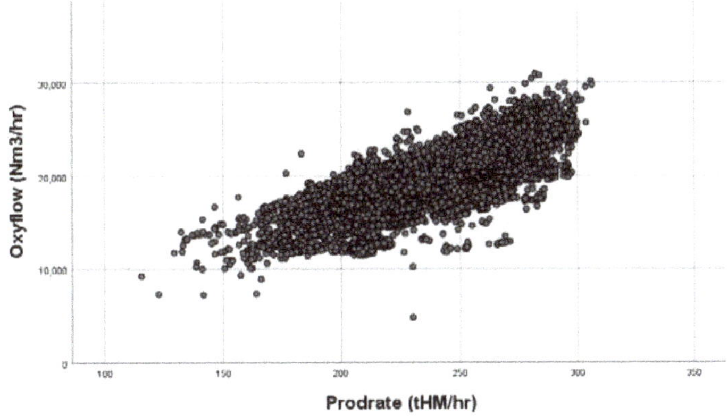

Figure 5. Scatter plot of the relationship between oxygen flow and the iron-production rate.

Table 2. Spearman's correlations of variables and measurements.

Dependent Variable	Independent Variable	Number of Data Points	Correlation Coefficient	Dependent Variable	Independent Variable	Number of Data Points	Correlation Coefficient
Gasification	Volatile matter	10856	−0.668 **	Production rate	Oxygen flow	17773	0.819
Sulphur	<125 μm particles	4708	0.669 **	Production rate	Blast pressure	17773	0.800
Burnout	Sulphur	7787	0.669 **	Production rate	Top pressure	17773	0.833
Burnout	<125 μm particles	4179	0.673 **	Production rate	Total volume	17773	0.903
Burnout	Gasification	11642	−0.643 **	Production rate	Blast volume	17773	0.844
Gasification	Coke rate	11915	−0.685 **	Coke rate	Coal rate	17773	0.842
Gasification	Coal rate	11915	0.600 **	Coke rate	Carbon dioxide	17773	−0.842
Oxygen enrichment	Oxygen flow	17773	0.730	Coal rate	Hydrogen	17773	−0.818
Oxygen enrichment	Nitrogen	17773	−0.636 **	Coal rate	Nitrogen	17773	−0.805
Oxygen flow	Blast pressure	17773	0.713	Blast pressure	Top pressure	17773	0.893
Oxygen flow	Top pressure	17773	0.641	Blast pressure	Total volume	17773	0.883
Oxygen flow	Total gas volume	17773	0.701	Blast pressure	Blast volume	17773	0.853
Oxygen flow	Nitrogen	17773	0.668 **	Top pressure	Total volume	17773	0.865
Coal rate	Carbon dioxide	17773	0.693 **	Top pressure	Blast volume	17773	0.862
Carbon dioxide	Nitrogen	17773	−0.690 **	Total gas volume	Blast volume	17773	0.966
Hydrogen	Nitrogen	17773	−0.750 **	-	-	-	-

Note: ** means significant at the 0.01 level (99%).

The raceway, formed by the hot blast directed into the furnace, is typically between 0.5–1.5 m in length, and the blast velocity is typically in the region of 180 m/s, so the residence time of coals injected into this region is short [16]. This means that the injected coal does not have long to burnout, and it is expected that partially burnt char particulates exit the raceway and ascend the furnace, where the reverse Boudouard reaction can take place between carbon dioxide and carbon to produce carbon monoxide. As a proxy for this, the reactivity of the partially burnt char residue from the incomplete coal combustion in the drop tube furnace was measured using a TGA. Increases in the measured char reactivity time ($t_{0.5}$) indicate slower char reaction with CO_2 and a less reactive char, whereas decreases in the char reactivity time indicate a more reactive char.

The moderate correlation of the char reactivity (formed in the laboratory DTF) with the coal rate (measured during blast furnace production) suggests a relationship between lower reactivity chars and higher blast furnace coal injection rates. Correspondingly, for the coke rate, the opposite was observed, where less reactive chars correlate with lower coke rates in the furnace.

It is understood that the reverse Boudouard reaction is endothermic, so more reactive chars could result in a larger cooling effect on the thermal reserve temperature of the blast furnace. This could be particularly problematic if the char retention by the burden is localised in certain regions due to retention by the raw material burden, leading to reduced blast furnace stability (temperature, pressure, and burden descent) and reduced coal rates.

Closer analysis of the scatter plot diagrams is necessary to interrogate the data more closely and verify that the data are not skewed or exhibiting bias. In the case of Figure 6, the data on the scatter plot are consistent with the −0.685 ** correlation of lower coke rates associated with lower char reactivates (indicated as longer gasification times).

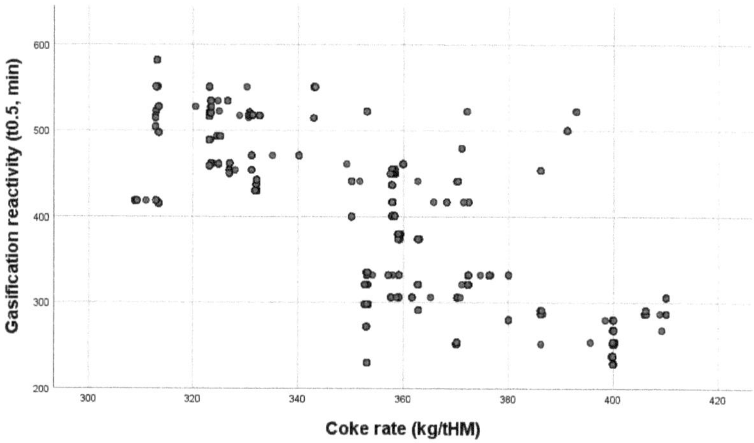

Figure 6. Scatter plot of the relationship between the gasification reactivity of the injectant coal char and the coke rate during blast furnace production.

The overall reductant charge to the blast furnace consists of both coke and coal and a closer look at the data in Figure 7 indicates that lower coke rates are associated with higher coal rates (high correlation 0.842 **), which is to be expected, as increased coal injection is used as a means to substitute the more expensive coke with its associated environmental emissions.

However, it should be noted that the scatter plot shows a stratification of the results into layers; this is due to wider variability in coal addition rates (injected into the bottom of the furnace) compared to the coke addition rates (added to the top of the furnace), which are added at a much more consistent rate. Coke is produced by a batch process, whereas coal is milled and injected as a continuous process, so additions are often subject to issues, such as blockages or variations in the milling process.

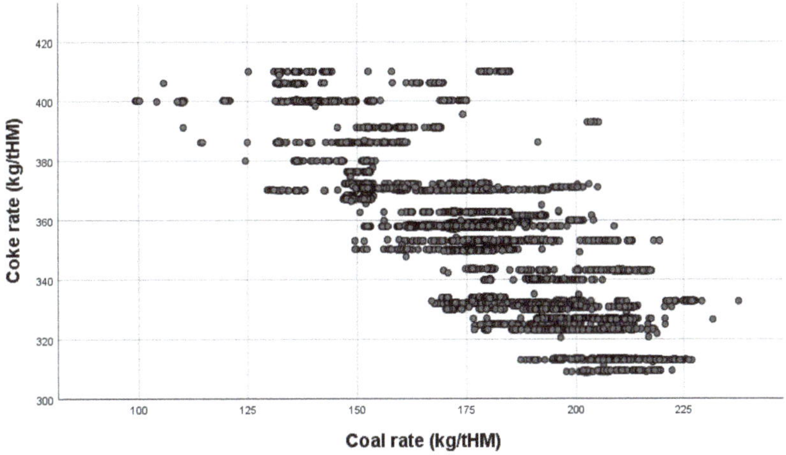

Figure 7. Scatter plot of the relationship between the injectant coal and the coke rate during blast furnace production.

The scatter plot for the char gasification against coal rate shown in Figure 8 is also consistent with the Spearman's correlation of 0.600.

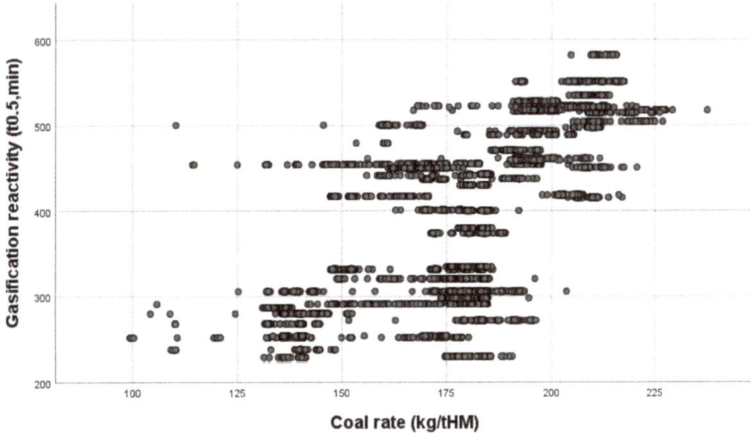

Figure 8. Scatter plot of the relationship between the gasification reactivity of the injectant coal and the coal injection rate during blast furnace production.

Figure 8 also shows some data stratified in layers. This is a reflection of the spread of values for the rate at which coal is injected (see Figure 9) during iron production, whereas the laboratory test results represent the average value of the coal sample collected on either the day or night shift (a composite of 12 h production).

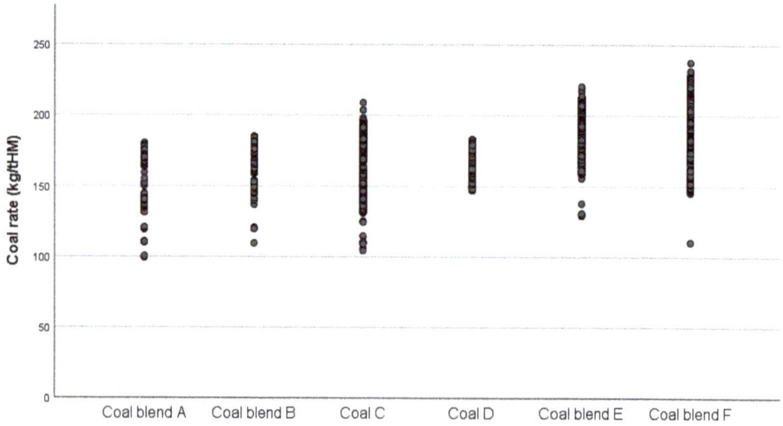

Figure 9. The range of the coal injection rates for each of the coals/coal blends used during the trial period of blast furnace production.

Comparing the gasification to the DTF burnout in Figure 10 and the coal volatile matter content in Figure 11 explains some of this relationship that higher char reactivities (lower gasification $t_{0.5}$ times) show a correlation with higher coal burnout (0.643) and higher volatile matter content (0.668); in both cases, these types of coals would combust to a greater extent, requiring more oxygen to maintain the injection rates of the respective injection coals.

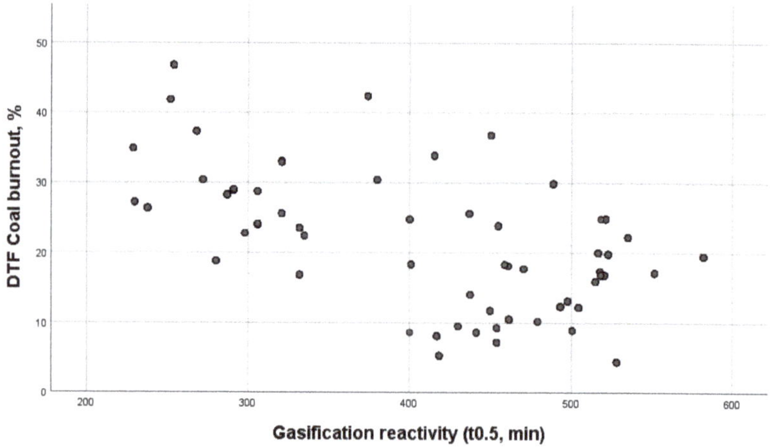

Figure 10. Scatter plot of the DTF burnouts of the injection coals versus their char gasification reactivities.

The large number of dependent and independent variables associated with iron making makes it challenging to draw reliable strong correlations, as the manual and automatic process control systems make adjustments to the process to compensate for changes associated with the different properties of the raw materials. It is also likely that there is considerable variation dependent on the coal milling process, sample properties, and coal blending.

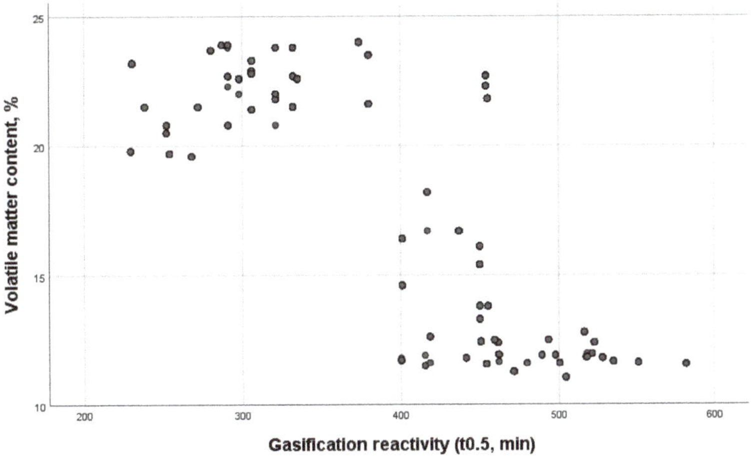

Figure 11. Scatter plot of the volatile matter content of the injectant coals versus their char gasification reactivities.

3.2. Box Plot Comparisons of Data

Another way to scrutinise the data is to use box plots; these give information on the spread of the results, their median, and the range. The box shows the range of results between the 25th percentile to the 75th percentile; inside of the box, the bold horizontal line shows the median; and finally, the whiskers that extend out of the box represent the minimum and maximum of the dataset (discounting any outliers which appear as circles).

All the box plots have been arranged in ascending order for the coal injection rate (Figure 12) where the box plots for the six different coal/coal blends used during the correlation trials are arranged from low to high coal rates. For Figure 12, the averages range from a minimum of 140 kg/tHM to a maximum of 200 kg/tHM. In contrast, Figure 13 is a plot of the coke addition rate for the time periods corresponding to the addition of different coals/coal blends during blast furnace production. The coke rate indicates a reverse trend, as less coke reductant is required with increasing coal injection rates.

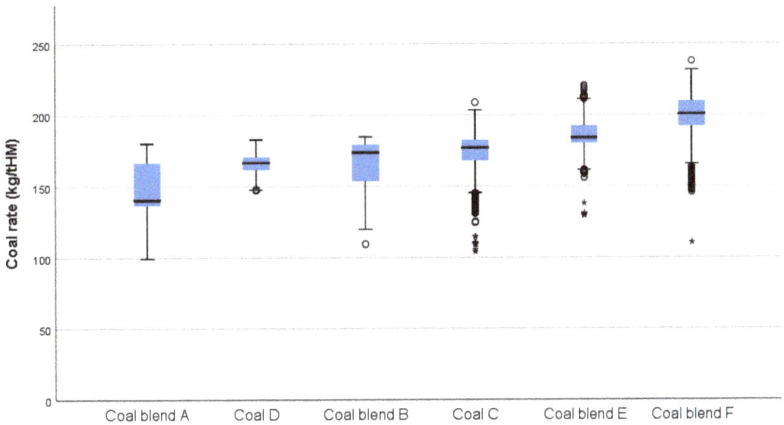

Figure 12. Box plots of the range of coal injection rates for each of the coals/coal blends used during blast furnace production (o represents outliers and * extreme outliers).

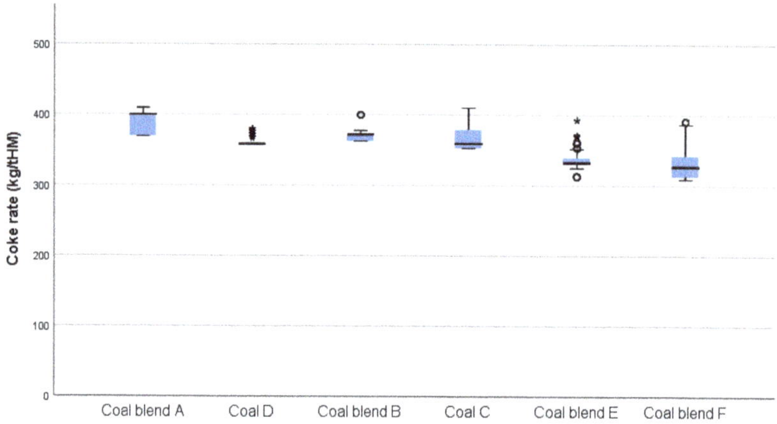

Figure 13. Box plots of the range of coke addition rates for the time periods corresponding to the addition of different coals/coal blends during blast furnace production (○ = high/low potential outliers and * = high/low extreme values).

In Figure 14 the data also show a similar ascending pattern that is in agreement with the Spearman's correlation for the gasification reactivity against coal rate (Figure 8, Table 2), indicating that lower DTF char reactivity is associated with higher blast furnace coal injection rates.

However, although five of the blends follow this trend, it is noted that the coal D does not. The proximate analysis of samples of this coal, taken during the six days that this coal was used for injection on production, had a wide variation (13–24%) about the mean for the volatile matter content (see Figure 1) and drop tube furnace burnout reactivity (see Figure 15) compared to the other blends.

Coal D was added to the blast furnace as a single coal injectant (not as blend with another coal), so a narrower variability might be expected, as the process of blending would not be such an issue. However, the wide range suggests the possibility of supply variation or contamination with other coals on the stockyard. In contrast, coal A had a very narrow gasification range over the 4 days that this coal was used for injection during production; the asterisks either side of the bar represent outlier results.

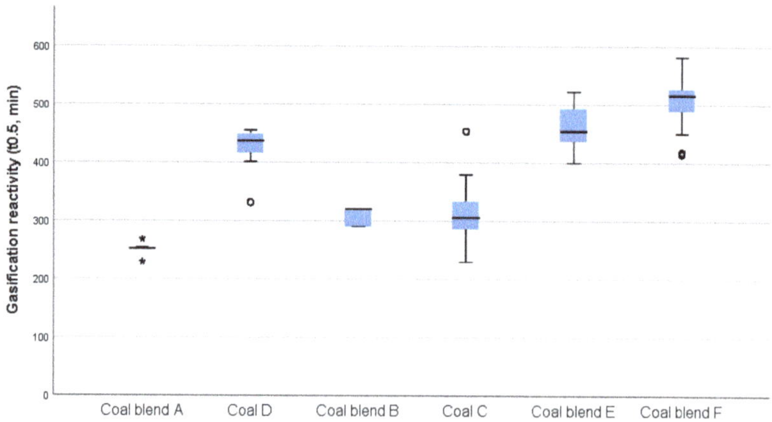

Figure 14. Box plots of the range of char gasification reactivity for each of the coals/coal blends used during blast furnace production (○ = high/low potential outliers and * = high/low extreme values).

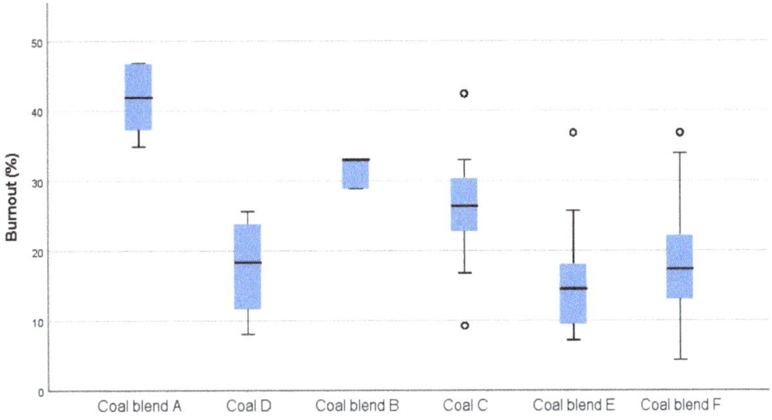

Figure 15. Range of coal burnout for each of the coals/coal blends used during observed blast furnace production (o = high/low potential outliers).

The coal burnout results shown in Figure 15 mirror the gasification results trend in reverse, indicating an association between lower DTF burnouts and higher coal rates in the blast furnace, except for Coal D.

3.3. Multiple Regression Modelling

Multiple regression modelling is used to give a more complete assessment of the process by comparing all the measured variables and parameters. In this way, it is used to predict the value of dependent variables from the multiple variables obtained from the blast furnace process information. This technique is used to determine the overall fit of the model and the relative contribution of each of the predictors to the total variance.

The model fitting for the blast furnace coal injection rate shown Figure 16 indicates that the data explains the relationship with a 95% accuracy, also shown by the relationship between the predicted versus actual coal rates in the scatter plot.

The model indicates that the most important predictor of the coal rate was the type of coal used, which is also observed from anecdotal blast furnace operator experience. However, the model did not indicate any relationship with the measured coal properties, such as volatile matter content or DTF burnout, etc. In contrast to the previous approaches, the multiple regression showed only a small relationship with char gasification.

However, in conjunction with the box plots, there is an indicative relationship of higher injection coal rates, lower burnouts, and lower char gasification rates.

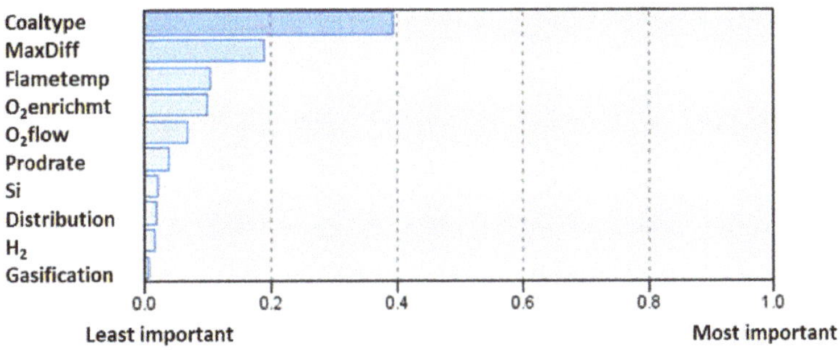

Figure 16. *Cont.*

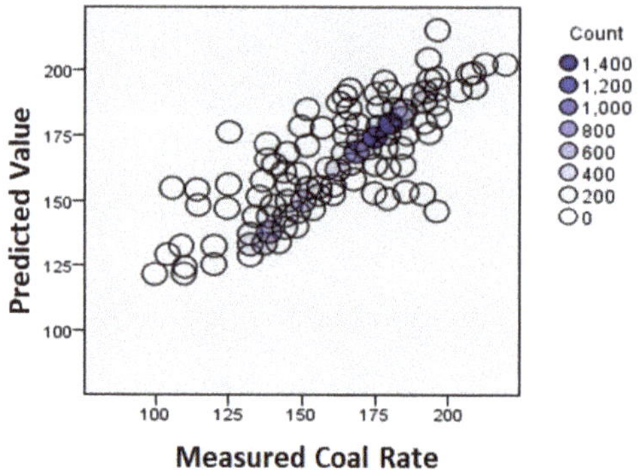

Figure 16. Relationship between the measured and predicted injection coal rate in the multiple linear regression model.

The reductant charged to the blast furnace is split between coke and coal; as one increases, the other is expected to decrease. Consequently, the type of injection coal is also the strongest predictor of the coke rate, too, and the model shown in Figure 17 explains the data with 87% accuracy. In the case of the coke addition rate, the maximum pressure difference was the highest predictor of coke rate, followed closely in second place by the injection coal type. Although this does not tell us specifically what coal property is important, in conjunction with the box plots in Figures 14 and 15, we can deduce that the lower burnouts and lower char gasification reactivities that are determined in the laboratory are important to the lower coke rates. This is particularly relevant, as coking coal is more expensive than injection coal and involves the additional process of coking, which can result in additional environmental emissions. Lower coke rates also relate to higher coal injection rates, which are linked to higher production efficiency.

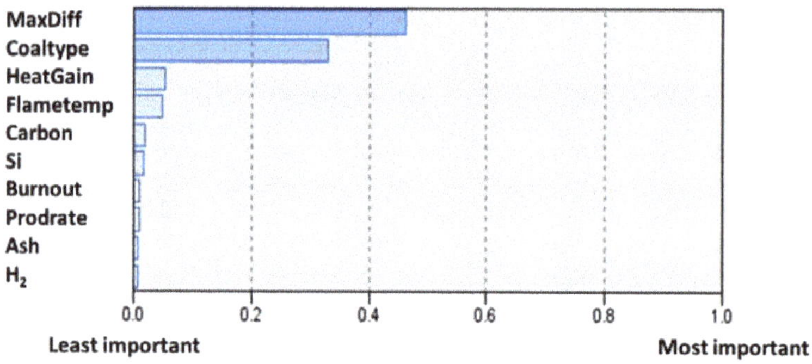

Figure 17. *Cont.*

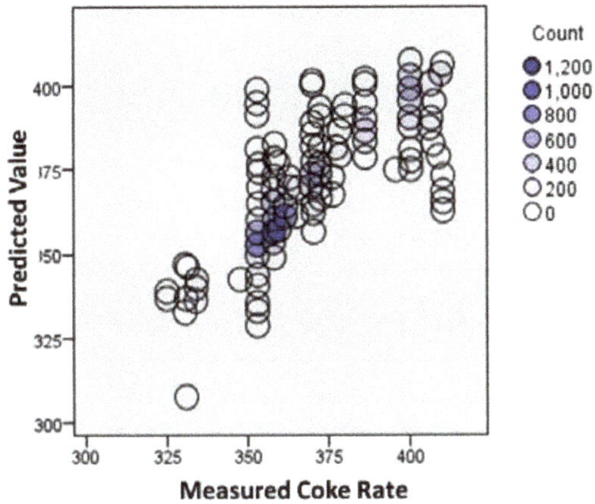

Figure 17. Relationship between the measured and predicted coke rate in the multiple linear regression model.

4. Conclusions

The aim of the study was to understand if laboratory test results could be related to the variables observed in a blast furnace during a production study; in this way, the laboratory testing of injection coals might be used as a predictor for the blast furnace performance of that process. Using the SPSS® statistical package, different statistical data analysis techniques were used to compare laboratory test results for the coals used over a two-month trial period against the process measurements for that production period. The main findings and conclusions from the study are summarised as follows:

- Strong correlations between the production variables, such as blast volume and oxygen enrichment, etc., make it difficult to differentiate more subtle relationships with the laboratory measured coal properties, as both operators and control systems make real-time adjustments for process variations.
- Combining the findings of the Spearman's correlation obtained from the scatter plots (Figures 6–8, Table 2), box plots (Figures 12–14), and multiple regression models (Figures 16 and 17) indicate that the coal injection rates through the blast furnace were higher for the chars (formed from the production coals tested in the drop tube furnace) with a lower gasification reactivity. Because efficient blast furnace operation aims to maintain consistent reductant additions, increased coal injection rates correspond with decreased coke rates accordingly so that there appears to be a relationship between lower gasification reactivity and lower coke rate.

Because the reverse Boudouard reaction is endothermic, it is thought that char with higher reactivity will consume more thermal energy and potentially reduce the thermal reserve of the blast furnace. This could result in a cooling effect, which could impact the stability of the process and the injection rates of coal obtainable with stable operation. This could be particularly problematic if the char deposition is localised in the furnace leading to blast furnace instabilities in the burden descent, pressure, and temperature variation.

- Both the box plot (Figure 12) and the multiple regression model (Figure 16) highlighted the importance of the type of coal used in the blast furnace compared to the coal rate. It was hoped that the variation in the laboratory analyses and char burnouts could be used to predict the addition rates; however, a consistent relationship was not observable across all the injection coals, suggesting the possibility of other non-tested variables, insufficient sampling frequency, or possibly sampling or testing inconsistencies.
- The composite coal samples were taken over a 12 h shift period. Because of this, the plotted results were sometimes stratified on the scatter plots, as the laboratory tests

referred to process variability over that production time period. This is a limitation, and more frequent sampling would have been preferable. Ideally, this study would have been carried out on single coal injection; however, coal blends were a limitation, as the practicalities of blending are known to result in variability that would not be accounted for.

Author Contributions: J.S. carried out the laboratory testing, including the drop tube furnace testing and char preparation; data analysis and interpretation; writing; and funding acquisition. R.M. contributed to conceputalisation; project planning; data interpretation; and funding acquisition. M.G. contributed to the project and production planning; coal sampling and acquisition; process data acquisition; production liaison and advice; and funding acquisition. All authors have read and agreed to the published version of the manuscript.

Funding: The authors would like to acknowledge TATA Steel UK in providing financial support and making samples and production data available for this study.

Institutional Review Board Statement: Not applicable.

Informed Consent Statement: Not applicable.

Data Availability Statement: Not applicable.

Conflicts of Interest: The authors declare no conflict of interest.

References

1. Carpenter, A.M. *Use of PCI in Blast Furnaces*; IEA Clean Coal Centre: London, UK, 2006.
2. Babich, A. Blast furnace injection for minimizing the coke rate and CO_2 emissions. *Ironmak. Steelmak.* **2021**, *48*, 728–741. [CrossRef]
3. Hutny, W.; Lee, G.; Price, J. Fundamentals of coal combustion during injection into a blast furnace. *Prog. Energy Combust. Sci.* **1991**, *17*, 373–395. [CrossRef]
4. Geerdes, M.; Toxopeus, H.; Van der Vilet, C. *Modern Blast Furnace Ironmaking: An Introduction*; IOS Press: Amsterdam, The Netherlands, 2009.
5. Ishii, K. *Advanced Pulverised Coal Injection and Blast Furnace Operation*; Elsevier: Amsterdam, The Netherlands, 2000.
6. Schott, R.; Kuttner, G.; Co, K.G. Optimisation strategies for pulverised coal injection into the blast furnace. In Proceedings of the European Steel Technology and Application Days (METEC & 2nd ESTAD 2015), Düsseldorf, Germany, 15–19 June 2015.
7. Ariyama, T.; Sato, M.; Nouchi, T.; Takahashi, K. Evolution of Blast Furnace Process toward Reductant Flexibility and Carbon Dioxide Mitigation in Steel Works. *ISIJ Int.* **2016**, *56*, 1681–1696. [CrossRef]
8. Born, S.; Babich, A.; Van Der Stel, J.; Ho, H.T.; Sert, D.; Ansseau, O.; Plancq, C.; Pierret, J.-C.; Geyer, R.; Senk, D.; et al. Char Formation by Coal Injection and Its Behavior in the Blast Furnace. *Steel Res. Int.* **2020**, *91*, 2000038. [CrossRef]
9. Puttinger, S.; Stocker, H. Toward a Better Understanding of Blast Furnace Raceway Block-ages. *Steel Res. Int.* **2020**, *91*, 2000227. [CrossRef]
10. Xing, X.; Rogers, H.; Zulli, P.; Hockings, K.; Ostrovski, O. Effect of coal properties on the strength of coke under simulated blast furnace conditions. *Fuel* **2018**, *237*, 775–785. [CrossRef]
11. Tiwari, H.P.; Das, A.; Singh, U. Novel technique for assessing the burnout potential of pulverised coals/coal blends for blast furnace injection. *Appl. Therm. Eng.* **2018**, *130*, 1279–1289. [CrossRef]
12. Cavaliere, P. *Clean Ironmaking and Steelmaking Processes-Efficient Technologies for Green-House Emissions Abatement*; Springer: Berlin/Heidelberg, Germany, 2019; ISBN 978-3-030-21208-7.
13. Oliveira, A.; Mahowald, P.; Muller, B.; Kinzel, K.-P.; Oliveira, V. pulverized coal injection for high injection rates. In Proceedings of the 46th Reduction/17th Iron Ore/4th Agglomeration Seminar, Rio de Janeiro, Brazil, 27–29 September 2016. [CrossRef]
14. Bortz, S.; Flament, G. Experiments on pulverized-coal combustion under conditions simulating blast furnace environments. *Ironmak. Steelmak.* **1982**, *10*, 222–229.
15. Suzuki, T.; Hirose, R.; Morimoto, K.; Abe, T. Pulverized coal combustion in high temperature furnaces for steelmaking. *JSME* **1984**, *27*, 2803–2810. [CrossRef]
16. Zhang, S.F.; Bai, C.G.; Wen, L.Y.; Qiu, G.B.; Lu, X.W. Gas particle flow and combustion charateristics of pulverised coal injection in blast furnace raceway. *J. Iron Steel Res.* **2010**, *17*, 8–12. [CrossRef]
17. Du, S.W.; Chen, W.H.; Lucas, J. Pulverised coal burnout in blast furnace simulated by a drop tube furnace. *Energy Fuels* **2010**, *35*, 576–581.
18. Li, H.; Elliott, L.; Rogers, H.; Austin, P.; Jin, Y.; Wall, T. Reactivity study of two coal chars produced in a drop tube furnace and a pulverised coal injection rig. *Energy Fuels* **2012**, *26*, 4690–4695. [CrossRef]
19. Ballantyne, T.R.; Ashman, P.J.; Mullinger, P.J. A new method for determining the conversion of low-ash coals using synthetic ash as a tracer. *Fuel* **2005**, *84*, 1980–1985. [CrossRef]

metals MDPI

Article

Dephosphorization in Double Slag Converter Steelmaking Process at Different Temperatures by Industrial Experiments

Han Sun [1], Jian Yang [1,*], Xinwu Lu [2], Wanshan Liu [2], Gefan Ye [1], Runhao Zhang [1] and Wenkui Yang [1]

1 State Key Laboratory of Advanced Special Steel, School of Materials Science and Engineering, Shanghai University, Shanghai 200444, China; sun_han@shu.edu.cn (H.S.); ye_gefan@shu.edu.cn (G.Y.); zhangrunhao@shu.edu.cn (R.Z.); yangwenkui@shu.edu.cn (W.Y.)
2 Ningbo Iron and Steel Group Co., Ltd., Ningbo 315807, China; luxinwu@ningbosteel.com (X.L.); liuwanshan@ningbosteel.com (W.L.)
* Correspondence: yang_jian@t.shu.edu.cn; Tel.: +86-21-6613-6580

Abstract: In the present work, the effect of dephosphorization endpoint temperature on the dephosphorization of hot metal was studied for the double slag converter steelmaking process under the conditions of low temperature and low basicity by the industrial experiments. In the temperature range of 1350–1450 °C, with an increasing dephosphorization endpoint temperature, the dephosphorization ratio and phosphorus distribution ratio first increase and then decrease. The phosphorus content in hot metal first decreases and then increases at the end of dephosphorization. At the dephosphorization temperature range of 1385–1410 °C, the dephosphorization ratio is higher than 55%, the P_2O_5 content in the dephosphorization slag is 3.93–4.17%, the $\log L_P$ value is 1.76–2.09, the value of P_{CO}^{P-C} of the selective oxidation reaction of carbon and phosphorus is 53–80 Pa, and the a_{FeO} value is 0.284–0.312. The path of phosphorus in hot metal entering the P-rich phase of dephosphorization slag can be reasonably inferred as: hot metal → Fe-rich phase → P-rich phase. Under the present industrial experimental conditions, the dephosphorization and rephosphorization reactions are in dynamic equilibrium at 1413 °C. Considering the experimental results and thermodynamic calculation results of industrial experiments by the double slag dephosphorization process, the optimal temperature range for intermediate deslagging is about 1400–1420 °C.

Keywords: double slag converter steelmaking process; hot metal dephosphorization; dephosphorization endpoint temperature; dephosphorization ratio; phosphorus distribution ratio; optimum temperature of intermediate deslagging

Citation: Sun, H.; Yang, J.; Lu, X.; Liu, W.; Ye, G.; Zhang, R.; Yang, W. Dephosphorization in Double Slag Converter Steelmaking Process at Different Temperatures by Industrial Experiments. *Metals* **2021**, *11*, 1030. https://doi.org/10.3390/met11071030

Academic Editor: Pasquale Cavaliere

Received: 17 May 2021
Accepted: 21 June 2021
Published: 26 June 2021

1. Introduction

The double slag converter steelmaking process developed by Nippon Steel in 2001 is the most representative, which is called Multi-Refining Converter (MURC) [1]. In this process, the desiliconization and dephosphorization operations are performed in the converter firstly, and the dephosphorization slag is poured out by tilting furnace. Then, the decarburization operation is performed in the same converter and the decarburization slag is left in the converter for reusing in the next charge.

Some scholars have carried out researches on the double slag converter steelmaking process by industrial experiments. Yang Wang et al. [2]. conducted the double slag converter steelmaking process by using a low oxygen lance height in the dephosphorization stage of the industrial converter. The stirring of molten steel is strengthened, and the phosphorus content is reduced at the end of decarburization from an average value of 0.018% to an average value of 0.011%, and the dephosphorization ratio is increased by more than 6%. He Wu et al. [3]. conducted the industrial experiments and showed that the average dephosphorization ratio of 62.2% and the average phosphorus content of 0.048% can be achieved under the process conditions of the lower temperature of 1330–1350 °C, the bottom blowing stirring intensity of 0.03–0.04 $m^3/t{\cdot}min$, the medium slag basicity of 2.0–3.0

and the high iron oxide content of 20–25%. Our previous experimental results [4] showed that with increasing the basicity of dephosphorization slag, the phosphorus distribution ratio and dephosphorization ratio increase, and the area fraction of phosphorus rich phase in dephosphorization slag increases with the increasing area fraction of iron-rich phase.

In the hot metal dephosphorization reaction, the slag is in the state of solid–liquid multiphase coexistence. While lime dissolves, $2CaO{\cdot}SiO_2$ (C_2S) is formed [5]. C_2S tends to combine with $3CaO{\cdot}P_2O_5$ (C_3P) in liquid slag to form solid solution $2CaO{\cdot}SiO_2$-$3CaO{\cdot}P_2O_5$ (C_2S-C_3P) [6,7]. Kitamura et al. [8] found that the mass transfer of CaO and SiO_2 occurs simultaneously with the mass transfer of P_2O_5. Xie et al. [9] found that the diffusion of phosphorus from the C_2S–C_3P solid solution reaction layer to the internal encapsulated C_2S is the rate-controlling step of phosphorus transport. Kakimoto et al. [10] found that the thickness of the FeO-CaO layer is related to the thickness of the C_2S phase. Wu et al. [11] found that almost all the phosphorus in CaO-FeO-SiO_2-MgO-MnO-Al_2O_3-P_2O_5-TiO_2 slag with the basicity close to 3 is combined with C_2S but not with $3CaO{\cdot}SiO2$ (C_3S), and C_2S crystal grows uniformly into a spherical shape. Pahlevani et al. [12] found that the distribution ratio of phosphorus in solid solution and liquid slag is determined by the CaO content in liquid slag and activity coefficient of P_2O_5 in solid solution.

The effect of temperature on dephosphorization in the double slag converter steelmaking process was one of the most important points for a highly efficient implementation of this process. Tian et al. [13] found that when the Fe^{3+} content in slag is between 2.33 and 5.98%, the basicity is between 1.72 and 3.14, and the temperature is between 1600 and 1700 °C, the dephosphorization ratio gradually decreases with increasing temperature. Yang et al. [14] found that in the temperature range of 1600–1660 °C, with increasing temperature, the phosphorus distribution ratio gradually decreases, the endpoint phosphorus content increases, and the dephosphorization ratio decreases. Zhou et al. [15] found that in the temperature range of 1590–1640 °C, the total rephosphorization content of molten steel first decreases and then increases. From the above literature, it can be seen that the effect of temperature on converter dephosphorization mainly focused on the endpoint temperature of converter decarburization, ranging from 1590 to 1700 °C; there were few reports on the temperature range of 1350–1450 °C.

The effect of the oxygen potentials at the different reaction interfaces on dephosphorization was also studied. Han et al. [16] concluded that when the dissolved oxygen content in the endpoint molten steel is between 0.03% and 0.045%, the [P]-[O] equilibrium phosphorus content is 21–38 times that of [P]-(FeO) equilibrium phosphorus content. Kitamura et al. [17] found that the oxygen activity of the FeO phase in dephosphorization slag is two orders of magnitude larger than that of equilibrium oxygen activity in hot metal containing more than 4% of carbon.

In addition, some scholars have studied the optimum deslagging temperature of dephosphorization slag in the double slag converter steelmaking process. Based on the selective oxidation reaction of carbon and phosphorus, Wu et al. [4] calculated the carbon–phosphorus selective oxidation transition temperature to be 1342 °C according to the dephosphorization slag compositions and semi-steel control target, and they took it as the guiding temperature for dephosphorization slag deslagging. Chaogang Zhou et al. [18] calculated the optimal deslagging temperature of dephosphorization slag and obtained that the equilibrium temperature of dephosphorization in the early stage of converter blowing is 1435 °C.

According to the above literatures, it can be seen that in the research of the double slag converter steelmaking process, the effect of dephosphorization endpoint temperature in the range of 1350–1450 °C on hot metal dephosphorization is rarely studied. Since the lower temperature range of 1350–1450 °C and the lower basicity of range about 1.5–1.6 are the main process range for the dephosphorization in the double slag converter steelmaking process, it is necessary to clarify the effect of the dephosphorization endpoint temperature on hot metal dephosphorization.

In this paper, through the 180 t double slag converter steelmaking industrial experiments, the effect of the dephosphorization endpoint temperature on the dephosphorization was studied at the low basicity slag in the relatively low temperature range of 1350–1450 °C. In addition, XRD and SEM-EDS were used to study the phase species and phase compositions of dephosphorization slag under the optimal dephosphorization ratio, and the enrichment path of phosphorus in dephosphorization slag was reasonably speculated. The effect of dephosphorization endpoint temperature on the oxygen activity at different reaction interfaces was analyzed. The phosphorus capacity and phosphorus distribution ratio of slag were calculated and compared with the empirical formula of the previous scholars. The optimum deslagging temperature of dephosphorization slag was further analyzed.

2. Experimental Procedure

2.1. Double Slag Converter Steelmaking Experiments

The industrial experiment was carried out with a 180 t converter in the steelmaking plant of Ningbo Iron and Steel Co. Ltd., Ningbo, China. Figure 1 shows the flow chart of the double slag converter steelmaking process, and the double slag converter steelmaking process is mainly divided into eight steps.

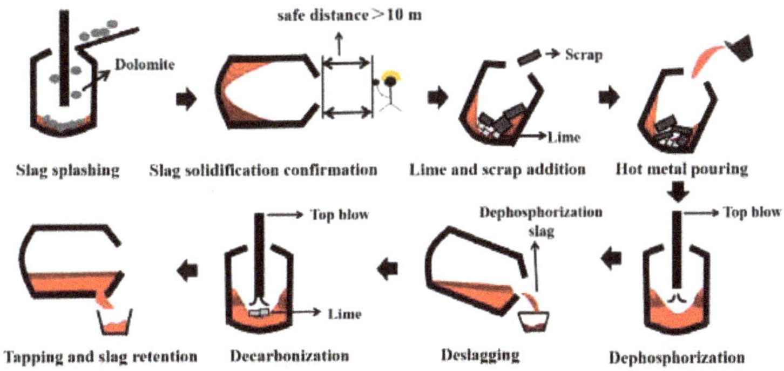

Figure 1. Flow chart of double slag converter steelmaking process.

After adding light burned dolomite as a slag-adjusting agent, slag splashing was carried out to protect the furnace wall. The added auxiliary materials included lime, magnesite, sinter, OG (Oxygen converter Gas recovery) slag pellets, etc. The OG slag pellets contained 55% T.Fe, 12.4% CaO, and 3.8% SiO_2, and they mainly include the mud produced by oxygen converter gas recovery. "T.Fe" refers to the total content of iron element in slag, (mass%). In addition, about 20 mass% of iron oxide scale, 13 mass% of lime, 2 mass% of binder, and 65 mass% of OG mud are molded by a high-pressure ball pressing machine and then dried at low temperature to form OG slag pellets, which can regulate converter temperature and promote rapid slag formation.

After confirming the solidification of the slag, the active lime and the scrap steel is added at first; then, the hot metal is poured into the furnace. Finally, an oxygen lance is ignited for dephosphorization and a low–high–low lance position is used in the dephosphorization stage. The initial low lance position aims to promote the ignition of the oxygen blowing. The subsequent high lance position is to promote the formation of molten slag in the early stage of dephosphorization and to increase the T.Fe content in slag. At the later stage of dephosphorization, the low lance position is used to promote dephosphorization. After the dephosphorization stage is completed, the dephosphorization slag is poured out by tilting the furnace, and then lime, dolomite, and other auxiliary materials are added for decarburization. After the decarburization, the molten steel is tapped out, and the decarburization slag is left in the furnace for the dephosphorization in the next charge.

2.2. Compositions of the Initial Hot Metal, the Hot Metal at the Endpoint of Dephosphorization, and Dephosphorization Slag

The compositions and temperatures of the initial hot metal in the converter are shown in Table 1. The number of experimental heats is T1350~T1450, which is used as the number of subsequent hot metal samples at the endpoint of dephosphorization and the dephosphorization slag samples. The hot metal after desulfurization is used for dephosphorization, and the temperature of the initial hot metal charged into the furnace is in the range of 1326–1366 °C. The added amount of the initial hot metal is 190–195 t, and the added amount of scrap steel is 20–30 t.

Table 1. The compositions and temperatures of the initial hot metal in converter (mass%).

Sample	C	Si	Mn	S	P	*T*/°C
T1350	4.65	0.29	0.18	0.03	0.145	1330
T1360	4.65	0.36	0.25	0.01	0.162	1326
T1370	4.65	0.34	0.29	0.03	0.148	1350
T1385	4.65	0.32	0.29	0.01	0.157	1350
T1395	4.65	0.41	0.28	0.01	0.141	1365
T1405	4.65	0.29	0.27	0.01	0.129	1366
T1410	4.65	0.31	0.25	0.02	0.136	1357
T1420	4.65	0.40	0.31	0.04	0.135	1355
T1450	4.65	0.39	0.29	0.03	0.158	1365

After the dephosphorization reaction, the slag sample is taken by sticking onto an iron rod and cooling in air. After grinding, the slag sample was passed through a 200-mesh sieve to remove the residual iron particles remaining in the slag. The compositions of slag samples were analyzed with an M4 TORNADO fluorescence spectrometer of Bruker Company in Germany. The temperature was measured, the hot metal was sampled with a sub-gun, and the compositions of hot metal were analyzed by SPECTROMAXx direct reading spectrometer of German SPECTRO Company. Before the analyses, the spectrometer was calibrated with the standard samples of the similar compositions.

Table 2 shows hot metal compositions, dephosphorization ratios, and temperatures at the endpoint of dephosphorization. Table 3 shows dephosphorization slag compositions and basicities at the endpoint of dephosphorization. It can be seen from Table 2 that the temperature of the dephosphorization endpoint is in the range of 1350–1450 °C. Basicity B is calculated with Equation (1), and η_P represents the dephosphorization ratio, which is calculated by Equation (2) [19]:

$$B = \frac{(\%\mathrm{CaO})}{(\%\mathrm{SiO2})} \quad (1)$$

$$\eta_P = \frac{[\%P]_0 - [\%P]_e}{[\%P]_0} \times 100\% \quad (2)$$

(%CaO) and (%SiO_2) are the mass percentages of CaO and SiO_2 in the dephosphorization slag respectively, $[\%P]_0$ is the initial phosphorus content in initial hot metal, and $[\%P]_e$ is the phosphorus content in the hot metal at the end of dephosphorization.

The phase analysis of dephosphorization slag was carried out by a D8 Advance X-ray powder diffractometer (XRD) of Bruker company in Germany. In the range of $2\theta = 10$–$90°$ and step size of $0.04° s^{-1}$, XRD data were collected by Cu-Kα radiation. A small amount of massive dephosphorization slag was embedded in epoxy resin. After solidification, it was ground and polished by an automatic grinding and polishing machine of the PRESI company in France. Then, the surface was carbonized by a magnetron sputtering MC1000 of the HITACHI company in Japan. The morphology of the P-rich phase of dephosphorization slag was observed by a Zeiss EVO 18 electronic scanning microscope of Zeiss company of Germany, and the chemical composition of the P-rich phase in dephosphorization slag was analyzed by an X-MaxN large area energy-dispersive spectrometer (SEM-EDS) of Oxford Instruments Company in the UK.

Table 2. Hot metal compositions, dephosphorization ratios, and temperatures at the end of dephosphorization (mass%).

Sample	C	Si	Mn	S	P	η_P/%	T/°C
T1350	3.076	0.010	0.060	0.018	0.073	49.7	1350
T1360	3.081	0.025	0.089	0.010	0.080	50.6	1360
T1370	3.241	0.015	0.097	0.009	0.071	52.0	1370
T1385	3.039	0.010	0.092	0.010	0.069	56.1	1385
T1395	3.069	0.022	0.117	0.018	0.050	64.5	1395
T1405	2.942	0.009	0.167	0.018	0.034	73.6	1405
T1410	2.943	0.012	0.143	0.017	0.051	62.5	1410
T1420	3.253	0.017	0.138	0.012	0.069	48.9	1420
T1450	3.488	0.015	0.147	0.019	0.103	34.8	1450

Table 3. Compositions and basicities of dephosphorization slag (mass%).

Sample	CaO	SiO_2	MgO	MnO	Al_2O_3	P_2O_5	FeO	B/-
T1350	35.96	22.77	7.24	9.37	2.52	3.17	18.97	1.58
T1360	34.87	21.53	9.05	9.76	2.56	3.85	18.38	1.62
T1370	34.66	22.51	8.15	9.25	2.83	3.73	18.87	1.54
T1385	35.22	21.60	8.58	10.01	3.59	3.96	17.04	1.63
T1395	34.88	22.48	8.98	9.92	2.86	3.93	16.95	1.55
T1405	37.30	22.71	8.23	8.49	2.27	4.17	16.83	1.64
T1410	35.73	22.91	8.94	8.29	3.05	3.95	17.13	1.56
T1420	35.51	23.71	9.18	8.80	2.85	3.57	16.38	1.50
T1450	35.28	23.84	9.35	8.59	2.23	3.17	17.54	1.48

3. Results

3.1. Effect of Dephosphorization Endpoint Temperature on Liquidus Region of Dephosphorization Slag

Figure 2 shows the effect of dephosphorization endpoint temperature on liquidus regions of the CaO-SiO_2-FeO-10%MgO-10%MnO-5%P_2O_5 dephosphorization slag. The liquidus projection section of dephosphorization slag at 1340–1460 °C was drawn by a phase diagram in Factsage7.3. Figure 2 shows that with increasing dephosphorization endpoint temperature, the liquid phase area of dephosphorization slag increases continuously, and the slag can be melted better. At the same time, the dephosphorization slag selected in the present work is marked in Figure 2 by using the ternary phase diagram of CaO, SiO_2, and FeO. Since the selected dephosphorization slags are from the different heats of the double slag converter steelmaking experiment, the basicities of dephosphorization slag are slightly different. Nine groups of dephosphorization slag components all located in red quadrilateral area indicate that the difference in the basicities of these dephosphorization slags is quite small. Therefore, it can be considered that the experimental results are mainly affected by the dephosphorization endpoint temperature.

3.2. Effect of Dephosphorization Endpoint Temperature on Dephosphorization

In industrial experiments, it is difficult to ensure that the other process parameters are the same to investigate the effect of dephosphorization endpoint temperature on hot metal dephosphorization. In the heats selected in the present work, the slag basicities ranged from 1.48 to 1.64, and the FeO contents in slag were from 16.38% to 18.97%. The MgO contents in slag from 7.24% to 9.35% increased with the increasing dephosphorization endpoint temperature, which was mainly caused by the erosion of the converter wall. The contents of CaO, SiO_2, MnO, and Al_2O_3 in the dephosphorization slag changed very little.

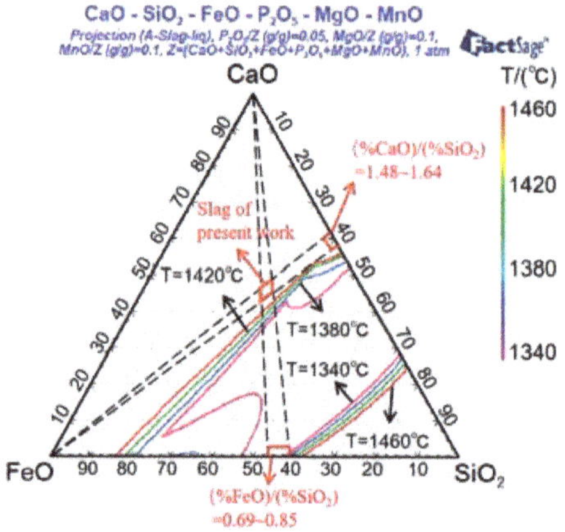

Figure 2. Effect of dephosphorization endpoint temperature on liquidus regions of the CaO-SiO_2-FeO-10%MgO-10%MnO-5%P_2O_5 dephosphorization slag.

Figure 3 shows the effect of dephosphorization endpoint temperature on the dephosphorization ratio, phosphorus content in hot metal, and P_2O_5 content in dephosphorization slag at the endpoint of dephosphorization. When the dephosphorization endpoint temperature is 1350–1385 °C, the dephosphorization ratio is between 49.7 and 56.1%. The phosphorus content and dephosphorization ratio in hot metal have no obvious change in this range. When the dephosphorization endpoint temperature is increased to 1385–1405 °C, the dephosphorization ratio increases significantly. At 1405 °C, the dephosphorization ratio reaches the highest value of 73.64% and the phosphorus content in hot metal reduces to the lowest value of 0.034%. This is because when the dephosphorization endpoint temperature is relatively low in the temperature range of 1350–1385 °C, although the thermodynamic condition is beneficial to dephosphorization, the lime cannot be well melted during the very short time period of about 5 min during the dephosphorization period in the double slag converter steelmaking process. Therefore, the poor kinetic conditions hindered the dephosphorization reaction. When the temperature is in the range of 1385–1405 °C, the auxiliary materials in the converter are well melted, the slagging is improved, and the dephosphorization thermodynamic conditions can still promote the dephosphorization of the converter, so the dephosphorization ratio is significantly increased in this range.

In the dephosphorization endpoint temperature range of 1350–1405 °C, the content of P_2O_5 in the dephosphorization slag increases with the increasing dephosphorization endpoint temperature and the content of P_2O_5 in the dephosphorization slag is relatively high, which is from 3.17% to 4.17%. This is because with the increasing dephosphorization endpoint temperature, the lime begins to melt, and the kinetic conditions of the reaction between molten slag and hot metal are improved, so that the initial dephosphorization capacity increases. When the dephosphorization endpoint temperature is 1405 °C, the P_2O_5 content reaches the maximum value of 4.17%.

When the dephosphorization endpoint temperature is further increased to higher than 1405 °C, the dephosphorization ratio of hot metal is decreased, the phosphorus content at the endpoint of dephosphorization is increased, and the content of P_2O_5 in the dephosphorization slag decreases. The reason is that the dephosphorization reaction is exothermic, and the high temperature will inhibit the dephosphorization reaction. Although the slagging is promoted and the kinetic conditions of dephosphorization in the converter are improved, too high temperature has a greatly adverse effect on the thermodynamic conditions of dephosphorization, leading to a great decrease in the equilibrium constant of dephosphorization reaction. Therefore, the thermodynamic and kinetic conditions of dephosphorization jointly determine the dephosphorization ratio.

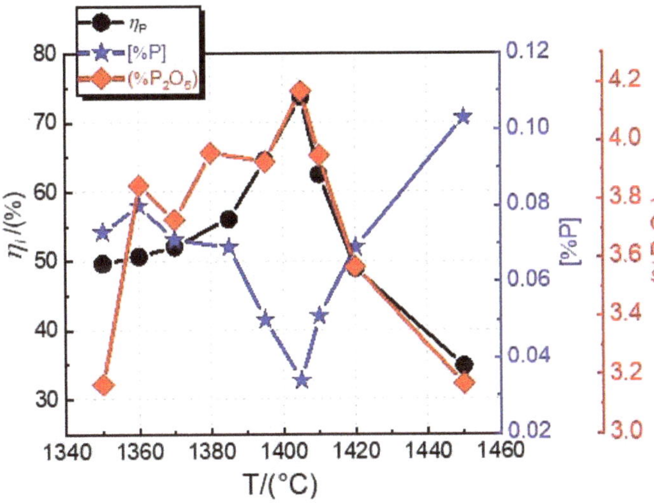

Figure 3. Effect of dephosphorization endpoint temperature on the dephosphorization ratio, phosphorus content in hot metal, and P_2O_5 content in dephosphorization slag at the endpoint of dephosphorization.

According to the results of Figure 3, when the dephosphorization endpoint temperature is between 1385 and 1410 °C, the dephosphorization ratio at the endpoint of dephosphorization rises to be above 55%, and the content of P_2O_5 in the dephosphorization slag is above 3.90%, with the excellent results of dephosphorization of hot metal.

3.3. Analysis of Dephosphorization Slag at the Optimum Dephosphorization Temperature of 1405 °C

Figure 4 shows the XRD analysis results of dephosphorization slag at 1405 °C. Figure 5 shows the images of dephosphorization slag by SEM-EDS at 1405 °C, with magnification of (a) 200 times and (b) 500 times, respectively. As shown in Figure 4, the dephosphorization slag is composed of dicalcium silicate Ca_2SiO_4, tricalcium silicate Ca_3SiO_4, phosphorus containing solid solution $6C_2S$-C_3P, silicate phase $Ca_3Mg(SiO_4)_2$, calcium ferrite phase $Ca_2Fe_2O_5$, and oxide phase RO. It can be seen from Figure 5a that there is dark gray phase 1, gray phase 2, light gray phase 3, white phase 4, gray phase 5 and black phase 6 in the dephosphorization slag at the dephosphorization endpoint temperature of 1405 °C with the basicity of 1.64. Furthermore, SEM-EDS point analysis and map scanning were used to analyze the phase composition. Table 4 shows the elements compositions of different phases in dephosphorization slag by SEM-EDS point analysis in Figure 5a, and Figure 6 shows the map scanning results of dephosphorization slag at 1405 °C under the magnification of 200 times.

It can be seen from Table 4 that the phosphorus content in the dark gray phase 1-1, 1-2, and 1-3 is the highest, with an average value of 4.34%, followed by the phosphorus content in the gray phase 2-1, 2-2, and 2-3, with an average value of 3.44%, and the phosphorus content in the light gray phase 3, with the value of 2.20%. The contents of phosphorus in white phase 4, gray white phase 5, and black phase 6 were less than 1.5%. According to the XRD analysis results in Figure 4 and the map scanning results in Figure 6, it can be judged that the dark gray phase 1, the gray phase 2, and the light gray phase 3 are all phosphorus rich phases (P-rich phase), but their phosphorus enrichment degrees are different, and the order of phosphorus enrichment degree is the dark gray phase 1 > the gray phase 2 > the light gray phase 3. The content of Fe in the white phase 4 is very high, so it can be judged that the white phase 4 is the mixed iron particles. The gray white phase 5 has high contents of Fe and Ca, which can be identified as $Ca_2Fe_2O_5$ by XRD analysis, which is named the Fe-rich phase. The black phase 6 contains a high content of Ca, Si, and a small amount of phosphorus. Therefore, the black phase 6 is considered as the matrix phase, which is known as the calcium silicate phase. From the SEM image of dephosphorization slag in Figure 5b, it is found that there are two kinds of Fe-rich phases in the dephosphorization slag at the dephosphorization endpoint temperature of 1405 °C with the basicity of 1.64, which are petal-like and strip-like respectively. There are large P-rich phases in the slag with the size of more than 40 μm, and the calcium silicate phases are dispersed in small pieces between the P-rich phase and the Fe-rich phase.

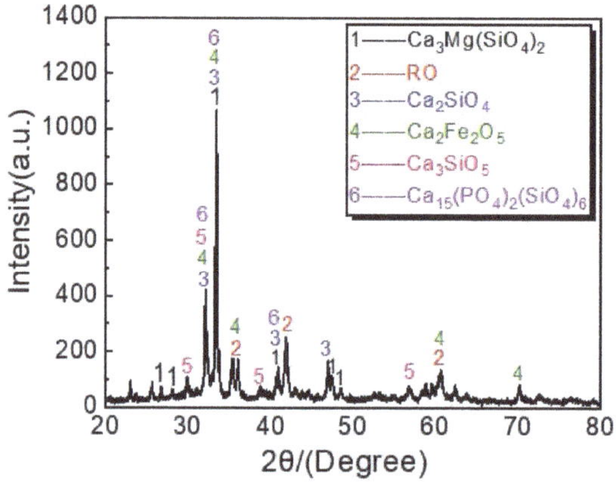

Figure 4. XRD analysis result of dephosphorization slag at 1405 °C.

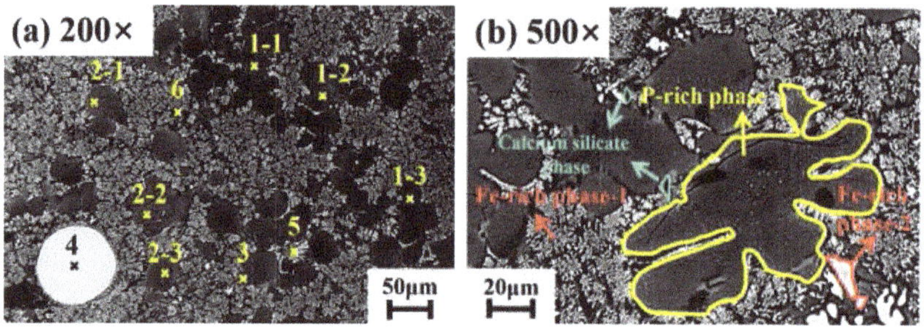

Figure 5. Images of dephosphorization slag by SEM-EDS at 1405 °C of (**a**) 200 times; (**b**) 500 times.

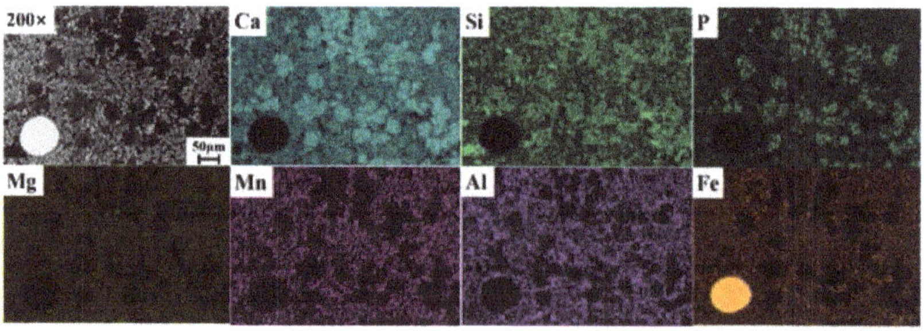

Figure 6. Element distribution results of dephosphorization slag at 1405 °C.

Table 4. Element compositions of different phases in dephosphorization slag by SEM-EDS point analysis (mass%).

Position	Ca	Si	Mn	Mg	P	Al	Fe	O
1-1	34.77	12.94	2.70	2.03	4.34	0.16	1.62	41.44
1-2	35.98	12.37	2.90	2.08	4.25	0.12	1.26	41.04
1-3	37.00	13.93	3.63	1.36	4.42	0.01	2.22	37.43
2-1	29.68	13.23	3.39	2.94	3.61	1.65	2.65	42.85
2-2	35.25	13.80	2.18	1.99	3.51	0.14	1.58	41.55
2-3	29.06	13.27	3.78	2.19	3.20	2.13	3.16	43.21
3	30.56	12.55	6.02	2.49	2.20	0.85	5.80	39.53
4	2.32	0.92	3.55	1.35	0.17	0.43	82.14	9.12
5	20.36	10.22	7.33	2.92	1.14	2.03	24.06	31.94
6	27.08	17.79	3.19	3.33	1.44	2.56	4.15	40.46

3.4. Analysis of the P-Rich Phase in Dephosphorization Slag at 1405 °C

SEM-EDS was used to further analyze the boundary of the P-rich phase. Figure 7a,b show the P-rich phase in dephosphorization slag at 1405 °C with the magnifications of 1000 times and 2000 times, Figure 7c shows the schematic diagram of the P-rich phase boundary, and Figure 7d shows the schematic diagram of formation process of the P-rich phase. In Figure 7a,b, the gray massive phase is the P-rich phase, the gray white phase is the Fe-rich phase and the black phase is calcium silicate in dephosphorization slag. At the same time, it is found that the gray P-rich phase is surrounded by the black calcium silicate phase, and the outermost part is surrounded by a layer of the gray white Fe-rich phase. Furthermore, the P-rich phase of dephosphorization slag was observed at 2000 times magnification, and the size of different phases was measured by Image Pro Plus image software. It can be seen that the size of the P-rich phase is 45.31 μm, the width of the black calcium silicate phase wrapped outside is about 1.15 μm, and the width of the outermost gray white Fe-rich phase is about 0.74 μm.

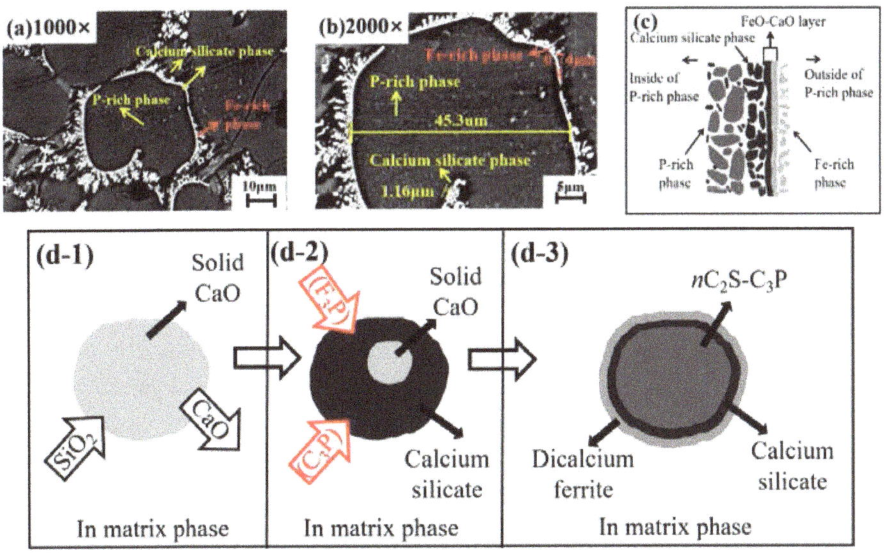

Figure 7. Images of the P-rich phase in dephosphorization slag at 1405 °C of (**a**) 1000 times; (**b**) 2000 times; (**c**) Schematic diagram of P-rich phase boundary; (**d**) Schematic diagram of formation process of the P-rich phase.

From Table 4, the content of phosphorus in the P-rich phase is higher than that in the calcium silicate phase, and the content of phosphorus in the calcium silicate phase is higher than that in the Fe-rich phase. Therefore, the enrichment path of phosphorus in hot metal can be reasonably inferred from the content gradient of phosphorus in the different phases, and the formation process of the P-rich phase was speculated, as shown in Figure 7d. The process from Figure 7(d-1) to (d-2) shows that the solid CaO particles in matrix phase will first form C_2S at the surface of CaO [20–22]. The process from Figure 7(d-2) to (d-3) shows that the phosphorus in hot metal is oxidized by iron oxide to form P_2O_5 and enters the slag, and FeO and CaO react with P_2O_5 to form $3FeO{\cdot}P_2O_5(F_3P)$ and C_3P, respectively, which initially fixes phosphorus in dephosphorization slag. Since the F_3P is extremely unstable especially in the high temperature, CaO in slag will further react with F_3P to form the stable C_3P phase [23]. Finally, phosphorus exists in the P-rich phase in the form of nC_2S-C_3P solid solution. Since the dephosphorization slag is a multiphase slag with solid–liquid coexistence between 1350 and 1450 °C, there is undissolved calcium silicate solid phase in the slag. With the progress of the dephosphorization reaction, the phosphorus containing solid solution grows into the massive P-rich phase. The calcium silicate phase wrapped outside shrinks, and the corresponding FeO-CaO layer shrinks, finally forming the P-rich phase boundary, as shown in Figure 7c. Therefore, the path of phosphorus in hot metal entering the P-rich phase of dephosphorization slag can be reasonably inferred as: hot metal → Fe-rich phase → P-rich phase in dephosphorization slag.

4. Discussion

4.1. Effect of Dephosphorization Endpoint Temperature on a_{FeO} and P_{CO}^{P-C}

In the dephosphorization reaction, the iron oxide in slag acts as the main oxidant for dephosphorization. Therefore, the content of iron oxide in dephosphorization slag has a significant effect on the dephosphorization ability of dephosphorization slag. In this study, it is assumed that all iron oxides in slag exist in the form of FeO, so as to analyze the effect of dephosphorization endpoint temperature on a_{FeO}. It should be noted that the activity of each substance in the dephosphorization slag adopts pure substances as the standard state, and a_{FeO} in the dephosphorization slag is expressed as Equation (3) [24].

$$a_{FeO} = \gamma_{FeO} \times X_{FeO} \tag{3}$$

X_i is the mole fraction of oxide i in dephosphorization slag, and γ_i is the activity coefficient of oxide i. The percent work uses the regular solution model to calculate γ_{FeO}, because this model can avoid the problem that the structure of the silicate anion changes with the number and type of cations and the temperature. In this model, it is assumed that various cations, such as Ca^{2+}, Fe^{2+}, Si^{4+}, and P^{5+}, are randomly distributed in the O^{2-} matrix, and O^{2-} is the common anion of various cations in the melt. According to the interaction energy between cations summarized by Ban-ya [25], γ_{FeO} can be expressed as Equation (4).

$$\begin{aligned} RT\ln\gamma_{FeO} = {} & 7110X_{MnO}^2 - 31,380X_{CaO}^2 + 33,470X_{MgO}^2 - 41,840X_{SiO2}^2 - 31,380X_{P2O5}^2 \\ & -41,000X_{Al2O3}^2 + 67,780X_{MnO}X_{CaO} - 21,410X_{MnO}X_{MgO} + 40,580X_{MnO}X_{SiO2} \\ & +60,670X_{MnO}X_{P2O5} + 49,790X_{MnO}X_{Al2O3} + 102,510X_{CaO}X_{MgO} + 60,670X_{CaO}X_{SiO2} \\ & +188,280X_{CaO}X_{P2O5} + 82,430X_{CaO}X_{Al2O3} + 58,570X_{MgO}X_{SiO2} + 39,750X_{MgO}X_{P2O5} \\ & +63,600X_{MgO}X_{Al2O3} - 156,900X_{SiO2}X_{P2O5} + 45,130X_{SiO2}X_{Al2O3} + 189,120X_{P2O5}X_{Al2O3} \end{aligned} \tag{4}$$

During the dephosphorization process of the converter, as the temperature of the hot metal rises and the dephosphorization reaction proceeds, carbon and phosphorus will undergo a selective oxidation transformation. The selective oxidation reaction of carbon and phosphorus is shown in Equation (5). Namely, when the temperature is lower than the selective oxidation temperature of carbon and phosphorus, the phosphorus in hot metal is preferentially oxidized. At the higher temperature, carbon is preferentially oxidized. The value of P_{CO} in a selective oxidation reaction of carbon and phosphorus is directly related to the selective oxidation temperature of carbon and phosphorus. Therefore, it is significant to study the relationship between the partial pressure of CO and dephosphorization endpoint temperature in the selective oxidation of carbon and phosphorus for the dephosphorization of hot metal. In this paper, P_{CO}^{P-C} is used to represent the partial pressure of CO in the selective oxidation of carbon and phosphorus.

$$2[P] + 5CO = (P2O5) + 5[C] \tag{5}$$

$$\Delta G^{\theta} = -594,505 + 748.25T \tag{6}$$

$$K^{\Theta} = \frac{a_{P2O5} \times a_{[C]}^5}{P_{CO}^5 \times a_{[P]}^2} \tag{7}$$

Equation (6) is the Gibbs free energy of the carbon and phosphorus selective oxidation reaction (J). Equation (7) is the equilibrium constant expression of the reaction. P_2O_5 in dephosphorization slag adopts pure substance as the standard state. $a_{[C]}$ and $a_{[P]}$ represent the activities of carbon and phosphorus in hot metal, respectively. The standard state of the activity of $a_{[C]}$ and $a_{[P]}$ in hot metal is 1% mass fraction solution, which conforms to Henry's law. Therefore, $a_{[C]}$ and $a_{[P]}$ can be expressed by Equations (8) and (9). a_{P2O5} is expressed as Equation (10).

$$a_{[C]} = f_{[C]} \times [\%C] \tag{8}$$

$$a_{[P]} = f_{[P]} \times [\%P] \tag{9}$$

$$a_{P2O5} = \gamma_{P2O5} \times X_{P2O5} \tag{10}$$

It should be noted that γ_{P2O5} is calculated by empirical Equation (11) [26]. The reason for not using regular solution calculation is that the value of a_{P2O5} in the actual slag is too small, usually smaller than 10^{-17}. [27] There is an error in the process of converting $a_{PO2.5}$ to a_{P2O5} obtained by the regular solution model, and the obtained a_{P2O5} is far from the calculated values in previous studies. [28–30] Based on the above reasons, the empirical formula summarized by Turkdogan and Pearson is selected to calculate γ_{P2O5} [26], and the activity coefficients of phosphorus and carbon in hot metal can be calculated by Equations (12) and (13).

$$\log \gamma_{P2O5} = -1.12(22X_{CaO} + 15X_{MgO} + 13X_{MnO} + 12X_{FeO} - 2X_{SiO_2}) - \frac{42,000}{T} + 23.58 \tag{11}$$

$$\log f_{[P]} = e_P^C[\%C] + e_P^{Si}[\%Si] + e_P^{Mn}[\%Mn] + e_P^S[\%S] + e_P^P[\%P] \tag{12}$$

$$\log f_{[C]} - e_C^C[\%C] + e_C^{Si}[\%Si] + e_C^{Mn}[\%Mn] + e_C^S[\%S] + e_C^P[\%P] \tag{13}$$

e_P^i and e_C^i are the interaction coefficients of solute element i to phosphorus and carbon dissolved in hot metal, respectively. With the progress of the dephosphorization reaction, a part of phosphorus in hot metal is fixed in dephosphorization slag and removed from hot metal. At the endpoint of dephosphorization stage in the double slag converter steelmaking process, it can be regarded that the dephosphorization quasi-equilibrium state is reached under the conditions of lower temperature and lower basicity. The value of a_{FeO} can be calculated by Equations (3) and (4), and the value of P_{CO}^{P-C} can be calculated by Equations (6)–(13). Considering that a_{FeO} is mainly affected by the basic oxide contents in slag and the dephosphorization endpoint temperature, through the multiple linear regression on a_{FeO}, Equation (14) can be obtained with the regression coefficient R^2 of is 0.97. Figure 8a shows the comparison between the predicted value based on multiple regression and the calculated value of a_{FeO} from the regular solution model of Equations (3) and (4). The predicted value is well consistent with the calculated result.

$$a_{FeO} = 0.0096 \times (\%FeO) - 0.004 \times (\%CaO) + 0.0026 \times (\%MnO) + 0.0134 \times (\%MgO) - 2.8923 \times 10^{-4} \times T + 0.5442 \tag{14}$$

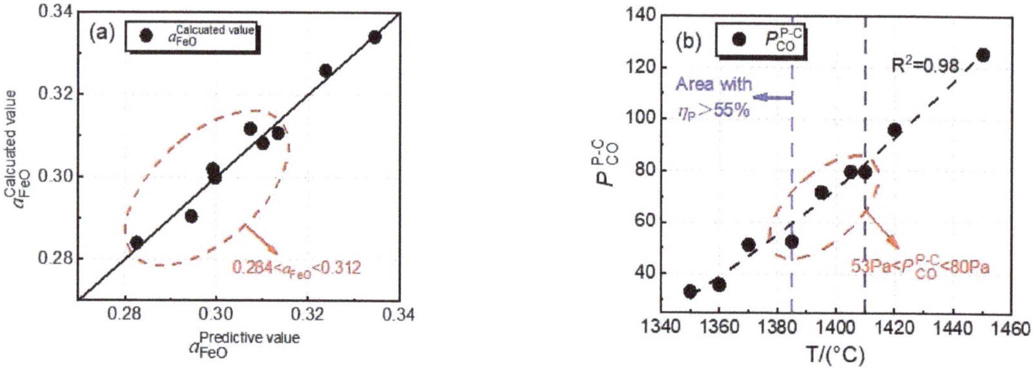

Figure 8. (**a**) Comparison between the predicted value based on multiple regression and the calculated value of a_{FeO} from the regular solution model; (**b**) Effect of dephosphorization endpoint temperature on P_{CO}^{P-C} of selective oxidation reaction of carbon and phosphorus.

Figure 8b shows the effect of dephosphorization endpoint temperature on P_{CO}^{P-C} of a selective oxidation reaction of carbon and phosphorus. In Figure 8b, P_{CO}^{P-C} data were fitted by an exponential curve, and the regression coefficient, R^2, is 0.98, which indicates that there is a good exponential relationship between the endpoint temperature of dephosphorization and P_{CO}^{P-C}; the fitting equation is shown in Equation (15). With increasing dephosphorization endpoint temperature, P_{CO}^{P-C} shows an exponential upward trend.

$$P_{CO}^{P-C} = 4.11 \times \exp(\frac{T - 576.82}{211.98}) - 126.41 \tag{15}$$

The blue dashed line area in Figure 8b shows that the dephosphorization ratio is greater than 55%, and the temperature range is 1385–1410 °C. In this range, the dephosphorization result is preferential, the P_{CO}^{P-C} value is 53–80 Pa, and a_{FeO} is 0.284–0.312. These two ranges are conducive to the dephosphorization of the converter.

4.2. Effect of Dephosphorization Endpoint Temperature on the Oxygen Activity on Hot Metal Surface, the Phosphorus Oxidation Equilibrium Oxygen Activity, and the Iron Oxidation Equilibrium Oxygen Activity at Slag–Hot Metal Interface

In the converter dephosphorization process, the oxygen in hot metal is mainly used to remove elements such as silicon, manganese, phosphorus, and carbon. Before dephosphorization reaches quasi equilibrium, with the decarburization reaction going on, the carbon in hot metal decreases continuously, and the carbon–oxygen reaction is in the state of excess oxygen. The equilibrium oxygen content in hot metal mainly depends on the carbon content in hot metal [16]. The carbon–oxygen reaction in the hot metal is shown in Equation (16), the Gibbs free energy of the reaction is expressed by Equation (17), and the equilibrium constant is given by Equation (18). It should be noted that P_{CO} in Equation (16) is the partial pressure of CO in the converter, and it is assumed to the value of 1 atm. In this paper, P_{CO}^{C-O} is used to represent the partial pressure of CO in the converter.

$$[C] + [O] = CO \tag{16}$$

$$\Delta G^{\theta} = -22,200 - 38.34T \tag{17}$$

$$K^{\Theta} = \frac{P_{CO}^{C-O}}{a_{[C]} \times a_{[O]}} \tag{18}$$

Combined with Equations (16)–(18), the theoretical oxygen activity on the surface of hot metal of $a_{[O]}^{iron}$ can be deduced as shown in Equation (19).

$$\log a_{[O]}^{iron} = -\frac{1160}{T} - 2.002 + \log P_{CO}^{C-O} - \log a_{[C]} \tag{19}$$

At the slag–hot metal reaction interface, phosphorus in hot metal is oxidized to P_2O_5, which enters slag and exists in the form of C_2S-C_3P solid solution. The oxidative dephosphorization reaction occurs as Equation (20). Equation (21) is the equilibrium constant of dephosphorization reaction obtained by Turkdogan and Pearson [26], and Equation (22) is the expression of the equilibrium constant.

$$2[P] + 5[O] = (P2O5) \tag{20}$$

$$\log K = \frac{37,160}{T} - 29.67 \tag{21}$$

$$K^{\ominus} = \frac{a_{P2O5}}{a_{[P]}^2 \times a_{[O]}^5} \tag{22}$$

When the dephosphorization reaction of slag and hot metal reaches quasi equilibrium, by combining Equations (20)–(22), the oxygen activity of phosphorus oxidation at the slag–hot metal interface of $a_{[O]}^{de[P]}$ is derived as Equation (23).

$$\log a_{[O]}^{de[P]} = (-\frac{37,160}{T} + 29.67 + \log a_{P2O5} - 2\log a_{[P]})/5 \tag{23}$$

The oxygen activity in the slag is mainly determined by the oxygen in the form of iron oxides. This article assumes that all iron oxides exist in the form of FeO, and the slag–hot metal interface reaches the quasi-equilibrium state of oxidation. Therefore, the oxygen activity of iron oxidation at the slag–hot metal interface can be expressed by the iron oxide reaction, as shown in Equation (24). The Gibbs free energy of reaction is shown in Equation (25) [24], and the equilibrium constant is shown in Equation (26).

$$[Fe] + [O] = (FeO) \tag{24}$$

$$\Delta G^{\theta} = -128,090 + 57.990T \tag{25}$$

$$K^{\ominus} = \frac{a_{FeO}}{a_{[O]} \times a_{[Fe]}} \tag{26}$$

The concentration of [Fe] in hot metal is very high, and the value of $a_{[Fe]}$ is taken as 1. When the iron oxidation reaction at the slag–hot metal interface reaches quasi equilibrium, by combining Equations (3), (4), and (26), the oxygen activity of the iron oxidation at the slag–hot metal interface of $a_{[O]}^{[Fe]-[O]}$ is obtained as shown in Equation (27).

$$\log a_{[O]}^{[Fe]-[O]} = -\frac{6689.77}{T} + 3.029 + \log a_{FeO} \tag{27}$$

The effect of the dephosphorization endpoint temperature on oxygen activity at different reaction interfaces in the dephosphorization stage of the converter is calculated by using Equations (19), (23), and (27), as shown in Figure 9a. It can be seen from Figure 9a that $a_{[O]}^{de[P]}$ and $a_{[O]}^{[Fe]-[O]}$ both increase with increasing dephosphorization endpoint temperature as a whole, while $a_{[O]}^{iron}$ first increases and then decreases. These results can be explained by the thermodynamic principle. The main decarburization reaction in steelmaking is Equation (16), which is an exothermic reaction. With dephosphorization endpoint temperature from 1350 to 1410 °C, the carbon–oxygen product increases; under the condition of little change in carbon content, the $a_{[O]}^{iron}$ and $a_{[O]}^{de[P]}$ increase rapidly [31]. When the dephosphorization endpoint temperature is 1410–1450 °C, the dephosphorization reaction is weakened, the oxygen content of reaction with phosphorus is reduced, the phosphorus content increases from 0.051% to 0.103%, and the carbon content increases from 2.94% to 3.49% in hot metal, so the $a_{[O]}^{iron}$ decreases, and $a_{[O]}^{de[P]}$ increases slowly. Since $a_{[O]}^{[Fe]-[O]}$ is mainly affected by the iron oxide content in slag and temperature, $a_{[O]}^{[Fe]-[O]}$ increases with the increase of temperature when the iron oxide content does not change obviously.

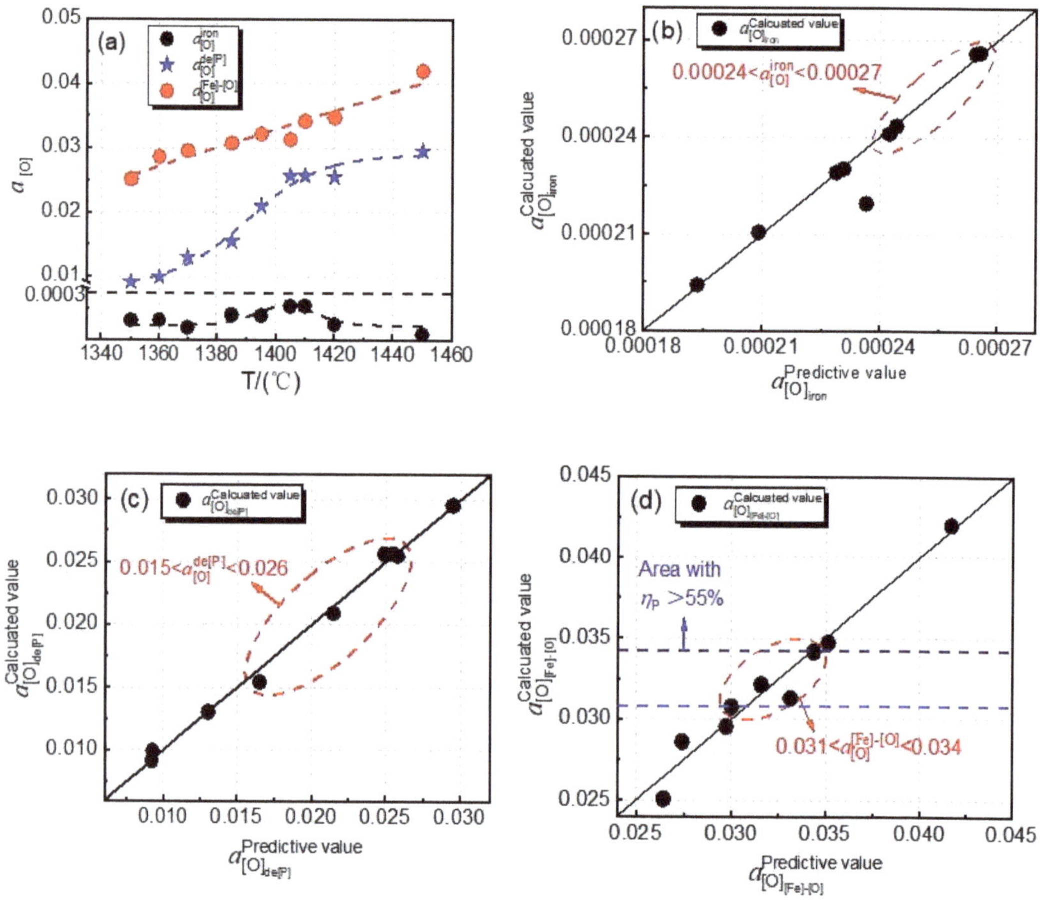

Figure 9. (**a**) Effect of dephosphorization endpoint temperature on $a_{[O]}^{iron}$, $a_{[O]}^{de[P]}$, and $a_{[O]}^{[Fe]-[O]}$; Comparison between predicted value and calculated value based on multiple regression, (**b**) $a_{[O]}^{iron}$; (**c**) $a_{[O]}^{de[P]}$; (**d**) $a_{[O]}^{[Fe]-[O]}$.

According to the different reaction interfaces of dephosphorization in the converter, the relationships between $a_{[O]}^{iron}$, $a_{[O]}^{de[P]}$, $a_{[O]}^{[Fe]-[O]}$, the dephosphorization endpoint temperature, and the compositions of hot metal and slag are fitted by multiple linear regression, which are expressed as Equations (28)–(30), respectively. It is found that the values of $a_{[O]}$ calculated by Equations (19), (23), and (27) all have a good multiple linear relationship with the regression coefficients, R^2, of 0.99, 0.97, and 0.96, respectively.

$$a_{[O]}^{iron} = -1.529 \times 10^{-4} \times [\%C] + 2.7657 \times 10^{-7} \times T + 3.258 \times 10^{-4} \quad (28)$$

$$a_{[O]}^{de[P]} = -0.00211 \times (\%P2O5) - 0.131 \times [\%P] + 2.423 \times 10^{-4} \times T - 0.3017 \quad (29)$$

$$a_{[O]}^{[Fe]-[O]} = 1.7046 \times 10^{-4} \times T + 0.00125 \times (\%FeO) - 0.2274 \quad (30)$$

It can be seen from Figure 9a that the increase trend of $a_{[O]}^{de[P]}$ is slightly faster, and the growth trends of $a_{[O]}^{iron}$ and $a_{[O]}^{[Fe]-[O]}$ are slightly slower in the temperature range of 1350–1405 °C, which indicates that the effect of temperature on the phosphorus oxygen equilibrium of the slag–hot metal interface is relatively large in the temperature range of 1350–1405 °C. The increase trend of $a_{[O]}^{[Fe]-[O]}$ is slightly faster in the temperature range of 1405–1450 °C, which indicates that the temperature has a greater effect on the iron oxygen equilibrium in the slag in temperature range of 1405–1450 °C.

In the range of 1350–1450 °C, $a_{[O]}^{[Fe]-[O]} > a_{[O]}^{de[P]} > a_{[O]}^{iron}$. The results show that $a_{[O]}^{[Fe]-[O]}$ is the highest, while $a_{[O]}^{iron}$ is the lowest, which is consistent with the research results of Kitamura. [17] Since there is a large amount of Fe in the dephosphorization process, the oxygen activity at the slag–hot metal interface mainly depends on the oxygen activity in equilibrium with iron. According to the analysis in Figure 9, $a_{[O]}^{[Fe]-[O]}$ plays a major role in the dephosphorization reaction in the range of 1350–1450 °C. Since the temperature range of the high dephosphorization ratio is 1385–1410 °C, $a_{[O]}^{[Fe]-[O]}$ should be between 0.031 and 0.034.

4.3. *Effect of Dephosphorization Endpoint Temperature on Phosphorus Capacity and Phosphorus Distribution Ratio of Dephosphorization Slag*

The solubility of phosphorus in molten slag can be expressed by phosphorus capacity, which is an important indicator of the potential dephosphorization capacity of slag. During the dephosphorization process of hot metal, the phosphorus that is oxidized and transferred to slag mainly exists in the form of phosphate ion PO_4^{3-}. According to the theory of ion dephosphorization, the formation reaction of PO_4^{3-} is Equation (31) [32].

$$[P] + \frac{5}{2}[O] + \frac{3}{2}(O^{2-}) = (PO_4^{3-})(PO_2 > 10^{-13}Pa) \quad (31)$$

Wagner proposed the concept of phosphate capacity $C_{PO_4^{3-}}$ based on the gas–slag equilibrium; the reaction expression is Equation (32), which is expressed as Equation (33) [33]. Yang et al. [32,34] defined the phosphorus capacity index, $C_{PO_4^{3-}index}$, which is based on slag–metal equilibrium. The reaction expression is Equation (31), with $C_{PO_4^{3-}index}$ expressed as Equation (34). The deduced relationship between $C_{PO_4^{3-}}$ and $C_{PO_4^{3-}index}$ can be expressed as Equation (35), and it can be applicable to any slag system [34].

$$\frac{1}{2}P_2(g) + \frac{5}{4}O_2(g) + \frac{3}{2}(O^{2-}) = (PO_4^{3-}) \quad (32)$$

$$C_{PO_4^{3-}} = \frac{\left(\%PO_4^{3-}\right)}{P_{P2}^{1/2} P_{O_2}^{5/4}} \quad (33)$$

$$C_{PO_4^{3-}index} = \frac{\left(\%PO_4^{3-}\right)}{a_{[P]} \times \left(a_{[O]}\right)^{5/2}} \quad (34)$$

$$\log C_{PO_4^{3-}} = \log C_{PO_4^{3-}index} + \frac{23,531.25}{T} + 0.1606 \quad (35)$$

In Equations (33) and (34), $(\%\mathrm{PO_4^{3-}})$ is the phosphate content in the slag, and the relationship between $(\%\mathrm{PO_4^{3-}})$ and $(\%\mathrm{P_2O_5})$ is shown in Equation (36) [32,34].

$$(\%\mathrm{PO_4^{3-}}) = \frac{2M\mathrm{PO_4^{3-}}}{M\mathrm{P_2O_5}}(\%\mathrm{P_2O_5}) = 1.3382(\%\mathrm{P_2O_5}) \tag{36}$$

M_i represents the molecular weight of oxide i. The expression of $\log C_{\mathrm{PO_4^{3-}}}$ is obtained by combining Equations (27) and (33)–(36), as shown in Equation (37).

$$\log C_{\mathrm{PO_4^{3-}}} = \log \frac{1.3382(\%\mathrm{P_2O_5})}{a_{[\mathrm{P}]} \times (a_{[\mathrm{O}]}^{[\mathrm{Fe}]-[\mathrm{O}]})^{5/2}} + \frac{23{,}531.25}{T} + 0.1606 \tag{37}$$

In Equation (37), $a_{[\mathrm{P}]}$ is calculated by Equation (11). Iron oxide is the main oxidant in the dephosphorization process. According to the discussion in Section 4.2, the oxygen activity in the slag is mainly determined by the oxygen activity of the iron oxidation equilibrium at slag–hot metal interface. Therefore, $a_{[\mathrm{O}]}$ in Equation (37) is calculated by using $a_{[\mathrm{O}]}^{[\mathrm{Fe}]-[\mathrm{O}]}$ in Equation (27).

The empirical formulas for calculating $\log C_{\mathrm{PO_4^{3-}}}$ are listed in Table 5, where Λ represents the optical basicity of the slag, which is expressed as Equation (38). x_i is the mole fraction of cations in oxide i, which is expressed as Equation (39). n_i is the number of oxygen atoms in oxide i and x'_i is the mole fraction of oxide i. Table 6 lists the optical basicities of various oxides [28].

$$\Lambda = \sum x_i \Lambda_i \tag{38}$$

$$x_i = \frac{n_i(\mathrm{O})x'_i}{\sum n_i(\mathrm{O})x'_i} \tag{39}$$

Table 5. Empirical formulas of $\log C_{\mathrm{PO_4^3}}$.

Scholar	Slag	Empirical Formulas
Yang [28]	$CaO\text{-}FeO\text{-}SiO_2\text{-}MgO\text{-}Al_2O_3$	$\log C_{\mathrm{PO_4^3}} = 19.4\Lambda + 6.74$
Maruoka [35]	$CaO\text{-}FeO\text{-}SiO_2\text{-}MgO\text{-}Al_2O_3$	$logCPO43 - -- = 20.67\Lambda + 66{,}204/T - - - 32.01$
Selin [36]	$CaO\text{-}SiO_2\text{-}CaF_2$	$\log C_{\mathrm{PO_4^3}} = 2.016\frac{(x\mathrm{CaO})}{(x\mathrm{SiO2})}0.34(\frac{(x\mathrm{CaO})}{(x\mathrm{SiO2})})^2 + 52{,}600/T$ 11.506
Sobandi [37]	$CaO\text{-}MnO\text{-}SiO_2\text{-}PO_{2.5}$(-MgO, Fe_tO)	$\log C_{\mathrm{PO_4^3}} = 2.60\{(\%\mathrm{CaO}) + 0.33(\%\mathrm{MnO}) + 0.55(\%\mathrm{MgO})$ $0.90(\%\mathrm{FetO})0.77(\%\mathrm{PO2.5})\}/(\%\mathrm{SiO2}) + 40{,}400/T6.48$

Table 6. Optical basicities of different oxides.

Oxide	CaO	SiO_2	MnO	P_2O_5	FeO	Fe_2O_3	MgO	Al_2O_3
Λ	1.00	0.46	0.59	0.40	0.51	0.48	0.78	0.60

Figure 10 shows the comparison of $\log C_{\mathrm{PO_4^{3-}}}$ between the calculated value of the empirical formulas and the calculated value in this paper at different dephosphorization endpoint temperatures. $\log C_{\mathrm{PO_4^{3-}}}$ in Figure 10 is calculated by Equation (37) and different empirical formulas in Table 5. It can be concluded that the calculated values of $\log C_{\mathrm{PO_4^{3-}}}$ from the empirical formulas of Maruoka et al. [35], Selin et al. [36], and Sobandi et al. [37] all decrease with increasing dephosphorization endpoint temperature, and the changing trend is similar, which is consistent with the thermodynamic result that the increase in temperature is not conducive to the dephosphorization of hot metal. The calculated values of $\log C_{\mathrm{PO_4^{3-}}}$ from Yang's empirical formula do not change with the increasing dephosphorization endpoint temperature significantly. This is because Yang et al. [28] mainly studied the effect of dephosphorization slag basicity on phosphorus capacity, and the optical basicity plays a decisive role in their empirical formula. The calculated values of $\log C_{PO_4^{3-}}$ in the present work show a trend of first decreasing, then increasing, and then decreasing again. The value of the calculated results at 1385–1410 °C in the present work are consistent to those of Yang et al. [24]. This is because the experimental temperature, dephosphorization slag basicity, and composition in the present work are similar to the experimental temperature of 1380 °C, slag basicity of 1.40–1.83, and slag compositions of CaO-MnO-MgO-SiO_2-P_2O_5-Fe_tO-Al_2O_3 in their paper, and they did the laboratorial experiments based on the double slag steelmaking process.

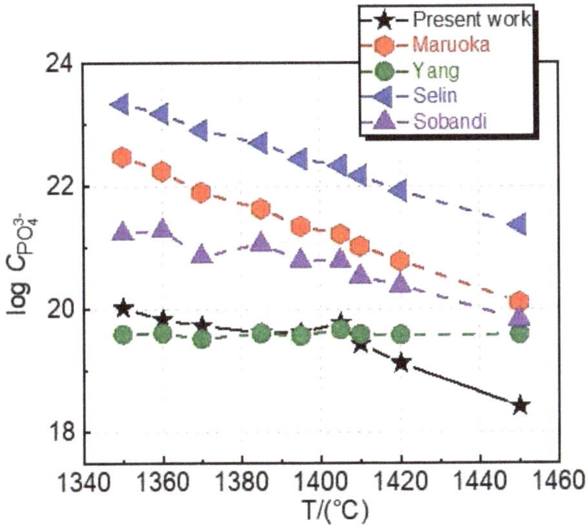

Figure 10. Comparison of $\log C_{PO_4^{3-}}$ between the calculated value of the empirical formulas and the calculated value in this paper at different dephosphorization endpoint temperatures.

The calculated results in the present work are smaller than those calculated by the empirical formulas, which become smaller when the temperature is increased. This is because most of the formulas in Table 6 are the results of laboratory equilibrium under ideal experimental conditions. However, the dephosphorization reaction has not reached equilibrium value due to the short time of dephosphorization in the industrial experiment, which is only about 5 min. Therefore, the $\log C_{PO_4^{3-}}$ value calculated in the percent work is smaller than those calculated by empirical formulas.

From Equations (12), (27), and (37), Equation (40) of the phosphorus distribution ratio $\log L_P$ based on the oxygen activity of iron oxidation equilibrium at the slag–hot metal interface can be deduced as follows:

$$\log L_P = \log C_{PO_4^{3-}} + \log f_P + \frac{5}{2}\log a_{[O]}^{[Fe]-[O]} - 23,531.25/T - 0.287 \quad (40)$$

Figure 11 shows the comparison of $\log L_P$ between the calculated values of the empirical formulas and the calculated values in this paper at different dephosphorization endpoint temperatures. The values of $\log L_P$ in Figure 11 are calculated by Equation (40) and the different empirical formulas in Table 7. According to Figure 11, all empirical formulas indicate that the values of $\log L_P$ show a downward trend at first, then an upward trend around 1405 °C, and then a downward trend again with increasing dephosphorization endpoint temperature. The $\log L_P$ value calculated by the Ogawa formula is the closest to the calculated results in the present paper. It is noticed that between 1395 and 1410 °C, the $\log L_P$ values calculated by the Ogawa formula are consistent with the calculated values in the present work. This is because the Ogawa formula is based on the formula summarized in the dephosphorization stage of the typical double slag converter steelmaking process—the MURC process. Therefore, the temperature and basicity used in the Ogawa formula are similar to those in the present paper. In addition, the effect of the higher [%C] content in hot metal on L_P is taken into consideration, so that it is suitable for the dephosphorization of hot metal with the higher carbon content in the dephosphorization stage of the present work. This is consistent with our previous results on the double slag converter steelmaking process [4].

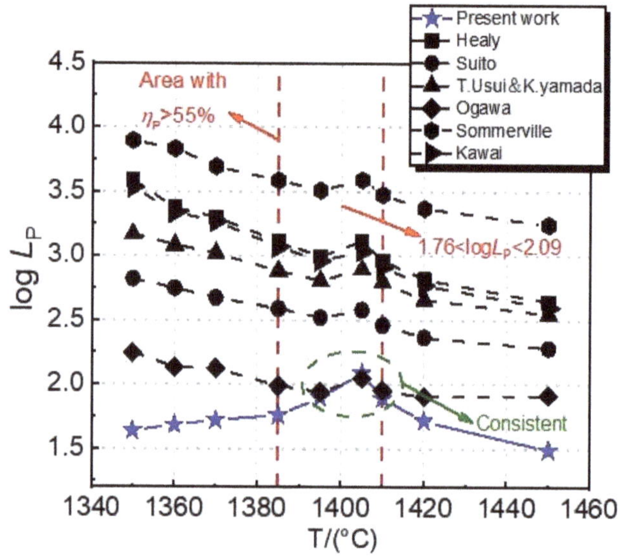

Figure 11. Comparison of logL_P between the calculated value of the empirical formulas and the calculated values in this paper at different dephosphorization endpoint temperatures.

Table 7. Empirical formulas of logL_P.

Scholar	Slag	Empirical Formulas
Healy [38]	$CaO\text{-}SiO_2\text{-}Fe_tO\text{-}P_2O_5$	$\log L_P = \log\frac{(\%P)}{[\%P]} = 22,350/T - 24.0 + 7\log(\%CaO) + 2.5\log(\%T\ Fe)$ $((\%CaO) \geq 30)$
		$\log L_P = \log\frac{(\%P)}{[\%P]} = 22,350/T - 16.0 + 0.08\,(\%CaO) + 2.5\log(\%T\ .Fe)$ $((\%CaO) < 30)$
Suito [39]	$CaO\text{-}SiO_2\text{-}Fe_tO\text{-}P_2O_5\text{-}MgO\text{-}MnO$	$\log L_P = \log\frac{(\%P)}{([\%P](\%T\ .Fe)^{5/2})} = 0.072[(\%CaO) + 0.3\,(\%MgO)$ $+0.6(\%P2O5) + 0.6(\%MnO)] + 2.5\lg(\%T\ .Fe) + 11,570/T - 10.52$
Kawai [40]	$CaO\text{-}SiO_2\text{-}Fe_tO\text{-}P_2O_5$	$\log L_P = \log\frac{(\%P)}{[\%P]} = 22,350/T - 21.876 + 5.6\log(\%CaO) + 2.5\log(\%T\ .Fe)$
T.Usui&K.Yamada [41]	$CaO\text{-}Fe_tO\text{-}SiO_2\text{-}P_2O_5\text{-}MgO\text{-}Al_2O_3\text{-}TiO_2$	$\log L_P = 5.60\log[(\%CaO) + 0.3(MgO) + 0.05(\%FetO)] + 14,800/T$ $18.038 + 0.5\log(\%P2O5) + 2.5\log(\%Fet)$
Ogawa [1]	$CaO\text{-}SiO_2\text{-}Fe_tO\text{-}P_2O_5\text{-}MgO\text{-}MnO$	$\log L_P = \log\frac{(\%P)}{[\%P]} = 2.5\log(\%T\ .Fe) + 0.0715\{(\%CaO) + 0.25\,(\%MgO)\}$ $+7710/T - 8.55 + (105.1\,/T + 0.0723)\ [\%C]$
Zhang [42]	$CaO\text{-}SiO_2\text{-}FeO\text{-}P_2O_5\text{-}MgO\text{-}MnO$	$\log L_P = \log\frac{(\%P2O5)}{[\%P]} = [162(\%CaO) + 127.5(\%MgO) + 28.5(\%MnO)]/T$ $+2.5\lg(\%FeO) - 6.287 \times 10^{-4} \times \log(\%SiO2)^2 + 11,000/T - 10.4$

It can be seen from Figure 11 that with the increasing dephosphorization endpoint temperature, the calculated results in the present work show logL_P first rising and then decreasing obviously, which is consistent with the changing trend of the dephosphorization ratio in Figure 3. When the dephosphorization endpoint temperature is in the range of 1350–1405 °C, with increasing temperature, the auxiliary materials such as lime and dolomite are easy to melt, so that the kinetic conditions of dephosphorization are improved. Although increasing temperature is unfavorable to the thermodynamic conditions of dephosphorization, the overall L_P still increases, and the dephosphorization result shows an upward trend. When the endpoint temperature of dephosphorization exceeds 1405 °C, with further increasing the temperature, the equilibrium constant of the dephosphorization reaction decreases greatly, and the dephosphorization ratio decreases, resulting in a decrease in L_P.

The calculated values of $\log L_P$ in the present work are closer to the empirical formula values of Ogawa et al. [1], but they are lower than the empirical formula values of the others, because the empirical formulas of Healy et al. [38], Suito et al. [39], Usui and Yamada et al. [41], and Zhang et al. [42] were from an equilibrium dephosphorization reaction under laboratorial experiments and not for the double slag process. The empirical formula of Kawai et al. [40] was from the industrial experiments but not for the double slag process. The empirical formula of Ogawa et al. [1] was for the double slag process. However, their dephosphorization times were up to 8 min, and their experiments were carried out in an 8 t converter so that the dephosphorization reaction could be carried out close to the equilibrium results. According to the results in Figure 11, the optimum dephosphorization temperature is in the range of 1385–1410 °C and the optimum $\log L_P$ value is in the range of 1.76–2.09.

4.4. Effect of Dephosphorization Endpoint Temperature on Dephosphorization, Rephosphorization, and the Optimum Temperature of Deslagging

According to previous scholars on the double slag converter steelmaking industrial experimental research, it is necessary to study the optimum temperature of intermediate deslagging theoretically [18,43,44]. As the desiliconization and dephosphorization reactions proceed, the converter temperature increases gradually. Since phosphorus in hot metal is oxidized by FeO and reduced by carbon at the same time, combining the dephosphorization reaction Equation (41) of oxidation of P by FeO and the rephosphorization reaction Equation (42) of the reduction of P_2O_5 by C, the coupling dephosphorization reaction among FeO, C, and P can be obtained as Equation (43). The Gibbs free energy expression of the reaction is taken from the literature [18]. Equation (46) is obtained from Equations (44) and (45).

$$2[\mathrm{P}] + 5(\mathrm{FeO}) = (\mathrm{P2O5}) + 5[\mathrm{Fe}] \quad (41)$$

$$5[\mathrm{C}] + (\mathrm{P_2O_5}) = 2[\mathrm{P}] + 5\mathrm{CO} \quad (42)$$

$$4[\mathrm{P}] + 5(\mathrm{FeO}) + 5\mathrm{CO} = 2(\mathrm{P_2O_5}) + 5[\mathrm{Fe}] + 5[\mathrm{C}] \quad (43)$$

$$\Delta G_{\mathrm{FeO-P}} = -27{,}763.36 + 255.23T + RT\ln\left(\frac{a_{\mathrm{P_2O_5}}}{a_{\mathrm{FeO}}^{5} \times a_{[\mathrm{P}]}^{2}}\right) \quad (44)$$

$$\Delta G_{\mathrm{C-P_2O_5}} = 521{,}758.34 - 709.02T + RT\ln\left(\frac{\left(P_{\mathrm{CO}}^{\mathrm{P-C}}\right)^{5} \times a_{[\mathrm{P}]}^{2}}{a_{\mathrm{P_2O_5}} \times a_{[\mathrm{C}]}^{5}}\right) \quad (45)$$

$$\Delta G_{\mathrm{FeO-CO-P}} = -549{,}521.71 + 964.26T + RT\ln\left(\frac{a_{\mathrm{P_2O_5}}^{2} \times a_{[\mathrm{C}]}^{5}}{P_{\mathrm{CO}}^{5} \times a_{[\mathrm{P}]}^{4} \times a_{\mathrm{FeO}}^{5}}\right) \quad (46)$$

Figure 12 shows the effect of dephosphorization endpoint temperature on the actual Gibbs free energy (ΔG) changes of FeO-P, C-P_2O_5, and FeO-CO-P reactions. The ΔG values of the three reactions of FeO-P, C-P_2O_5, and FeO-CO-P are calculated from Equations (44)–(46), where $a_{P_2O_5}$ is calculated from Equations (10) and (11), P_{CO}^{P-C} is calculated by Equations (6)–(13), $a_{[C]}$, $a_{[P]}$ and a_{FeO} are calculated by Equations (8) and (9), and Equations (3) and (4), respectively.

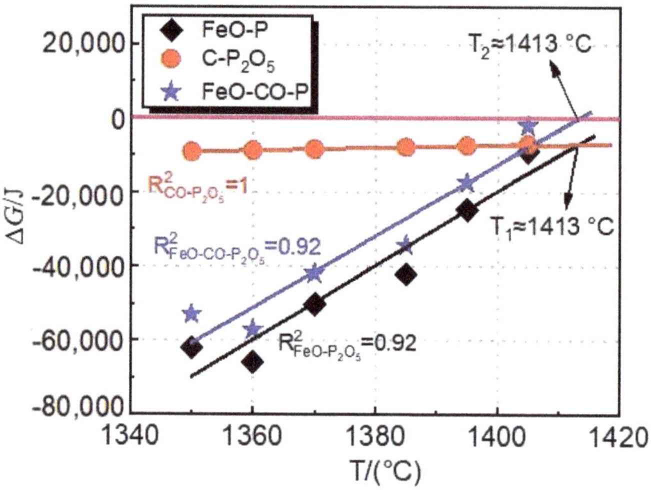

Figure 12. Effect of dephosphorization endpoint temperature on the actual Gibbs free energy changes of FeO-P, C-P_2O_5, and FeO-CO-P reactions.

It can be seen from Figure 12 that the ΔG values of FeO-P and C-P_2O_5 are both smaller than 0, which means that in the dephosphorization stage with the temperature in the range of 1350–1405 °C, both the dephosphorization reaction of phosphorus oxidation by FeO and rephosphorization reaction of reduction of P_2O_5 by carbon occur. With the increasing dephosphorization endpoint temperature, the ΔG value of the C-P_2O_5 reaction does not change significantly, which indicates that temperature has little effect on rephosphorization in 1350–1405 °C.

The ΔG values of the three reactions are linearly fitted with the dephosphorization endpoint temperature, with the regression coefficients, R^2, are 0.92, 0.99, and 0.92, respectively. The fitting line of the C-P_2O_5 ΔG value and the fitting line of the FeO-P ΔG value cross when T_1 is about 1413 °C, which means the oxidation of phosphorus by FeO and the reduction of P_2O_5 by carbon are in dynamic equilibrium at this temperature. At the temperature lower than 1413 °C, dephosphorization and rephosphorization reactions are carried out simultaneously in the converter, and the trend of dephosphorization reaction is greater than that of rephosphorization. When the temperature is higher than T_1, the trend of the rephosphorization reaction in the converter will be greater than that of the dephosphorization reaction, resulting in the rephosphorization of hot metal.

Figure 12 also shows the result of Equation (46) obtained by coupling Equations (44) and (45), as shown by the blue line in the Figure 12. The ΔG value of the FeO-CO-P reaction increases with the increasing dephosphorization endpoint temperature. When the temperature T_2 exceeds 1413 °C, the ΔG is greater than 0. The trend of rephosphorization in the FeO-CO-P coupling dephosphorization reaction is greater than that of dephosphorization.

It can be seen from Figure 12 that the temperature T_2 at which the ΔG value of the FeO-CO-P reaction is 0 is the same as T_1. It shows that the FeO-CO-P coupled dephosphorization reaction is equivalent to combining the dephosphorization reaction and the rephosphorization reaction. Considering the experimental results and thermodynamic calculation results of industrial experiments by the double slag dephosphorization process, the optimal temperature range for intermediate deslagging is about 1400–1420 °C.

5. Conclusions

In the present work, the industrial experiments of dephosphorization in the double slag converter steelmaking process under the conditions of low temperature and low basicity were carried out. The effect of dephosphorization endpoint temperature on the dephosphorization of hot metal was studied, and the following conclusions were drawn:

(1) In the temperature range of 1350–1450 °C, with increasing dephosphorization endpoint temperature, the dephosphorization ratio and phosphorus distribution ratio first increase and then decrease. The phosphorus content in hot metal at the end of dephosphorization first decreases and then increases. The optimum dephosphorization temperature is in the range of 1385–1410 °C, with the dephosphorization ratio higher than 55%, the P_2O_5 content in the dephosphorization slag of 3.93–4.17%, and the $\log L_P$ value of 1.76–2.09.

(2) Dephosphorization slag is mainly composed of the gray massive P-rich phase, gray white Fe-rich phase, and black calcium silicate phase. The path of phosphorus in hot metal entering the P-rich phase of dephosphorization slag can be reasonably inferred as: hot metal → Fe-rich phase → P-rich phase in dephosphorization slag.

(3) a_{FeO} is decided by the dephosphorization endpoint temperature and basic oxide content in dephosphorization slag, and a_{FeO} has multiple linear correlation with them. The P_{CO}^{P-C} of selective oxidation reaction of carbon and phosphorus has a good exponential relationship with the endpoint temperature of dephosphorization. It is beneficial to converter dephosphorization when the temperature is 1385–1410 °C, the value of P_{CO}^{P-C} is 53–80 Pa, and a_{FeO} value is 0.284–0.312.

(4) The oxygen activity on the surface of hot metal of $a_{[O]}^{iron}$, the oxygen activity of the phosphorus oxidation at the slag–hot metal interface of $a_{[O]}^{de[P]}$, and the oxygen activity of the iron oxidation at the slag–hot metal interface of $a_{[O]}^{[Fe]-[O]}$ all increase with increasing the dephosphorization endpoint temperature in the range of 1350–1405 °C. The temperature has the greater effect on the phosphorus oxygen equilibrium of the slag–hot metal interface in the temperature range of 1350–1405 °C, and the temperature has a greater effect on the iron oxygen equilibrium in the slag in the temperature range of 1405–1450 °C. In the temperature range of 1350–1450 °C, owing to $a_{[O]}^{[Fe]-[O]} > a_{[O]}^{de[P]} > a_{[O]}^{iron}$, $a_{[O]}^{[Fe]-[O]}$ plays the most important role on the dephosphorization in the double slag steelmaking process.

(5) Under the present industrial experimental conditions, when the temperature is 1413 °C, the dephosphorization and rephosphorization reactions are in dynamic equilibrium. Considering the experimental results and thermodynamic calculation results of industrial experiments by the double slag dephosphorization process, the optimal temperature range for intermediate deslagging is about 1400–1420 °C.

Author Contributions: Conceptualization, H.S. and J.Y.; methodology, J.Y.; validation, X.L. and W.L.; formal analysis, H.S.; investigation, H.S.; resources, X.L. and W.L.; data curation, W.Y., G.Y., R.Z. and H.S.; writing—original draft preparation, H.S.; writing—review and editing, J.Y.; visualization, H.S.; supervision, J.Y.; project administration, J.Y.; funding acquisition, J.Y. All authors have read and agreed to the published version of the manuscript.

Funding: This work is financially supported by the National Natural Science Foundation of China (Grant No. U1960202).

Data Availability Statement: Data supporting reported results can be found in this paper.

Acknowledgments: The authors would like to thank the National Natural Science Foundation of China (Grant No. U1960202). The equipment was partially provided by the State Key Laboratory of Advanced Special Steel, School of Materials Science and Engineering, Shanghai University.

Conflicts of Interest: The authors declare no conflict of interest.

References

1. Ogawa, Y.; Yano, M.; Kitamura, S.; Hirata, H. Development of the Continuous Dephosphorization and Decarburization Process Using BOF. *Tetsu-to-Hagane* **2001**, *87*, 21–28. [CrossRef]
2. Wang, Y.; Yang, S.; Li, J.; Feng, J.; Wang, F. Dephosphorization by Double-Slag Process in Converter Steelmaking. *High Temp. Mater. Process.* **2017**, *37*, 625–633. [CrossRef]
3. Wu, H.; Li, J.; Zhou, C.G.; Cai, K.S.; Wu, G.P.; Cai, Y.L. Practice on first deslagging process of double slag dephosphorization in 120 t top and bottom combined blown converter. *Spec. Steel* **2013**, *34*, 30–32.
4. Ye, G.-F.; Yang, J.; Zhang, R.-H.; Yang, W.-K.; Sun, H. Behavior of phosphorus enrichment in dephosphorization slag at low temperature and low basicity. *Int. J. Miner. Met. Mater.* **2021**, *28*, 66–75. [CrossRef]
5. Kobayashi, Y.; Yiming, T.; Takahashi, S.; Endo, R. Thermal Conductivity of $2CaO{\cdot}SiO_2$ Bearing Solid Solution. *ISIJ Int.* **2017**, *57*, 1698–1702. [CrossRef]
6. Kitamura, S.-Y.; Shibata, H.; Maruoka, N. Kinetic Model of Hot Metal Dephosphorization by Liquid and Solid Coexisting Slags. *Steel Res. Int.* **2008**, *79*, 586–590. [CrossRef]
7. Suito, H.; Inoue, R. Behavior of phosphorous transfer from $CaO\text{-}Fe_tO\text{-}P_2O_5(\text{-}SiO_2)$ slag to CaO particles. *ISIJ Int.* **2006**, *46*, 180–187. [CrossRef]
8. Kitamura, S.Y.; Saito, S.; Utagawa, K.; Shibata, H.; Robertson, D.G. Mass transfer of P2O5 between liquid slag and solid solution of $2CaO{\cdot}SiO_2$ and $3CaO{\cdot}P_2O_5$. *ISIJ Int.* **2009**, *49*, 1838–1844. [CrossRef]
9. Xie, S.; Wang, W.; Luo, Z.; Huang, D. Mass Transfer Behavior of Phosphorus from the Liquid Slag Phase to Solid $2CaO{\cdot}SiO_2$ in the Multiphase Dephosphorization Slag. *Met. Mater. Trans. A* **2016**, *47*, 1583–1593. [CrossRef]
10. Kakimoto, S.; Kiyose, A.; Murao, R. Influence of P_2O_5 on dissolution behavior of lime in molten slag. *ISIJ Int.* **2017**, *49*, 1710–1717. [CrossRef]
11. Wu, X.R.; Wang, P.; Li, L.S.; Wu, Z.J.; Chen, R.H. Distribution and enrichment of phosphorus in solidified BOF steelmaking slag. *Ironmak. Steelmak.* **2011**, *38*, 185–188. [CrossRef]
12. Pahlevani, F.; Kitamura, S.Y.; Shibata, H.; Maruoka, N. Distribution of P_2O_5 between solid solution of $2CaO{\cdot}SiO_2\text{-}3CaO{\cdot}P_2O_5$ and liquid phase. *ISIJ Int.* **2010**, *50*, 822–829. [CrossRef]
13. Tian, Z.H.; Li, B.H.; Zhang, X.M.; Jiang, Z.H. Double slag operation dephosphorization in BOF for producing low phosphorus steel. *J. Iron Steel Res. Int.* **2009**, *16*, 6–14. [CrossRef]
14. Yang, X.; Sun, F.-M.; Yang, J.-L.; Liu, F.; Cheng, K.-S.; Wang, J.-H. Optimization of Low Phosphorus Steel Production with Double Slag Process in BOF. *J. Iron Steel Res. Int.* **2013**, *20*, 41–47. [CrossRef]
15. Zhou, C.G.; Li, J.; Wu, H.; Yang, K.Z.; Cai, K.S.; Wu, G.P.; Cao, Y.L. Study of factors affecting liquid steel rephosphorization. *Iron Steel Vanadium Titan.* **2014**, *35*, 116–122.
16. Han, X.; Li, J.; Hu, X.G.; Zhou, C.G.; Zhang, H.Y.; Guo, C. Effect of oxidability of final slag in converter on dephosphorization. *Foundry Technol.* **2015**, *36*, 412–415.
17. Kitamura, S.Y.; Yonezawa, K.; Ogawa, Y.; Sasaki, N. Improvement of reaction efficiency in hot metal dephosphorization. *Ironmak. Steelmak.* **2002**, *29*, 121–124. [CrossRef]
18. Zhou, C.G.; Li, J.; Wu, H.; Cai, K.S.; Wu, G.P.; Cao, Y.L. Study on the temperature of first deslagging of double slag dephosphorization in converter. *Iron Steel* **2014**, *49*, 24–28.
19. Karbowniczek, M.; Cebula, E.K.; Reichel, J. Investigations of the dephosphorization of liquid iron solution containing chromium and nickel mater. *Met. Mater. Trans. B* **2012**, *43*, 554–561. [CrossRef]
20. Hamano, T.; Fukagai, S.; Tsukihashi, F. Reaction mechanism between solid CaO and $FeOx\text{-}CaO\text{-}SiO_2\text{-}P_2O_5$ slag at 1573 K. *ISIJ Int.* **2006**, *46*, 490–495. [CrossRef]
21. Yang, X.; Matsuura, H.; Tsukihashi, F. Condensation of P_2O_5 at the interface between 2CaO center dot SiO_2 and $CaO\text{-}SiO_2\text{-}FeO_x\text{-}P_2O_5$ slag. *ISIJ Int.* **2009**, *49*, 1298–1307. [CrossRef]
22. Yang, X.; Matsuura, H.; Tsukihashi, F. Reaction Behavior of P_2O_5 at the interface between solid 2CaO center dot SiO_2 and liquid $CaO\text{-}SiO_2\text{-}FeO_x\text{-}P_2O_5$ slags saturated with solid 5CaO center dot SiO_2 center dot P_2O_5 at 1573 K. *ISIJ Int.* **2010**, *50*, 702–711. [CrossRef]
23. Xie, S.; Wang, W.; Huang, D.; Li, H.; Du, Y. Clarification of the Dissolution of Solid CaO and the Phosphorus-Enrichment Capability of Calcium Silicates in the Multiphase Slag Based on the Ion and Molecule Coexistence Theory. *Steel Res. Int.* **2018**, *89*. [CrossRef]
24. Huang, X.H. *Principles of Iron and Steel Metallurgy*; Metallurgical Industry Press: Beijing, China, 2002; pp. 213–215.
25. Ban-Ya, S. Mathematical Expression of Slag-Metal Reactions in Steelmaking Process by Quadratic Formalism Based on the Regular Solution Model. *ISIJ Int.* **1993**, *33*, 2–11. [CrossRef]
26. Turkdogan, E.T.; Pearson, J. Activities of constituents of iron and steelmaking slags, Part 3-Phosphorus pentoxide. *Iron Steel* **1953**, *175*, 398–403.
27. Yang, X.M.; Duan, J.P.; Shi, C.B.; Zhang, M.; Zhang, Y.L.; Wang, J.C. A thermodynamic model of phosphate distribution ratio between $CaO\text{-}SiO_2\text{-}MgO\text{-}FeO\text{-}Fe_2O_3\text{-}MnO\text{-}Al_2O_3\text{-}P_2O_5$ slags and molten steel during a top–bottom combined blown converter steelmaking process based on the ion and molecule coexistence theory. *Met. Mater. Trans. B* **2011**, *42*, 738–770. [CrossRef]
28. Yang, W.; Yang, J.; Shi, Y.; Yang, Z.; Gao, F.; Zhang, R.; Ye, G. Effect of basicity on dephosphorization of hot metal with a low basicity slag at 1653 K. *Ironmak. Steelmak.* **2021**, *48*, 69–77. [CrossRef]

29. Turkdogan, E.T. Assessment of P2O5 Activity Coefficients in Molten Slags. *ISIJ Int.* **2000**, *40*, 964–970. [CrossRef]
30. Basu, S.; Lahiri, A.K.; Seetharaman, S. A Model for Activity Coefficient of P2O5 in BOF Slag and Phosphorus Distribution between Liquid Steel and Slag. *ISIJ Int.* **2007**, *47*, 1236–1238. [CrossRef]
31. Wu, X.; Bao, Y.P.; Yue, F.; Feng, J.; Tang, D.C. Study on factors to affect the product of carbon content and oxygen content at blowing end-point of BOF steelmaking. *Res. Iron Steel* **2010**, *38*, 26–29.
32. Yang, X.M.; Li, J.Y.; Chai, G.M.; Duan, D.P.; Zhang, J. A thermodynamic model for predicting phosphate capacity of CaO-based slags during hot metal dephosphorization pretreatment process. *Ironmak. Steelmak.* **2017**, *44*, 437–454. [CrossRef]
33. Wagner, C. The concept of the basicity of slags. *Met. Mater. Trans. A* **1975**, *6*, 405–409. [CrossRef]
34. Yang, X.M.; Shi, C.B.; Zhang, M.; Duan, J.P.; Zhang, J. A thermodynamic model of phosphate capacity for CaO-SiO_2-MgO-FeO-Fe_2O_3-MnO-Al_2O_3-P_2O_5 slags equilibrated with molten steel during a top–bottom combined blown converter steelmaking process based on the ion and molecule coexistence theory. *Met. Mater. Trans. B* **2011**, *42*, 951–976. [CrossRef]
35. Maruoka, N.; Ono, S.; Kitamura, S.Y. Equilibrium distribution ratio of phosphorus between solid iron and magnesio-wustite-saturated Al_2O_3-CaO-Fe_tO-MgO-SiO_2 slag at 1623K. *ISIJ Int.* **2013**, *53*, 1709–1714. [CrossRef]
36. Roger, S.; Dong, Y.C.; Wu, Q.A. Uses of limes-based fluxes for simultaneous removal of phosphorus and sulfur in hot metal pretreatment. *Scand. J. Met.* **1990**, *19*, 98–109.
37. Sobandi, A.; Katayama, H.G.; Momono, T. Activity of phosphorus oxide in CaO-MnO-SiO_2-$PO_{2.5}$(-MgO, Fe_tO) slags. *ISIJ Int.* **1998**, *38*, 781–788. [CrossRef]
38. Healy, G.W. A new look at phosphorus distribution. *Iron Steel* **1970**, *208*, 664–668.
39. Suito, H.; Inoue, R. Phosphorus distribution between MgO-saturated CaO-Fe_tO-SiO_2-P_2O_5-MnO slags and liquid iron. *Trans. ISIJ* **1984**, *24*, 40–46. [CrossRef]
40. Kawai, Y.; Takahashi, I.; Miyashita, Y.; Tachibana, K. For dephosphorization equilibrium between slag and molten steel in the converter furnace. *Tetsu-to-Hagané* **1977**, *63*, 156.
41. Usui, T.; Yamada, K.; Kawai, Y.; Inoue, S.; Ishikawa, H.; Nimura, Y. Experiment of phosphorus and oxygen distribution between CaO-SiO_2-MgO-Fe_tO slag and liquid steel and estimation of phosphorus content at end point of top and bottom blowing converter. *Tetsu-to-Hagané* **1991**, *77*, 1641–1648. [CrossRef]
42. Zhang, X.F.; Sommerville, I.D.; Toguri, J.M. Equation for the equilibrium distribution of phosphorus between basic slags and steel. *Trans. Iron Steel Soc.* **1985**, *6*, 29–35.
43. Zhou, C.G.; Li, J.; Luo, K.M.; Han, X.; Zhang, Z.M.; Liu, Z.M.; Deng, C.F. First deslagging practice of double slag process for dephosphorization. *Iron Steel Vanadium Titan.* **2016**, *37*, 119–126.
44. Zhang, T.X.; Song, J.S.; Zhou, D.D.; Liu, X.Y. Calculation and process test for dephosphorization parameter of 120t converter. *Spec. Steel* **2020**, *41*, 25–27.

metals

MDPI

Article

Forecasting Model of Silicon Content in Molten Iron Using Wavelet Decomposition and Artificial Neural Networks

Ana P. Miranda Diniz [1], Klaus Fabian Côco [1], Flávio S. Vitorino Gomes [2] and José L. Félix Salles [1,*]

[1] Electrical Engineering Department, Campus de Goiabeiras, Federal University of Espírito Santo, Vitória 29075-910, ES, Brazil; anapmdiniz@gmail.com (A.P.M.D.); klaus.coco@ufes.br (K.F.C.)
[2] Renewable Energy Engineering Department, Federal University of Paraíba, João Pessoa 58051-900, PB, Brazil; flavio@cear.ufpb.br
* Correspondence: jleandro@ele.ufes.br

Abstract: Silicon content forecasting models have been requested by the operational team to anticipate necessary actions during the blast furnace operation when producing molten iron, to control the quality of the product and reduce costs. This paper proposed a new algorithm to perform the silicon content time series up to 8 h ahead, immediately after the molten iron chemical analysis is delivered by the laboratory. Due to the delay of the laboratory when delivering the silicon content measurement, the proposed algorithm considers a minimum useful forecasting horizon of 3 h ahead. In a first step, it decomposes the silicon content time series into different subseries using the Maximal Overlap Discrete Wavelet Packet Transform (MODWPT). Next, all subseries forecasts were determined through Nonlinear Autoregressive (NAR) networks, and finally, these forecasts were summed to furnish the long-term forecast of silicon content. Using data from a real industry, we showed that the prediction error was within an acceptable range according to the blast furnace technical team.

Keywords: blast furnaces; silicon content; maximal overlap discrete wavelet packet; artificial neural network; forecasting; time series analysis

Citation: Diniz, A.P.M.; Côco, K.F.; Gomes, F.S.V.; Salles, J.L.F. Forecasting Model of Silicon Content in Molten Iron Using Wavelet Decomposition and Artificial Neural Networks. *Metals* **2021**, *11*, 1001. https://doi.org/10.3390/met11071001

Academic Editors: Pasquale Cavaliere and Alexander McLean

Received: 28 April 2021
Accepted: 17 June 2021
Published: 23 June 2021

1. Introduction

Due to the higher competitiveness in the globalized market, industries are always seeking to reduce their operational costs without affecting the quality of the final product through control techniques of the relevant variables involved in the manufacturing process. The iron making industry has played a significant role in the world economy, boosting the process of industrialization. This industry produces molten iron (or hot metal or pig iron) using blast furnaces and next directs it to be refined in the steel plant, resulting in steel. The stable and efficient operation of the blast furnace depends on the acquisition of a large amount of data to support operators in decision making through big data platforms and data-driven models [1].

The quality of the produced molten iron affects the cost of the production of the steel and defines for which ultimate purpose it can be sold [2]. However, the adverse conditions presented inside a blast furnace (high pressure, temperature, and erosive environment) may impair the direct measurements of some essential variables for quality control of the molten iron [3]. Thus, the laboratory analysis of some chemical variables of the molten iron samples is fundamental for monitoring the blast furnace operating state [4]. Notably, one of the most critical variables is the silicon content of the molten iron, due to its strong correlation with the thermal state of a blast furnace and the quality of the produced iron [2]. High values of the silicon content indicate an excess of coke in the blast furnace (i.e., a large amount of thermal energy), while lower values indicate the lack thereof. Thus, since the coke cost is prevalent in molten iron production, obvious economic benefits arise from more rigorous control of the silicon content [4–12]. In addition, better silicon control in molten iron can reduce the CO_2 emissions to the atmosphere due to the steelmaking time decrease in the pig iron refining stage and other steps of the production process.

Over the past two decades, several papers have proposed neural network models with different inputs to predict and even control the silicon content of molten iron and other parameters related to the quality of the blast furnace process. Due to the complexity and a large number of exogenous variables involved in the molten iron production process, References [4–12] stressed the difficulty when choosing an appropriate algorithm that can establish effective the input–output relationships of neural networks. Consequently, we can observe in the literature a significant quantitative and qualitative discrepancy among the chosen inputs. All these literature works performed forecasts of silicon content with forecasting horizons lower than 3 h ahead and also disregarded the delay between the collection and laboratory analysis results of the silicon content, which is typical in any blast furnace [11]. The more we anticipate future information on the silicon content, the more effective the actions of the operators will be, aiming at increasing the quality of the final product, as well as reducing the CO_2 emissions and costs associated with the production of iron.

Initially, Reference [4] used a Multilayer Perceptron (MLP) without performing a careful analysis of the most relevant input variables and their respective time delays. Next, Reference [5] chose the inputs of a Nonlinear Input-Output (NIO) model through the blast furnace operator experience, and References [4–8] applied a pruning algorithm [13]. The work [9] considered a Nonlinear Autoregressive (NARX) neural network, including the silicon content as one of the network inputs. The authors of [10] eliminated the exogenous inputs, by decomposing the silicon content time series into different subseries from Empirical Mode Decomposition (EMD). After, these authors added the subseries forecasts through NAR neural network models. However, Reference [10] did not provide a numerical analysis of the forecasting quality.

Recently, References [11,12] used Extreme Learning Machine (ELM)-based networks with up to thee hidden layers containing about one-thousand neurons. Unlike backpropagation-based algorithms, ELM chooses the input and hidden layer weights randomly, while the output layer weights are determined analytically. Thus, ELM performs faster training than backpropagation [14]. However, due to the absence of weight adjustment in the ELM neural network, it requires many more neurons, which may increase its computational cost [15]. Therefore, because of the hardware limitation of the existing supervisory distributed computer control systems in the current steel industry, the ELM network implementation in this hardware is much more complicated than the neural network trained with backpropagation.

In this paper, we proposed a long-term forecasting algorithm with horizons from 3 h ahead, because the silicon content measurement in the molten iron is available for operators through a chemical laboratory analysis updated every three hours approximately after being collected. We considered a parsimonious neural network model trained by backpropagation, so that with a smaller number of parameters, it was possible to model and even forecast the silicon content with superior performance to those NIO and NARX neural networks. The proposed forecasting algorithm decomposes the silicon content time series into different subseries using Maximum Overlap Discrete Wavelet Packet Transformation (MODWPT) [16]. We modeled each subseries decomposed by MODWPT through a NAR neural network, and we performed the silicon content forecast by adding each subseries' forecasts.

MODWPT is the third generation of wavelet transform, and just as the Discrete Wavelet Transform (DWT) and Discrete Wavelet Packet Transform (DWPT), MODWPT also applies lowpass and highpass filters to the input signal at each level. However, MODWPT overcomes the disadvantages inherent to DWT and DWPT as it not only presents a uniform frequency bandwidth, but is also time-invariant. Besides, using MODWPT avoids edge effects at the end of the time series and allows the addition of new samples without the need to recalculate the entire transform again [16]. These properties are worthwhile when dealing with real-time series applications focused on performing nonstationary time series forecasting. MODWPT decomposes the time series into stationary and nonlinear subseries through highpass and lowpass filters [17], characterized by frequency bands with different resolutions.

The work [18] applied the wavelet transform of the first generation in a neural model to reduce data noise and forecast the permeability index for blast furnaces. However, the analysis presented in [18] was entirely based on static data, i.e., the data were not processed sample-by-sample, which is inherent to real-time data processing. This inhibits the application of this model in industrial dynamic systems, whose forecasts must be processed for each sample, or a small set of samples collected and processed in real time, as we considered in this article.

There are some differences between the EMD decomposition technique used by [10] and MODWPT. EMD is an iterative procedure that extracts oscillatory-like features from the data in such a way that it uses adaptive bases that are directly generated by the data, i.e., EMD is a wavelet approach. EMD is an adaptive technique, and MODWPT is a math transform that lends itself to robust statistical analysis [19].

In Section 2, we briefly explain the molten iron production process; in Section 3, we apply the pruning algorithm to determine the neural network models with and without the MODWPT decomposition; in Section 4, using real data from a blast furnace located in Brazil, we compare the proposed algorithm, denoted by MODWPT-NAR, with the NAR, NARX, and NIO models without considering time series decomposition; finally, we conclude this article in Section 5.

2. The Blast Furnace Process

2.1. Background

The blast furnace is a big thermo-chemical reactor used for molten iron production, which is the main raw material for steel production. The operation of the blast furnace occurs following the countercurrent principle [2]. During the production process, synthetic ferrous materials (sinter and iron ore pellets) or natural raw materials (granulated iron ore) enter the upper part of the blast furnace, with one type of fuel (coke or charcoal). In addition, there may also be auxiliary injections of fuel oil, tar, pulverized coal, or natural gas. The objective is to produce an alloy, denominated molten iron, at a temperature of approximately 1500 °C, in the liquid state, composed of iron (between 92% and 95%), carbon (between 3.5% and 5%), and other elements such as silicon and manganese. In addition, slag and blast furnace gas are also generated as byproducts [2].

Molten iron is formed from the reduction of iron ore, while slag comes from the agglomeration of minerals and ashes resulting from burning fuels. Generally, it takes about six to eight hours for the raw materials to be processed in the form of molten iron and slag, which are accumulated at the base of the blast furnace (called blast furnace hearth). The molten iron remains on the bottom of the furnace hearth, while the slag remains on the upper part of it [2]. Ores and coke may contain many different kinds of impurities, and most of them are reduced in the blast furnace and eliminated in the form of slag. Some elements are fully incorporated into the molten iron (such as phosphorus and copper), while others are partially incorporated (such as sulfur, manganese, carbon, and silicon). From the point of view of thermal control, silicon is one of the most important elements [2].

The silicon in the blast furnace comes mainly from ores and coke ashes. The content of silicon depends not only on the quantity, but also on the types of raw materials used and on the method applied for producing the molten iron. The transfer of silicon to molten iron can occur either directly from slag to molten iron (through metal–slag reactions) or indirectly. For the latter, the silicon is transferred from slag (or coke ash) to the gaseous phase (SiO for example), which then reacts with molten iron–carbon, incorporating silicon [2].

The silicon content of the molten iron is an essential variable for the blast furnace process because it expresses not only the thermal state, but also the quality of it. The fluctuations of the silicon content reflect the thermal variations within the blast furnace.

This way, a low silicon content indicates that the furnace is cooling, which can compromise the operation, while high values indicate the waste of coke, shown by excessive heat generation. Thus, since the cost of coke is prevalent (usually more than 30% of the total amount) in molten iron production, the stricter the control of silicon content, the more significant the reduction in production costs is [6,8].

After passing through the blast furnace, the liquid molten iron is taken to the steelworks to transform it into steel. The thermal input of the molten iron is mainly responsible for the process heating supply. Besides, this material must show adequate physicochemical characteristics; otherwise, it is necessary to perform adjustments in the process, which can decrease productivity. Therefore, the high content of silicon can hinder the process of converting iron into steel and, consequently, generate even higher expenses. However, the presence of silicon in the steelworks is desirable, since it contributes to removing the oxygen present in the steel, through the oxidation of SiO_2 [2].

One of the major problems during the quality control of molten iron production is the long time spent from charging the raw material at the top of the blast furnace until the molten iron removal from the bottom of the blast furnace hearth [2]. Furthermore, in the real production process at hand, the laboratory analysis of the molten iron chemical composition takes about three hours to perform, as well as the silicon content amount to be available for the operators. Another problem is the nonregularity of the data collection periods of the silicon content, since the opening and closing times of the pig iron tapholes may vary according to the blast furnace production rate [20]. All these factors make it difficult to control the silicon content inside the blast furnace to reach adequate levels that guarantee the quality of the final product and the low cost of production.

2.2. The Dataset

In this study, we used data extracted from a blast furnace located in Brazil. This blast furnace has 4 tapholes, 34 tuyeres, 3617 m^3 of internal volume, and a nominal annual production capacity of 2.8 Mt. Typically, one taphole is opened at a time, although in some cases, it is necessary to open a second taphole to accelerate material extraction [3].

Regarding the blast furnace considered in this paper, twenty-eight variables may influence the silicon content. Such variables are described in Table 1.

It is noted that for each molten iron casting (time elapsed between an opening and a closing of the blast furnace taphole), three torpedo cars are filled, on average. A torpedo car is made with steel and coated with refractory, and its primary function is to transport molten iron from the blast furnace to the steelworks. Usually, the blast furnace operators collect a molten iron sample from each torpedo car loading for laboratory analysis. The result of such analysis, regarding the percentage of silicon, is then updated in the system. Because the acquisition time of the molten iron samples is not synchronous, presenting variations of around 50 min, it is crucial to resample it at regular times so that the neural network inputs can receive the past values of the silicon content time series. The molten iron temperature and electromotive force (input variables x_{27} and x_{28}) are also measured at each torpedo car loading and, therefore, also present an asynchronous sampling period, similar to the silicon content. The other input variables described in Table 1 have a sample period of 30 min. Thus, due to the interest of blast furnace operators in specifying the data at each 60 min interval, all input and output variables are resampled at this sample time.

Due to the harsh conditions of the blast furnace operation, process and measurement noises can contaminate the dataset. In this sense, noise reduction is required in data preprocessing [11]. All input and output data collected were processed to correct outliers and measurement errors, as well as to exclude significant periods of blast furnace shutdowns. In general, an abrupt decrease in the blast furnace production rate indicates a blast furnace shutdown (i.e., a period of strong process instability), and often, these moments are followed by outliers in the silicon content time series. Figure 1 shows an example of a plant shutdown identified during the analysis of the corresponding data. As the objective of this work was to forecast the silicon content during regular operational periods, the samples corresponding to the abnormalities must be detected [21] and excluded from the dataset.

Table 1. Input variables that may influence the silicon content.

Variable	Description
x_1	flame temperature (°C)
x_2	rate of CO/CO_2
x_3	H_2 gas efficiency (%)
x_4	molten iron production (tons/min)
x_5	coke rate (Kg/ton of molten iron)
x_6	pulverized coal injection rate (Kg/ton of molten iron)
x_7	direct reduction (%)
x_8	ore/coke ratio
x_9	thermal index of H0 (10^3 Kcal/ton of molten iron)
x_{10}	north central gaseous flow index
x_{11}	north peripheral gaseous flow index
x_{12}	south central gaseous flow index
x_{13}	south peripheral gaseous flow index
x_{14}	blowing air volume (Nm^3/min)
x_{15}	blowing air moisture (g/Nm^3)
x_{16}	blowing air temperature (°C)
x_{17}	blowing air pressure (Kg/cm^2)
x_{18}	top pressure (Kg/cm^2)
x_{19}	top temperature (°C)
x_{20}	composition of H_2 on the top gas (%)
x_{21}	composition of N_2 on the top gas (%)
x_{22}	composition of CO on the top gas (%)
x_{23}	composition of CO_2 on the top gas (%)
x_{24}	O_2 enrichment rate (%)
x_{25}	CO efficiency (%)
x_{26}	staves' thermal losses (10^4 Kcal/h)
x_{27}	molten iron temperature
x_{28}	electromotive force (%)

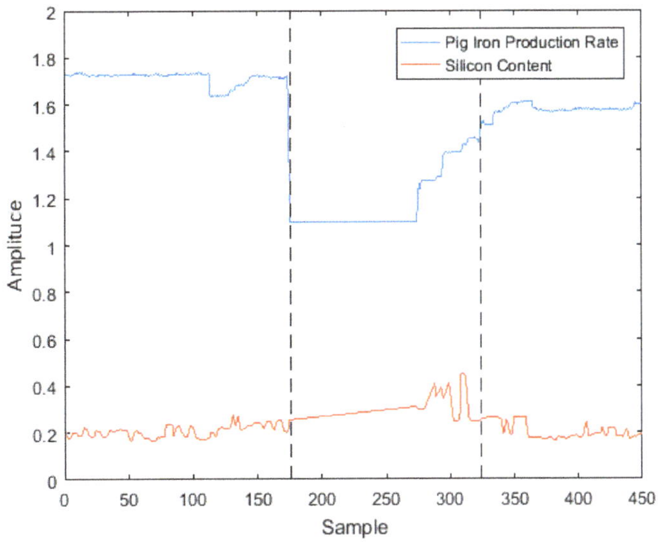

Figure 1. Abnormality period in the blast furnace.

After removing the data corresponding to instability periods in the blast furnace, we performed the stationarity analysis of the silicon content time series. First, we detected the nonstationary characteristic of this time series through the Kwiatkowski, Phillips, Schmidt, and Shin (KPSS) test [22]. Then, we used the Lilliefors [23] and Kruskal–Wallis [24] tests to confirm the existence of nonlinearities by splitting the time series into five packages of data; see Figure 2. Each one of the packages was composed of a 10-day measurement, to show that these samples were not from the same distribution.

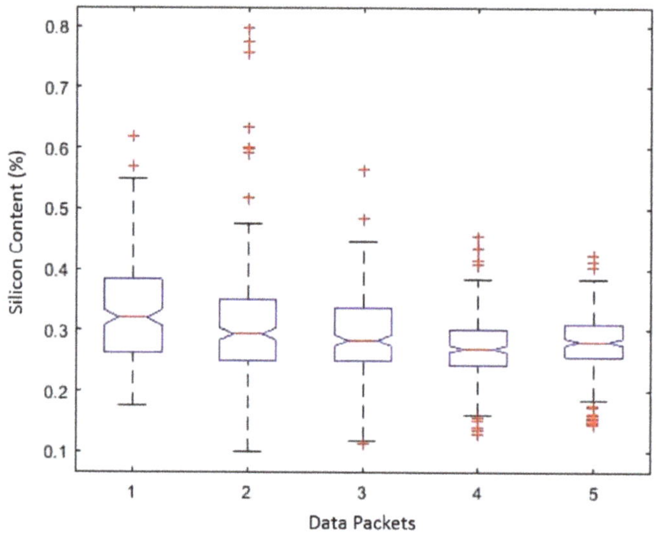

Figure 2. Box plot test of silicon content time series.

Based on these analyses, we concluded that the time series had nonlinear components due to the asymmetric dispersion concerning the normal distribution, as shown in Figure 3.

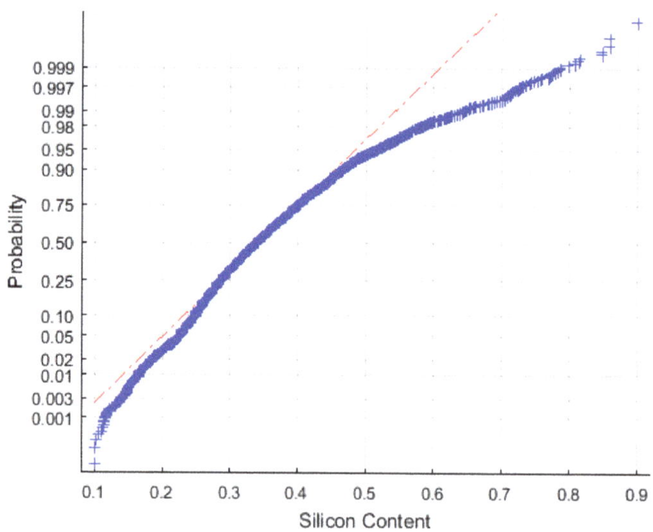

Figure 3. Q-Q plot of silicon content time series.

3. Silicon Content Time Series Modeling

Firstly, we present the ANN models without considering the decomposition of silicon content using MODWPT.

3.1. NIO, NARX, and NAR Neural Network Models

Inspired by the architecture and learning of the human brain, ANNs exhibit self-adaptive behavior and are very favorable to model and forecast nonlinear processes that have complex characteristics, in contrast to traditional deterministic methods. They work as processors made up of simple processing units—the neurons—that can store the knowledge acquired and can also adapt to a changing environment through a learning process. The links that connect neurons to carry out signal transmission are called "synapses," over which there is a corresponding weight. Each neuron receives multiple inputs from other neurons, computes them, and generates a single output, which can be propagated to several other neurons [25].

In this paper, we addressed the Time-Delay Neural Networks (TDNN). Such networks are extensions of Multilayer Perceptron (MLP) networks, with the introduction of one unit delay in the input variable, aiming at representing the dynamics (cause and effect relationships) of the nonlinear model [26]. Examples of TDNN are the NIO, NARX, and NAR neural networks. Their outputs at k-steps ahead ($\hat{y}(t+k)$) are described by (1)–(3), respectively, where n_a and n_b are the maximum delays of the output $y(t)$ (silicon content) and inputs $x(t) = (x_1(t), x_2(t), \ldots, x_{28}(t))$ (defined in Table 1), respectively; t is the moment immediately after the molten iron chemical analysis is delivered by the laboratory. $f()$ represents a nonlinear function that is parameterized through the synaptic weights calculated during the neural network training phase [25,26].

$$\hat{y}(t+k) = f(x(t), \ldots, x(t-n_b)) \tag{1}$$

$$\hat{y}(t+k) = f(y(t), \ldots, y(t-n_a), x(t), \ldots, x(t-n_b)) \tag{2}$$

$$\hat{y}(t+k) = f(y(t), \ldots, y(t-n_a)) \tag{3}$$

The NIO, NARX, and NAR neural network models were defined by the input and output sets $\{x(t-n), \ldots, x(t)\}$ and $\{y(t-n), \ldots, y(t)\}$, respectively, where n is the total of historical data samples. We split both datasets into two blocks. Regarding the first block, each time series had 2160 samples (90 days) used for training, and as for the second block, each time series had 720 samples (30 days), used for validation. Each time series was normalized within the range $[0.1, 0.9]$. Normalization was important to ensure non-zero values in the time series, thus allowing us to calculate the performance indices used to compare the neural network models. Furthermore, the normalization technique can provide a reduction of errors in the training of the neural networks [25]. High-frequency noises present in the input time series were removed by applying wavelet filters of the symlet type with order four [27].

The criteria used to define the ANN models were the k-steps ahead Mean Squared Error (MSE(k)) and k-steps ahead Mean Absolute Percentage Error (MAPE(k)), defined according to (4) and (5), respectively:

$$\text{MSE}(k) = \frac{1}{n} \sum_{t=1}^{n} (y(t+k) - \hat{y}(t+k))^2, \tag{4}$$

$$\text{MAPE}(k) = \frac{1}{n} \sum_{t=1}^{n} \left| \frac{y(t+k) - \hat{y}(t+k)}{y(t+k)} \right|, \tag{5}$$

We considered a NAR neural network topology with six inputs because the leaks (casting) through the tapholes occurred at a frequency of between eight- and fourteen-times a day, which corresponded to a leakage period of 1.5 to 3 h. Besides, the residence time of the load inside the blast furnace was 6 h approximately; therefore, the current sample of the silicon content depended on the samples collected up to 6 h before, i.e., up to six lags considering a sample time of one hour. For that reason, the NAR neural network must have six entries, which corresponded to the current sample plus five previous samples.

Regarding the NIO and NARX neural networks, the delays of the input variables $x_1 \ldots, x_{28}$ were, initially, equal to their respective average response times, i.e., the average times the production process responded to the variations of each input. In this sense, except the coke (x_5) and ore/coke (x_8) rates, which presented a slower response time (8 h), all input variables of the NARX network had a delay of three samples (3 h).

The NIO, NARX, and NAR neural networks have a hyperbolic tangent-type activation function for the hidden layer neurons and a linear activation function for the output layer neuron. We used a single hidden layer on each of the networks, which were defined by testing structures with five to two-hundred hidden neurons, adding five neurons per test. For each of these networks, we determined the MSE(0) errors after five consecutive training rounds performed by using the Levenberg–Marquardt learning algorithm. From these experiments, we concluded that the hidden layers for the NIO, NARX, and NAR neural models with 95, 90, and 100 neurons, respectively, provided the lowest MSE(0) errors in the training phase.

Due to a large number of variables involved in the process, it was challenging to choose an appropriate algorithm that could select the most relevant inputs and, consequently, to establish input–output relations effectively [6]. To remove redundant connections between the neurons of the input and hidden layers, without harming the model performance, we applied a pruning algorithm based on Hessian matrix, using the Optimal Brain Surgeon method (OBS) [28,29]. The pruning algorithm allowed removing regressors (input variables) that were not significant for the model, thus facilitating the choice of relevant variables. We verified that after applying the pruning algorithm, the number of hidden neurons for the NIO, NARX, and NAR neural networks reduced from 95 to 3, 90 to 6, and 100 to 13, respectively. The pruning algorithm also detected the significant regressors to be considered in the input layer of each neural network, as shown in Table 2. Each of the hidden layer neurons of these final ANN models was associated with more than one regressor, although the total number of regressors decreased significantly concerning the ANN models initially proposed. In Section 4, we compare the performances of the NIO, NARX, and NAR neural networks before and after applying the pruning algorithm.

Table 2. Significant regressors in the input layers of neural networks.

ANN Model	Regressors
NIO	$x_5(t-5),\ x_{24}(t-1),\ x_{27}(t-2)$
NARX	$x_2(t-2),\ x_3(t-3),\ x_4(t-3),\ x_5(t-k)$ for $k=1,2,3$ $x_7(t-3),\ x_8(t-1),\ x_{10}(t-k)$ for $k=1,2,$ $x_{12}(t-2),\ x_{14}(t-1),\ x_{15}(t-1),\ x_{19}(t-2),$ $x_{23}(t-2),\ x_{24}(t-3),\ x_{25}(t-2),\ x_{27}(t-k)$ for $k=1,2,$ $x_{28}(t-1),\ y(t-k)$ for $k=1,2,3,4,5.$
NAR	$y(t-k)$ for $k=1,2,3,4,5.$

3.2. *MODWPT-NAR Neural Network Model*

To increase the accuracy of prediction models using neural networks, we developed a hybrid algorithm based on MODWPT decomposition and ANN models to forecast the content of silicon in molten iron (see Figure 4). Firstly, the silicon content time series was decomposed into three levels through the MODWPT, using a four-order symlet mother wavelet. As a result, we obtain eight data subsets $\{X_{WP1}\}, \{X_{WP2}\}, \ldots, \{X_{WP8}\})$, which are illustrated in Figure 5.

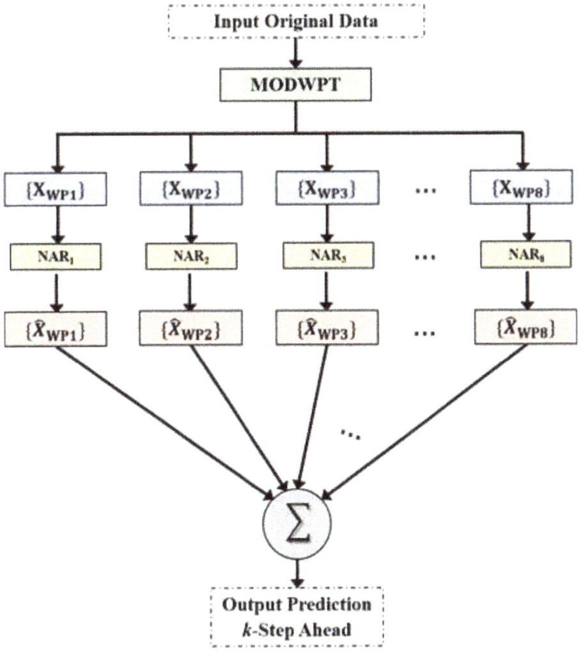

Figure 4. MODWPT-NAR neural network model.

After that, we used eight NAR networks to perform multistep forecasts of each one of the eight subseries. Finally, we added the subseries forecasts to obtain the original signal forecasting.

The Discrete Wavelet Transform (DWT) is a time and frequency multiresolution analysis technique that decomposes the original signal into several levels. According to [27], multiresolution analysis is a useful tool since many signals found in practical applications have high-frequency components for short time durations and low-frequency components for a long duration of time.

The implementation of DWT is an iterative process with successive decompositions of a signal, which uses two sets of functions called scale functions and displacement functions, associated with the lowpass and highpass filters of the wavelets. Employing convolution, each one of the lowpass and highpass filters yields signals in the approximation and detail components, respectively, of the original signal [27].

A single function $\Psi(t)$, called the mother wavelet, generated the wavelet functions. Mother wavelets can take many forms (Daubechies wavelet (Daubechies), Daubechies' least-asymmetric wavelet (symlet), and coiflet [27]) and, in general, are defined by [16]:

$$\Psi_{j,k}(t) = a_0^{-j/2}\Psi(a_0^{-j}t - kb_0) \tag{6}$$

where j and k are the expansion and translation coefficients, respectively, and a_0 and b_0 give the expansion variation and the translation step, respectively. We usually assumed binary expansion and unit translations, i.e., $a_0 = 1$ and $b_0 = 2$, resulting in the following expression:

$$\Psi_{j,k} = 2^{-j/2}\Psi(2^{-j}t - k) \tag{7}$$

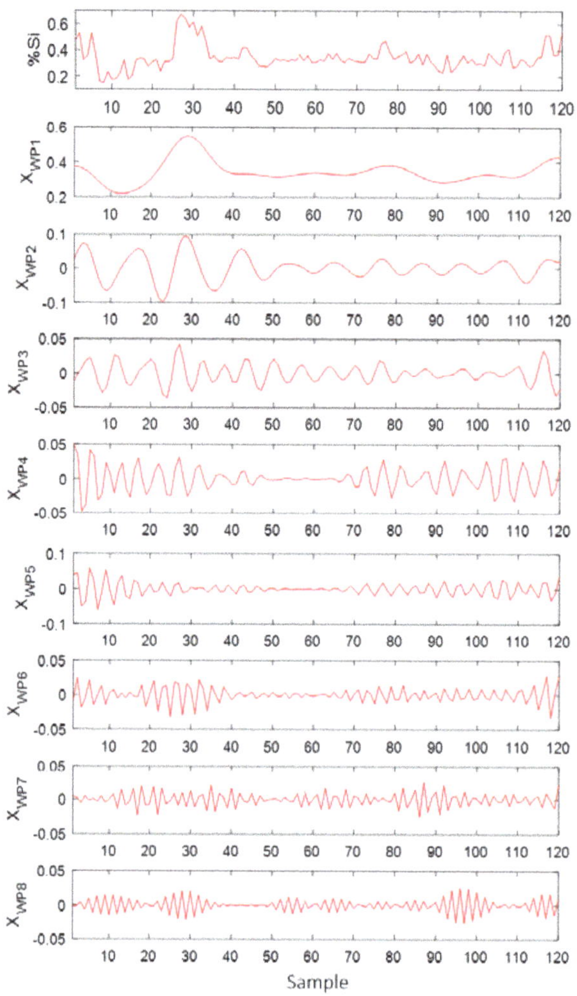

Figure 5. MODWT decomposition of silicon content time series.

However, at each DWT iteration, only the elements were divided into new approximation and detail components. In contrast to DWT, the Discrete Wavelet Packet Transform (DWPT) allows detail elements to be also divided into new approximation and detail components, resulting in uniform frequency bands. In addition, in the DWT, the approximation coefficients identify the frequency range $[0, \frac{f_s}{2^{(j+1)}}]$, while the detail coefficients at level j describe the frequency band $[\frac{f_s}{2^{(j+1)}}, \frac{f_s}{2^j}]$, where f_s represents the sampling frequency of the original signal. On the other hand, the DWPT allows that, at each level j, the frequency band is divided into 2^j partitions of equal length [16].

In the DWT and DWPT approaches, the signal length is limited to the power of two integers. Let L be the maximum decomposition level, int() the function that returns the nearest integer, and n the length of the input signal. We defined the maximum decomposition level by [27]:

$$L = \text{int}(log_2(n)) \tag{8}$$

In contrast to DWT and DWPT, the Maximal Overlap Discrete Wavelet Packet Transform (MODWPT) is invariant after adding new samples in the time series. It has the same number of coefficients at each level of decomposition since there is no signal decimation at each level. The MODWPT has often been used as an analytical tool to represent signals in the analysis of nonstationary time series, by dividing the variance of the data by the scale [27]. The main characteristic of such a transform is the lack of sensitivity to the time series starting point; consequently, the temporal analysis does not depend on the instant of time we take the samples [16].

The MODWPT is a function that allows the signal representation in frequency bands with different resolutions [16]. As the DWPT, it divides the frequency band into 2^j partitions of equal length for each level j [16].

Another feature of subseries resulting from the decomposition process via MODWPT is that the original signal is perfectly reconstructed since the decomposition coefficients pass through lowpass and reverse-highpass filters [16]. The MODWPT overcomes the disadvantages of the DWT and DWPT, i.e., it not only has uniform frequency bandwidth, but it is time-invariant. Furthermore, since the MODWPT components are nonorthogonal, there is a slight overlap of frequency bands enabling solving the edge effect problem found in time series forecasting. These properties allow the addition of new samples in the time series without the need to recalculate the entire transform again [16,17].

In the proposed hybrid MODWPT-NAR algorithm, the decomposed signals were modeled by the $NAR_1, \ldots, NAR_8$ neural networks. We performed two experiments to select the number of neurons in the hidden layer of each NAR model by using the learning algorithm Levenberg–Marquardt [25]. The first considered the same number of neurons in the hidden layer of all NAR_1 to NAR_8 networks. We varied the hidden neurons from one to thirty for all these networks, simultaneously, and we obtained 30 MODWPT-NAR models. The MODWPT-NAR model with seven neurons in the hidden layer of all NAR_1 to NAR_8 networks had the lowest training error, and its MSE(0) index was equal to 5.32×10^{-5}.

Regarding the second experiment, we considered different numbers of neurons in the hidden layer of each NAR_1 to NAR_8 network within the range of one to thirty neurons. This was necessary because we determined the number of neurons of the hidden layer of each neural network according to the complexity of the decomposed subseries. Next, we found the number of neurons in the hidden layer of each NAR_1 to NAR_8 network that provided the lowest MSE(0) error in the training phase, as shown in Table 3. We calculated this error from each decomposed subseries and corresponding forecast, without considering the original time series.

Table 3. Hidden layer size of networks for the second experiment.

Neural Network	Hidden Layer Size	MSE(0) ($\times 10^{-4}$)
NAR_1	27	0.0169
NAR_2	8	0.2642
NAR_3	7	0.1299
NAR_4	19	0.0487
NAR_5	21	0.0500
NAR_6	25	0.0442
NAR_7	23	0.0708
NAR_8	14	0.0047

The training error obtained from the MODWPT-NAR model after the second experiment had an MSE(0) index equal to 6.55×10^{-5}. Therefore, we chose the MODWPT-NAR model obtained from the first experiment as the final hybrid model, because it had a smaller hidden layer size for all NAR_1 to NAR_8 networks and provided a smaller MSE(0) training error.

4. Comparative Analysis of Forecasting Models

In this section, we compare the silicon content forecasting performances obtained from the NAR, NIO, and NARX neural networks and the MODWPT-NAR model. All numerical results were performed on a desktop PC i3-3110M, 2.4-GHZ CPU, and 4G RAM with the MATLAB (2018a) software, using real data obtained from a steel industry located in Brazil. We used the toolboxes provided by [29] to implement the neural networks and perform the pruning algorithm.

In addition to the MSE and MAPE indices, another important criterion used to compare the silicon forecasting models is the absolute error percentage of the k-steps ahead forecasting lower than ϵ ($\text{AEP}_\epsilon(k)$), given by [9]:

$$\text{AEP}_\epsilon(k) = \frac{1}{n}\sum_{t=1}^{n} N_E \times 100 \tag{9}$$

where

$$N_E = \begin{cases} 1 & \text{if } \ |y(t+k) - \hat{y}(t+k)| < \epsilon \\ 0 & \text{otherwise} \end{cases} \tag{10}$$

The $\text{AEP}_\epsilon(k)$ index is defined according to the blast furnace specialists' need. In this paper, we considered $\epsilon = 0.05$, i.e., the forecasting absolute errors at k-steps ahead were acceptable when they were lower than this range. Therefore, this index is better when the resulting value is higher.

To better understand the forecasting results presented in this section, let us define by t the time when the last molten iron chemical analysis is delivered by the laboratory. Since silicon content measurements were updated every 3 h and the molten iron samples were collected at intervals of 1 h, the collection time of the current molten iron sample was $t+3$. Therefore, the silicon content forecast at this time was $\hat{y}(t+3)$, which was obtained through neural models using the data stored until time t. In addition, the future silicon content samples would be collected at times $t+k, k = 4, 5, \ldots,$ and the respective forecasts would be $\hat{y}(t+k)$. In fact, to control the silicon content in the blast furnace and steelmaking plant, only forecasting horizons longer than or equal to 3 h ahead starting from instant t were adequate; due to the great inertia of the blast furnace, and the time spent in the steelmaking shop floor operations, the longer the forecasting horizon was, the better actions could be obtained in correcting the silicon content.

Tables 4 and 5 show the forecasting performances obtained by the NIO, NARX, and NAR networks before and after implementing the pruning algorithm.

From Tables 4 and 5, one can observe a significant improvement of the networks with exogenous inputs (NIO and NARX), in addition to a considerable decrease in the size of the hidden layer. For example, for the NARX network before applying the pruning algorithm, the number of hidden neurons was 90, the MSE(3) value 13.42×10^{-3} the MAPE(3) value 32.15%, and the $\text{AEP}_\epsilon(3)$ value 33.80%, for a three-hour-ahead forecasting horizon. In turn, after applying the pruning algorithm, the $\text{AEP}_\epsilon(3)$ value became 56.33% and the MAPE(3) value 17.98%, by considering only six neurons for the hidden layer.

Table 4. Performance of NIO, NARX, and NAR models before implementing the pruning algorithm.

Model	Horizon (k)	MSE(k) (10^{-3})	MAPE(k) (%)	AEP$_\epsilon$(k) (%)
NIO	3	21.15	38.85	27.32
	4	20.54	38.28	30.60
	5	20.29	38.05	30.90
	6	20.48	38.25	28.14
NARX	3	13.42	32.15	33.80
	4	11.65	29.55	36.53
	5	12.37	30.42	34.74
	6	11.87	29.04	36.63
NAR	3	3.05	12.88	73.73
	4	4.40	15.07	66.94
	5	9.07	21.76	53.80
	6	15.17	29.50	40.48

Table 5. Performance of NIO, NARX, and NAR models after implementing the pruning algorithm.

Model	Horizon (k)	MSE(k) (10^{-3})	MAPE(k) (%)	AEP$_\epsilon$(k) (%)
NIO	3	5.24	18.76	58.30
	4	5.28	18.76	58.53
	5	5.38	18.74	60.02
	6	5.45	18.76	59.83
NARX	3	5.31	17.98	56.33
	4	5.67	18.82	54.58
	5	6.04	19.55	51.41
	6	6.40	20.13	49.50
NAR	3	3.36	12.29	76.26
	4	5.88	16.49	66.24
	5	15.31	26.26	52.53
	6	21.52	31.76	40.76

We observed the same effect for the NIO network, which previously had 95 hidden neurons and presented an MSE(3) of 21.15×10^{-3}, a MAPE(3) of 38.85%, and an AEP$_\epsilon$(3) of 27.323%, for a forecasting horizon of 3 h ahead. After applying the pruning algorithm, on the other hand, this network could perform forecasts with superior performance (MSE(3) of 5.24×10^{-3}, MAPE(3) of 18.76%, and AEP$_\epsilon$(3) of 58.309%) using only three hidden neurons. Figure 6 illustrates the six-hour-ahead forecasts of silicon content obtained from the NIO and NARX models.

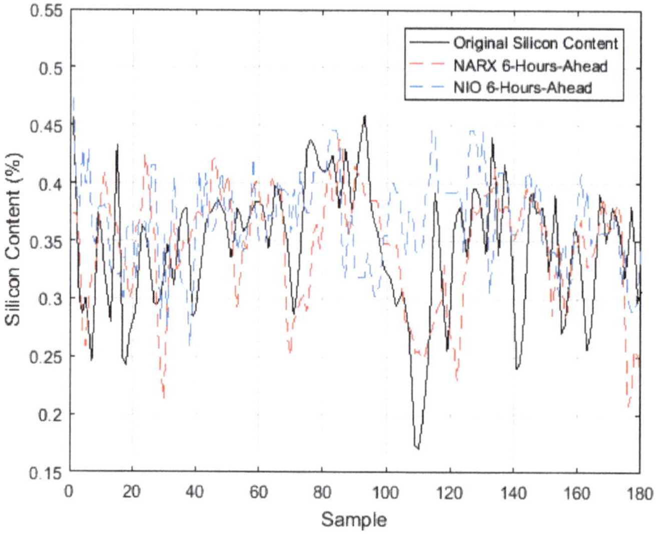

Figure 6. Silicon content and its six-hour-ahead forecasts from the NIO and NARX models.

The most significant effect brought by the pruning algorithm for the NAR network was that the size of the hidden layer was reduced from 100 neurons to 13 hidden neurons, without harming the performance. Notice in Table 5 the improvements in NIO and NARX networks due to the use of the pruning algorithm. Besides, observe in Table 5 that the NAR network still performed the best in forecasting for horizons until four steps ahead, and for longer horizons, the performance of NAR was worse than NIO and NARX. Thus, it was relevant to use the previous values of silicon content as the ANN input. However, the nonstationary behavior of this time series could impair the learning ability of the NAR and NARX networks for a longer forecasting horizon.

In Table 6, observe the performance of the MODWPT-NAR forecasts for the same data considered in Table 5.

Table 6. Performance of the MODWPT-NAR model.

Horizon (k)	MSE(k) (10^{-3})	MAPE(k) (%)	AEP_{ϵ} (%)
3	0.043	1.58	100
4	0.14	2.82	99.72
5	0.22	3.54	99.02
6	0.30	4.21	98.18
7	0.33	4.54	97.90
8	0.83	6.57	93.70

We noticed that this hybrid algorithm presented superior performance to the NAR model. As an example, the 3-h-ahead forecasting of the MODWPT-NAR model reached a MAPE(3) of 1.58%, while for the NAR model, this index was 12.29%. Furthermore, when the forecasting horizon increased, notice from Table 6 that the performance of the hybrid model did not decrease as much as in the model that used only the NAR neural network without considering MOPWPT decomposition. However, the proposed forecasting model became inaccurate for horizons longer than 8 h ahead, because the AEP index remained below 90%.

We also verified from Tables 5 and 6 that 40.76% of the six-hour-ahead forecasts obtained from the NAR model were inside the tolerance range, against the 98.18% obtained by the MODWPT-NAR hybrid model. We justified this result by the fact that the operators frequently take actions on the blast furnace input variables, aiming to correct instabilities to keep the silicon content within an acceptable range. These actions may increase the nonstationary behavior of the silicon content time series and hamper the learning ability of the NAR network when compared to the proposed MOPWT-NAR hybrid model. This is because MODWPT allows the decomposition of the nonstationary signal into stationary subseries. Figure 7 illustrates the six-hour-ahead forecasts of silicon content obtained from the NAR and MODWPT-NAR models.

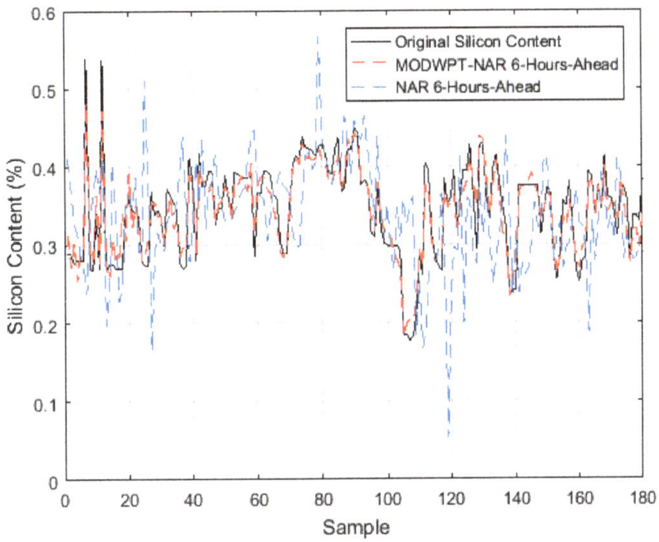

Figure 7. Silicon content and its six-hour-ahead forecasts from the NAR and MODWT-NAR models.

5. Conclusions

In this paper, we proposed a long-term forecasting model of silicon content, which performed the decomposition of the time series into additive components using MODWPT. We used NAR neural networks to model each decomposed signal, and the long-term silicon content forecast was performed by adding each subseries forecasting. In this case study, ninety-three-point-seven percent of the 8-h-ahead forecasting errors determined by MOPDWT NAR model were within the acceptable range of 0.05%, while at most 60% of the 6-h-ahead forecasting errors furnished by the ARX, NIO, and NAR neural network models without decomposition were within this acceptable range. Therefore, with this hybrid model, we obtained very accurate forecasts with a horizon of up to 8 h ahead from the moment the last molten iron chemical analysis was delivered by the laboratory.

The proposed model proved to be a promising tool for the prediction of pig iron silicon content, and we can extend it to other blast furnaces, with appropriate changes. To implement this forecasting system on industrial plants, an operating station accessing the process database to collect input data and send the forecasts in real time to the supervisory system is necessary. In the control room, the operator may observe the long-term silicon content forecasts while performing blast furnace operational changes. Thus, such a forecasting system can optimize the blast furnace raw materials, reduce the operational costs, and increase the quality of the pig iron. It also allows optimizing the steelmaking logistics since, currently, the technical team of the steelmaking plant receives the silicon measurements 3 h after the molten iron sample's collection. In this way, the anticipation of silicon content information can decrease the steelmaking time and, consequently, reduce the production costs and the CO_2 emissions to the atmosphere.

Author Contributions: Conceptualization, A.P.M.D., K.F.C., J.L.F.S., and F.S.V.G.; methodology, A.P.M.D., K.F.C., J.L.F.S., and F.S.V.G.; software, A.P.M.D. and K.F.C.; validation, A.P.M.D., J.L.F.S., and F.S.V.G.; formal analysis, A.P.M.D., K.F.C., J.L.F.S., and F.S.V.G.; investigation, A.P.M.D. and K.F.C.; writing—original draft preparation, A.P.M.D., K.F.C., and J.L.F.S.; writing—review and editing, A.P.M.D., K.F.C., J.L.F.S., and F.S.V.G.; visualization, A.P.M.D. and J.L.F.S.; supervision, K.F.C. and J.L.F.S.; project administration, J.L.F.S. and F.S.V.G. All authors read and agreed to the published version of the manuscript.

Funding: This research was funded by Coordenação de Aperfeiçoamento de Pessoal de Nível Superior—Brasil (CAPES), Grant Number 88882.385150/2019-01.

Data Availability Statement: Restrictions apply to the availability of these data. Data was obtained from ArcelorMittal Brazil S.A. and are available from the authors with the permission of ArcelorMittal Brazil S.A.

Acknowledgments: We thank the ArcelorMittal Brazil S.A. steel company for the technical support.

Conflicts of Interest: The authors declare no conflict of interest.The funders had no role in the design of the study; in the collection, analyses, or interpretation of data; in the writing of the manuscript, or in the decision to publish the results.

References

1. Li, H.; Bu, X.; Liu, X.; Li, X.; Li, H.; Liu, F.; Lyu, Q. Evaluation and Prediction of Blast Furnace Status Based on Big Data Platform of Ironmaking and Data Mining. *ISIJ Int.* **2021**, *61*, 108–118. [CrossRef]
2. Geerdes, M.D.; Toxopeus, H.; Vliet, C.; Young, G.O. *Modern Blast Furnace Ironmaking: An Introduction*; IOS Press BV: Amsterdam, The Netherlands, 2009.
3. Gomes, F.S.V.; Coco, K.F.; Salles, J.L.F. Multistep forecasting models of the liquid level in a blast furnace hearth. *IEEE Trans. Autom. Sci.* **2017**, *14*, 1286–1296. [CrossRef]
4. Singh, H.; Sridhar, N.V.; Deo, B. Artificial neural nets for prediction of silicon content of blast furnace hot metal.*Steel Res. Int.* **1996**, *67*, 521–527. [CrossRef]
5. Chen, J. A predictive system for blast furnaces by integrating a neural network with qualitative analysis. *Eng. Appl. Artif. Intell.* **2001**, *14*, 77–85. [CrossRef]
6. Saxén, H.; Pettersson, F. Nonlinear prediction of the hot metal silicon content in the blast furnace. *ISIJ Int.* **2007**, *47*, 1732–1737. [CrossRef]
7. Nurkkala, A.; Pettersson, F.; Saxén, H. Non linear modeling method applied to prediction of hot metal silicon in the ironmaking blast furnace. *Ind. Eng. Chem. Res.* **2011**, *50*, 9236–9248. [CrossRef]
8. Saxén, H.; Gao, C.; Gao, Z. Data-Driven Time Discrete Models for Dynamic Prediction of the Hot Metal Silicon Content in the Blast Furnace—A Review. *IEEE Trans. Ind. Inform.* **2013**, *9*, 2213–2225. [CrossRef]
9. Chen, W.; Wang, B.X.; Han, H.L. Prediction and control for silicon content in pig iron of blast furnace by integrating artificial neural network with genetic algorithm. *Ironmak. Steelmak.* **2010**, *37*, 458–463. [CrossRef]
10. Wang, H.; Yang, G.; Pan, C.; Gong, Q. Prediction of hot metal silicon content in blast furnace based on EMD and DNN. In Proceedings of the 34th Chinese Control Conference, Hangzhou, China, 28–30 July 2015; pp. 8214–8218.
11. Zhang, H.; Zhang, S.; Yin, Y.; Chen, X. Prediction of the hot metal silicon content in blast furnace based on extreme learning machine. *Int. J. Mach. Learn. Cybern.* **2017**,*9*, 697–1706. [CrossRef]
12. Su, X.; Zhang, S.; Yin, Y.; Hui, Y.; Xiao, W. Prediction of hot metal silicon content for blast furnace based on multi-layer online sequential extreme learning machine. In Proceedings of the 37th Chinese Control Conference (CCC), Wuhan, China, 25–27 July 2018; pp. 8025–8030.
13. Saxén, H.; Pettersson, F. Method for the selection of inputs and structure of feedforward neural networks. *Comput. Chem. Eng.* **2006**, *30*, 1038–1045. [CrossRef]
14. Cao, J.; Lin, Z.; Huang, G.B. Self-adaptive evolutionary extreme learning machine. *Neural Process. Lett.* **2012**, *36*, 285–305. [CrossRef]
15. Chazal, P.; Tapson, J.; Schaik, A.V. A comparison of extreme learning machines and backpropagation trained feed-forward networks processing the mnist database. In Proceedings of the 2015 IEEE International Conference on Acoustics, Speech and Signal Processing (ICASSP), South Brisbane, QLD, Australia, 19–24 April 2015; pp. 2165–2168.
16. Percival, D.B.; Walden, A.T. *Wavelet Methods for Time Series Analysis*; Cambridge University Press: Cambridge, UK, 2000.
17. Walden, A.T.; Cristan, A.C. The phase-corrected undecimated discrete wavelet packet transform and its application to interpreting the timing of events. *Proc. R. Soc. Lond.* **1998**, *454*, 2243–2266. [CrossRef]
18. Su, X.; Zhen, S.; Yin, Y.; Xiao, W. Prediction model of permeability index for blast furnace based on the improved multi-layer extreme learn machine and wavelet transform. *J. Frankl. Inst.* **2018**, *355*, 1663–1691. [CrossRef]
19. Labate, D.; Foresta, F.L.; Occhiuto, G.; Morabito, F.C.; Lay-Ekuakille, A.; Vergallo, P. EMD vs. DWT for the extraction of respiratory signal from single-channel ECG: A comparison. *IEEE Sens. J.* **2013**, *13*, 2666–2674. [CrossRef]

20. Saxén, H. Short-term prediction of silicon content in pig iron. *Can. Metall. Q.* **1994**, *33*, 319–326. [CrossRef]
21. Zhou, B.; Ye, H.; Zhang, H.; Li, M. Process monotoring of iron-making process with PCA-based methods. *Control Eng. Pract.* **2016**, *47*, 1–14. [CrossRef]
22. Kwiatkowski, D.; Phillips, P.C.B.; Schmidt, P.; Shin, Y. Testing the null hypothesis of stationarity against the alternative of a unit root. *J. Econom.* **1982**, *54*, 159–178. [CrossRef]
23. Lilliefors, H.W. On the Kolmogorov–Smirnov test for normality with mean and variance unknown. *J. Am. Stat. Assoc.* **1967**, *62*, 399–402. [CrossRef]
24. Kruskal, W.; Wallis, W.A. Use of ranks in one-criterion variance analysis. *J. Am. Stat. Assoc.* **1952**, *47*, 583–621. [CrossRef]
25. Haykin, S. *Neural Networks and Learning Machines*; Pearson Education, Inc.: Upper Saddle River, NJ, USA, 2009.
26. Hu, Y.H.; Hwang, J. *Handbook of Neural Network Signal Processing*; CRC Press: Boca Raton, FL, USA, 2002.
27. Mallat, S. *A Wavelet Tour of Signal Processing*; Academic Press: Burlington, MA, USA, 2008.
28. Hassibi, B.; Stork, D.G.; Wolff, G. Optimal Brain Surgeon and general network pruning. In Proceedings of the IEEE International Conference on Neural Networks, San Francisco, CA, USA, 28 March–1 April 1993; pp. 293–299.
29. The NNSYSID Toolbox-a MATLAB(R) Toolbox for System Identification with Neural Networks. Available online: https://www.mathworks.com/matlabcentral/fileexchange/87-nnsysid (accessed on 18 May 2021).

metals

MDPI

Article

Numerical Simulation of Bottom-Blowing Stirring in Different Smelting Stages of Electric Arc Furnace Steelmaking

Hang Hu, Lingzhi Yang *, Yufeng Guo, Feng Chen, Shuai Wang, Fuqiang Zheng and Bo Li

School of Minerals Processing and Bioengineering, Central South University, Changsha 410083, China; csu-huhang@csu.edu.cn (H.H.); yfguo@csu.edu.cn (Y.G.); csuchenf@csu.edu.cn (F.C.); wang_shuai@csu.edu.cn (S.W.); f.q.zheng@csu.edu.cn (F.Z.); lee0109@csu.edu.cn (B.L.)
* Correspondence: yanglingzhi@csu.edu.cn

Abstract: Electric arc furnace (EAF) steel bottom-blowing can effectively improve the temperature and composition uniformity of the molten pool during smelting process. To explore the effect of molten-steel characteristics on bottom-blowing at various stages of smelting, we divided the smelting process of the EAF into four stages: the melting stage, the early decarburization stage, the intermediate smelting stage, and the ending smelting stage. The numerical simulation software ANSYS Fluent 18.2 was used to simulate the velocity field of molten steel under the condition of bottom-blowing stirring in different stages in EAF steelmaking process. The properties of bottom-blowing and the kinetic conditions of the steel-slag at this interface were investigated. Our results showed that at a bottom-blowing gas flow rate of 100 L/min, the average flow rates of the four stages were v1 = 0.0081 m/s, v2 = 0.0069 m/s, v3 = 0.0063 m/s, and v4 = 0.0053 m/s. The physical model verification confirmed the results, that is, the viscosity of molten steel decreased as the smelting progressed, and the flow velocity of the molten steel caused by the agitation of bottom-blowing also decreased, the effect of bottom-blowing decreased. Based on these results, a theoretical basis was provided for the development of the bottom-blowing process.

Keywords: electric arc furnace steelmaking; bottom-stirring; different smelting time; molten steel flow; numerical simulation

Citation: Hu, H.; Yang, L.; Guo, Y.; Chen, F.; Wang, S.; Zheng, F.; Li, B. Numerical Simulation of Bottom-Blowing Stirring in Different Smelting Stages of Electric Arc Furnace Steelmaking. *Metals* **2021**, *11*, 799. https://doi.org/10.3390/met11050799

Academic Editor: Pasquale Cavaliere

Received: 25 March 2021
Accepted: 8 May 2021
Published: 14 May 2021

1. Introduction

Electric arc furnace (EAF) steelmaking, as the core of the short-process steelmaking method, has the characteristics of having a short process, low energy consumption, diversified charge, and diversified product structure [1,2]. The flow characteristics of molten steel in electric arc furnace (EAF) steelmaking remarkably influence heat transfer, momentum transfer, mass transfer, and chemical reactions among different phases in the smelting process [3–5]. The appropriate flow characteristics depend on the temperature and chemical composition of molten steel and determine the success of the smelting process. Disadvantages such as small free space, weak stirring ability, and uneven distribution of composition and temperature in the arc furnace can be improved by applying bottom-blowing technology in molten pool stirring. Blowing Ar, N_2, CO_2, and other gases into the molten pool through the gas-supply component arranged at the bottom of the arc furnace can strengthen mixing in the molten pool, increase the reaction contact area, promote the steel-slag reaction, and accelerate metallurgical speed. Thus, it is of great significance to improve the quality of molten steel and metal yield [6–10].

The flow characteristics of molten steel under bottom-blowing technology in EAF smelting have been extensively studied. The disadvantages of low mass and heat transfer rates caused by poor stirring in the molten pool during EAF steelmaking have been improved after the adoption of bottom-blowing technology [11]. Ramírez [12] established a mathematical model to describe the flow in the molten pool of a direct current (DC) electric arc furnace and found that the maximum temperature of liquid steel rose from 1966 to

1999 K after the application of bottom-blowing, and the maximum flow velocity of liquid steel rose from 0.8 to 4 m/s, which significantly improved the flow of the molten pool and smelting effect. Detailed analyses on the effect of the bottom-blowing system, through numerical simulation, experimental research, and industrial applications, were conducted by the Zhu group at the University of Science and Technology Beijing [13–17]. Dong [14] studied the application effect of bottom-blowing gas at a flow rate of 0.3−0.9 N m^3/h in 70 t EAF and revealed decarburization speed increased from 0.04 to 0.10 %/min, and lime consumption reduced by 10.4 kg/t. The bath blended time is only 1.7−2.5 min. Wei [15] studied the velocity distribution in the molten bath at different bottom-blowing gas flow rates and revealed the velocity of molten steel increased when the bottom-blowing gas rates increased, the content of phosphorus in the molten steel was decreased by 0.005 wt.%, the contents of FeO and T. Fe in endpoint slag were, respectively, reduced by 4.1 and 4.7 wt.%, the dephosphorization and decarburization rate were, respectively, increased by 12.1 and 11.8%, and the endpoint carbon–oxygen equilibrium of the molten steel was improved by 0.0024. Liu [16] researched stirring effects of six kinds of bottom-blowing arrangements on the molten bath in a 75 t EAF, the results showed increasing flow rate, weakening impeding force of sidewall, and improving stirring effect on the molten bath in eccentric bottom-tapping (EBT) region would decrease mixing time and improved stirring ability. Ma [17] studied the flow field with bottom-blowing in a 70 t EAF, and the numerical simulation studies showed that the turbulent kinetic energy of the EAF molten pool was increased by 87.9% and metal velocity was increased by 98%. The dead area decreased by 79%. When using the combined blowing technology, it can significantly improve the stirring intensity and the mixing effect of liquid steel, reducing smelting cost.

The above-mentioned research results preliminarily show that bottom-blowing stirring can enhance the flow characteristics of molten steel and promote smooth smelting. Because EAF smelting is performed at high temperatures, the dynamic characteristics of the molten pool in different smelting stages have not been well understood, and studies concerned the molten pool state in a single stage. In the smelting process, the bottom- blowing flow rate is usually constant, while the composition, temperature, and physical characteristics of molten steel change. The influences of molten steel with different physical characteristics affected the molten pool differently under the same bottom-blowing conditions. Therefore, at a certain stage of smelting, it is easy to cause insufficient stirring intensity due to insufficient bottom-blowing gas or the scouring of the furnace lining and waste of bottom-blowing gas.

Thus, determining the main flow properties of molten steel in different stages and studying the effect of bottom-blowing stirring on molten steel can help understand the bottom-blowing effect in different stages of the EAF smelting process, which is conducive to improving the utilization rate of the bottom-blowing gas and forming a better smelting effect. In this paper, numerical simulation and model verification were used to study the influence of molten steel characteristics in different smelting periods on the flow rate of molten steel under the same bottom-blowing conditions.

2. Flow Characteristics of Molten Steel in Different Smelting Periods of EAF

Modern electric arc furnace smelting retains the main melting, heating, and necessary refining processes. Dephosphorization and partial decarburization are advanced to the early stage of smelting as far as possible. In the middle and late stages of smelting, only the carbon content and temperature of steel and slag are controlled, and the tap-to-tap time is shortened. In a short period, the steel scrap is melted, and the temperature of the molten steel is adjusted to meet the requirements of tapping. Due to the current use of eccentric bottom-tapping (EBT) tapping and retained steel operation, the molten pool already exists when the steel scrap melts. Combined with enhanced oxygen jet technology and bottom-blowing stirring technology, the EAF offers favorable conditions for the metallurgical reaction.

2.1. Factors Influencing Liquid Steel Flow Characteristics

Decarburization, dephosphorization, and metal oxidation reactions occur in EAF. The related operation system and the corresponding temperature and composition changes of molten steel and slag in the steelmaking process change the physical characteristics of molten steel and, consequently, the gas stirring effect. Density and viscosity are the main physical characteristics of molten steel, and its main influencing factors include chemical composition and temperature.

Density is a basic variable to explain the behavior and properties of liquid metals, and it is a basic parameter to reflect and analyze the melt structure. Systematic studies on the density of various liquid metals have been performed [18–23]. Viscosity is the momentum of fluid molecules moving from the current liquid layer to another liquid layer perpendicular to the streamline direction. Viscosity is one of the important physical properties of liquid metals, determining their hydrodynamic characteristics [24]. It also has a great influence on heat, momentum, and mass transfer and chemical reaction between phases in EAF steelmaking process. The Roscoe formula modified by Iida has high reliability and accuracy in calculating the viscosities of liquid metals, such as molten steel and liquid iron [25–27].

Studies on the density and viscosity of liquid iron and molten steel were mainly based on experimental design and theoretical analysis, to determine a relative value, without coupling the influences of multi-fluid and multi-phase interactions in the steelmaking process. Stirring methods commonly used in EAF steelmaking process, such as oxygen jet [28], bottom-blowing stirring, and electromagnetic stirring [29], can strengthen the momentum transfer and heat and mass exchange of molten steel in the furnace by inputting material and energy and improve the fluidity of molten pool to a certain extent. However, the molten pool flow is ultimately determined by the flow characteristics of molten steel. Excessive external input is inefficient when the internal molten steel viscosity only slightly changes, causing material and energy consumption. Therefore, analyzing the influencing factors of the liquid steel viscosity in EAF steelmaking requires understanding the effect of bottom-blowing stirring to improve smelting efficiency.

The viscosity of molten steel is affected by the temperature and chemical elements content of molten steel, especially the temperature. With the temperature range from melting point to 1923 K, the relationship between the viscosity and temperature [24] is shown as follows:

$$lg\eta = \frac{1951}{T} - 3.327 \tag{1}$$

where η is the viscosity of molten steel, Pa·s; T is the temperature of the molten steel, K; with the temperature increase, the viscosity of molten steel decreases. When the temperature increases, the momentum of atoms in molten steel jumping into the adjacent liquid layer increases, providing enough energy for particles to move so that the number of particles with the viscous flow activation energy increases, and the viscosity of the melt decreases.

2.2. Feature Point Selection in Different Smelting Periods

EAF steelmaking is accompanied by many physical changes and chemical reactions. In the initial stage of smelting, the scrap melts gradually and then forms the molten pool, and silicon and manganese in the molten steel are first oxidized and then floated to the molten steel surface or suspended in the molten steel. The dephosphorization agent, CaO, carries oxygen or argon and disperses it in the molten steel in the form of powder particles for the dephosphorization reaction. In this process, the power of electrode heating is larger, heating the molten pool and increasing the temperature of it with the maximum power, which is conducive to the dephosphorization reaction. In the middle of smelting, the molten pool begins to skim slag, slowing down the heating rate, increasing the alkalinity of steel slag, and increasing the oxygen supply rate, to improve dephosphorization conditions. After the dephosphorization reaction stops, arc heating is stopped, and the supply of oxygen is reduced. At the same time, many phosphorus-containing slags are stripped to prevent phosphorus recovery. During the dephosphorization reaction, the decarburization reaction also accompanies it, and the iron element in the molten steel is oxidized. At a later stage of smelting, oxygen is supplied at a medium rate, and the electric arc furnace is used for only high-power heating to facilitate the decarburization reaction and adjust the temperature. At the end of smelting, the oxygen supply and heating are stopped, the composition and temperature of molten steel are stabilized, and the steel is prepared for reduction refining.

According to the various characteristics of elements and temperature in the EAF steelmaking process, four liquid steel components and temperatures were selected in this work as the characteristic points of the four smelting stages. The theoretical values of the liquid steel density and viscosity were calculated based on Equation (1) and Reference [30]. Before the EAF steelmaking process, the chemical composition of hot metal and scrap would be tested and analyzed. Then, the initial carbon and silicon contents of the bath could be obtained by dividing the total mass of carbon and silicon by the total weight of molten steel, respectively. The specific parameters are shown in Table 1.

Table 1. Relevant parameters of characteristic points in different smelting periods.

Physical Properties	Temperature (°C)	Carbon Content (%)	Silicon Content (%)	Viscosity (Pa·s)	Density (g/cm^3)
Point 1 (Melting stage)	1140	2.581	0.3165	0.0113	6.900
Point 2 (Early decarburization stage)	1400	2.3259	0.0593	0.0069	6.662
Point 3 (intermediate smelting stage)	1500	1.132	0	0.0059	7.003
Point 4 (Ending smelting stage)	1650	0.1035	0	0.0049	6.912

3. Numerical Simulation of Bottom-Blowing

3.1. Governing Equation

The molten steel, slag, and argon need to satisfy not only the mass, energy, and momentum conservation equations but also the control equation of the finite element model and turbulence control equation. To study the effect of bottom-blowing gas agitation on the molten steel flow rate in the EAF steelmaking process, a fluid volume function (VOF) model calculation was introduced during the establishment of the CFD model. In the simulation process, different fluid equation components share a set of momentum equations. The effective density ρ_e of the argon–slag–steel liquid (three-phase) system explored in this study was derived from the following equation:

$$\rho_e = \alpha_{Ar}\rho_{Ar} + \alpha_{Sl}\rho_{Sl} + \alpha_{St}\rho_{St} \tag{2}$$

where α_{Ar}, α_{Sl}, and α_{St} (%), ρ_{Ar}, ρ_{Sl}, and ρ_{St} (kg/m^3) are the volume fractions and the densities of argon, slag, and molten steel, respectively.

The energy equation in the calculation domain is shared by all phases, and the expression is as follows:

$$\frac{\partial(\rho E)}{\partial t} + \nabla\cdot(u(\rho E + p)) = \nabla\cdot\left(k_{eff}\nabla T\right) + S_h \tag{3}$$

where ρ is the density of the gas, kg/m^3; E is the element energy, J; t is the time, s; p is the static pressure of the fluid, MPa; T is the temperature, K; k_{eff} is the effective thermal conductivity, W/(m·K); ρ_e and k_{eff} are shared with all phases; and the source term S_h is provided by radiant heat transfer and other volumetric heat sources.

In the VOF model, the energy E can be obtained by the mass weighted average method which is described by the following formula.

$$E = \sum_{i=1}^{n} \alpha_i \rho_i E_i \Big/ \sum_{i=1}^{n} \alpha_i \rho_i \tag{4}$$

where α_i (%) and ρ_i (kg/m^3) are the volume fraction and density of phase i, respectively; and E_i is based on the specific heat and shared temperature for each phase. The standard $k-\varepsilon$ turbulence model was used in this study. The turbulent flow energy k (m^2/s^{-2}) and the dissipation rate ε (m^2/s^3) were determined by the following transfer equations, respectively.

$$\frac{\partial(\rho k)}{\partial t} + \frac{\partial(\rho k v_i)}{\partial x_i} = \frac{\partial}{\partial x_j}\left[\left(\mu + \frac{\mu_t}{\sigma_k}\right)\cdot\frac{\partial k}{\partial x_i}\right] + G_k + G_b - \rho\varepsilon - Y_M + S_k \tag{5}$$

$$\frac{\partial(\rho \varepsilon)}{\partial t} + \frac{\partial(\rho \varepsilon v_i)}{\partial x_i} = \frac{\partial}{\partial x_j}\left[\left(\mu + \frac{\mu_t}{\sigma_\varepsilon}\right)\cdot\frac{\partial \varepsilon}{\partial x_i}\right] + C_{1\varepsilon}\frac{\varepsilon}{k}(G_k + C_{3\varepsilon}G_b) + C_{2\varepsilon}\rho\frac{\varepsilon^2}{k} + S_\varepsilon \tag{6}$$

In these equations, μ represents the dynamic viscosity, Pa·s; and x_i and x_j represent the coordinates along the coordinate axis i and j directions, respectively;

G_k and G_b are turbulent flow energies generated by the average fluid velocity and buoyancy, J, respectively; Y_M is the turbulent dissipation rate generated by a compressible turbulent pulsation; S_k and S_ε are custom source terms. The expression for calculating the turbulent viscosity μ_t (Pa·s) using k and ε is as follows:

$$\mu_t = \rho C_\mu \frac{\varepsilon^2}{k} \tag{7}$$

where $C_{1\varepsilon}$, $C_{2\varepsilon}$, $C_{3\varepsilon}$, σ_k, σ_ε, and C_μ are the constant terms of the $k-\varepsilon$ model, and their values were provided by Launder, as 1.44, 1.92, 0.8, 1.0, 0.9, and 0.009, respectively [31]. The continuity equation is as follows:

$$\frac{1}{\rho_i}\left[\frac{\partial}{\partial t}(\alpha_i\rho_i) + \nabla\cdot(\alpha_i\rho_i v_i)\right] = S_{\alpha_i} + \sum_{i=1}^{n}(m_{ji} - m_{ij}) \tag{8}$$

where v_i is the velocity component in direction i, m/s; m_{ij} is the mass transmitted from phase i to phase j, kg; m_{ji} is the mass transmitted from phase j to phase i, kg; and S_{α_i} is a custom source item. The momentum equation is as follows:

$$\frac{\partial}{\partial t}\left(\rho\vec{v}\right) + \nabla\cdot\left(\rho\vec{v}\vec{v}\right) = -\nabla p + \nabla\cdot\left[\mu\left(\nabla\vec{v} + \nabla\vec{v}^T\right)\right] + \rho\vec{g} + F \tag{9}$$

where $\vec{v}$ is the instantaneous velocity of the fluid, m/s; p is the static pressure, MPa; g is the acceleration of gravity, m/s^2; and F is the other force that the control body receives, N.

3.2. Grid Model

A 100 t industrial top-charge EAF in a steel plant was selected to study the physical and chemical properties of molten steel in the EAF steelmaking process. To ensure the reliability of the results, a grid calculation model with a ratio of 1:1 was established using numerical simulation software. Based on the previous research results, the optimized bottom-blowing layout was set in this study. The geometry and bottom-blowing layout of the 100 t EAF are shown in Figure 1.

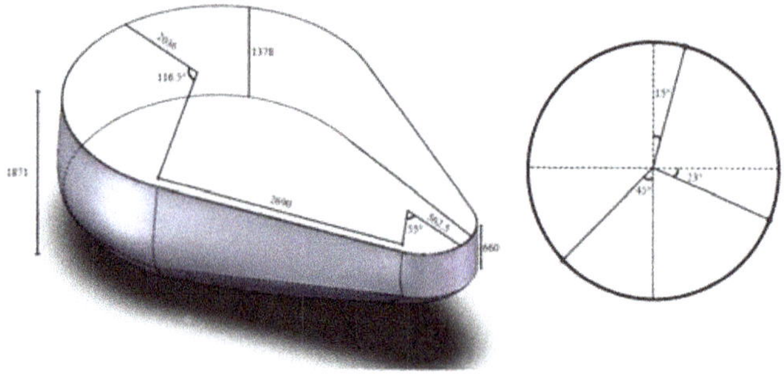

Figure 1. Geometric parameters and bottom blowhole arrangement of the Electric arc furnace (EAF).

In the geometric model, the diameter of the reference circle on the top surface of the eccentric area was 1374 mm, the diameter of the reference circle on the top surface of the main body of the EAF was 4072 mm, the diameter of the bottom circle of the main body of the EAF was 2122 mm, and the furnace length of the EAF was 6355 mm. The length of the hypotenuse was 2690 mm, and the actual fan angles at the top of the EAF model were 233° and 110°, respectively. The bottom-blowing hole has a hydraulic radius of 10 mm and is arranged on a circumference of 1001.1 mm from the center; the hydraulic radius of the molten pool was 5481 mm, the depth was 1650 mm, the internal molten steel depth was 1430 mm, and the thickness of the slag phase was 165 mm. The number of meshes in the mesh model established based on the geometric model was 55,924, and the number of nodes was 540,395 [31] (Figure 2).

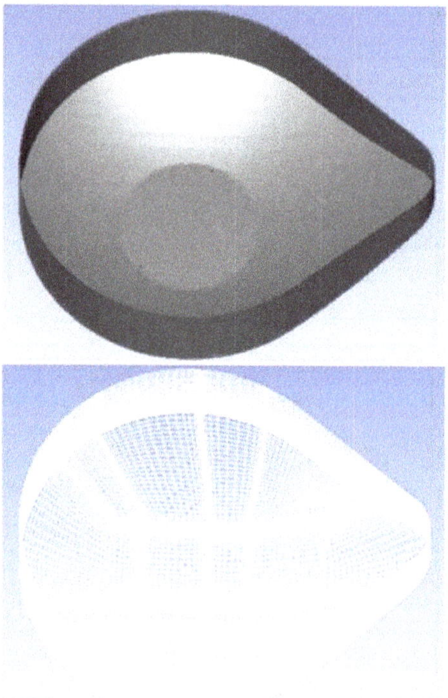

Figure 2. Simulation and grid diagram of the EAF.

3.3. Calculation Hypothesis

(1) The slag, molten steel, and gas phases formed a transient non-isothermal three-phase flow fluid, and the three phases were immiscible.;

(2) The gas phase was regarded as a compressible Newtonian fluid, and the slag and steel liquid phases were regarded as incompressible Newtonian fluids;

(3) All the walls during the simulation were non-smooth wall models, and the average velocity of the fluid close to the wall was simulated by the standard wall model;

(4) The simulation process did not consider the chemical reactions occurring between the components.

3.4. Calculation Settings

The materials involved in this work include argon, molten steel, and slag. Table 1 involves the related parameters of molten steel, and we set the viscosity of the steel slag to 0.35 Pa·s. The specific heat of argon gas does not change significantly in the temperature range of 298-2000 K; therefore, the values in the built-in database were used. The material properties of steel slag and argon are shown in Table 2.

Table 2. Parameter settings of materials in the fluid domain.

Materials	Slag	Argon
Density (kg/m^3)	3000	Ideal-gas
Specific Heat (J/(kg·K))	1200	520.64
Thermal Conductivity (W/m·K)	1.2	0.0158
Viscosity (kg/m·s)	0.35 [15,31]	2.125×10^{-5} (298.15 K)
Molar Mass (kg/mol)	31.996	39.948

In the simulation calculation, three mass flow inlet boundaries and a pressure outlet boundary were used to simulate the bottom-blowing flow (100 L/min) of the three bottom-blowing holes and the gas flow and recirculation at the top of the EAF, respectively. The other walls were adiabatic and non-smooth, which were used to simulate the lining and wall parts of the EAF; in addition, a fluid condition was set to simulate the distribution of the molten steel, slag, and argon in the EAF.

To couple pressure and velocity, the most widely used SIMPLE algorithm was adopted. The calculation area was divided into structured grids, and the gradient was calculated based on the least square method of the smallest unit. The pressure calculation was performed using the PRESTO algorithm. The momentum, turbulent kinetic energy, and turbulent dissipation rate and energy were solved by the second-order upwind scheme.

4. Results and Discussion

4.1. Numerical Simulation Results

The numerical results of each step were calculated, and the ANSYS Fluent 18.2 (version: 18.2, ANSYS Inc., Canonsburg, PA, USA) [32] statistical calculation was outputted in the form of a file. Each step corresponds to the average speed of molten steel. The statistical data greatly fluctuated at the beginning of the calculation, and the data tended to be flat in the later period; thus, the second half of the data that tended to be flat after 200,000 steps was selected as the basis. The average flow rate of this part represents the average flow rate of molten steel after its smooth blowing [31].

Each smelting stage corresponded to the state of the molten pool after the calculation was completed. Since the velocity cloud at the bottom of the arc furnace fluctuated greatly, only the average value of the flow rate of the molten steel in the upper part was used to represent the average flow rate of the molten steel in the entire arc furnace. The molten steel was divided into six planes. At each stage, the velocity clouds of 0.1, 0.2, 0.3, 0.4, 0.5, and 0.6 m were taken from the slag–steel contact surface, as shown in Figure 3. The calculation of the average flow rate of the molten steel at the end of the simulation requires statistical calculation. A single cloud image was imported into Photoshop 2018, and the color range function was used with the default setting to count the area of the color distribution corresponding to each speed. The average speed was calculated with the area as the weight, which was the average flow rate of the molten steel in that section, and the average value of the corresponding flow velocity of the six planes was calculated as the average value of the molten steel (Table 3).

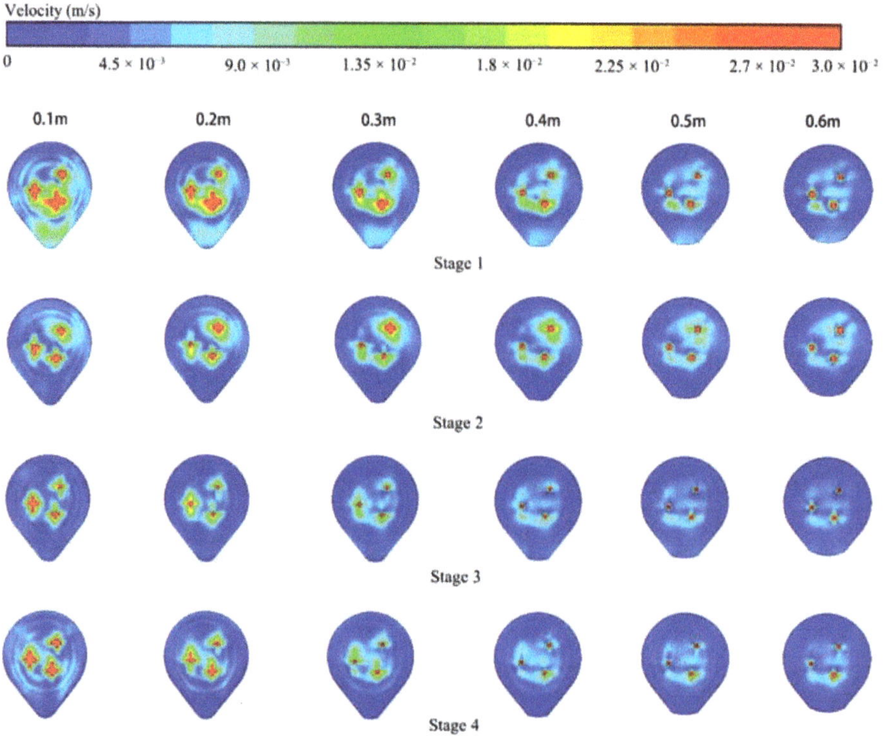

Figure 3. Speed cloud map of each interface in different smelting stages.

Table 3. Average flow velocity of molten steel at different smelting stages.

Number of Smelting Stages	Average Flow Velocity of the Liquid Surface at the Corresponding Depth from the Slag–Steel Interface ($\times 10^{-3}$ m/s)						Average value ($\times 10^{-3}$ m/s)
	0.1 m	0.2 m	0.3 m	0.4 m	0.5 m	0.6 m	
1	10.970	9.422	8.448	7.550	6.913	6.499	8.107
2	7.517	7.392	7.277	6.815	6.554	6.306	6.941
3	7.797	7.078	6.452	5.915	5.659	5.367	6.266
4	7.018	5.883	5.383	4.961	4.799	4.518	5.336

When the bottom-blowing gas flow rate was 100 L/min, the average flow rates of the four stages were $v_1 = 0.0081$ m/s, $v_2 = 0.0069$ m/s, $v_3 = 0.0063$ m/s, and $v_4 = 0.0053$ m/s. With the progress of smelting, the flow rate of the stirring intensity of the molten steel under the same bottom-blowing conditions gradually decreased.

The average flow rates of the molten steel at different stages and distances from the steel-slag contact surface at different stages are shown in Figure 4. Under the condition that the bottom-blowing gas flow rate remained unchanged, the average molten steel flow rate decreased significantly as the EAF smelting process proceeded. Simultaneously, the average flow velocity of the molten steel decreased with the distance farther away from the contact surface of the slag–steel.

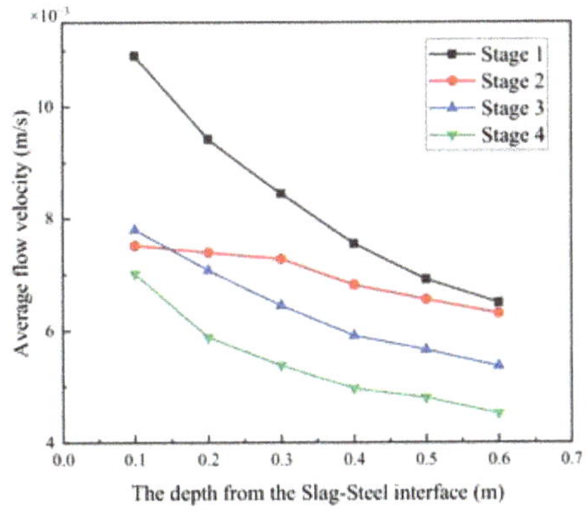

Figure 4. Speed distribution at each interface.

According to the simulation data, the viscosity of the molten steel and the average speed were plotted, as shown in Figure 5. At the same time, the curve was linearly fitted to obtain the following equation of molten steel flow velocity and viscosity:

$$v = -85048.18\eta^2 + 1808.75\eta - 1.473 \tag{10}$$

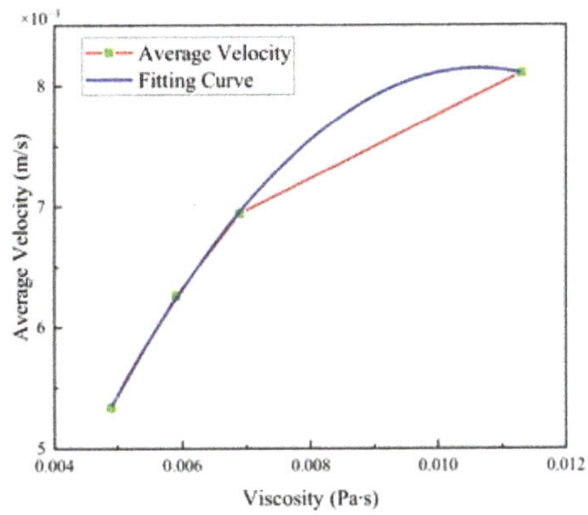

Figure 5. Average velocity versus molten steel viscosity diagram and fitting curve.

The sum of squared residuals of the fitted curve was 0.00122, and the range of the squared range was 0.9997, meeting the statistical fitting error requirements. In the formula, v is the flow velocity of the molten steel, 10^{-3} m/s; and η is the viscosity of the molten steel, Pa·s.

Under the same bottom-blowing gas flow rate, when the viscosity of the molten steel was high in the early stage of smelting, the acceleration effect of the bottom-blowing stirring on the molten steel was better, and the molten steel flow rate was faster. With the progress of smelting, the viscosity of the molten steel decreased, the acceleration effect of the bottom-blowing stirring on the molten steel worsened, and the flow rate of the molten steel decreased.

4.2. Physical Model Verification

The numerical simulation results showed that the ongoing smelting process made the molten steel more difficult to be stirred. However, artificial experience and industrial production practice on EAF steelmaking suggest that the viscosity of molten steel is reduced, the velocity of molten steel is faster, and the stirring effect of bottom-blowing is better under the same bottom-blowing conditions, which is contrary to the results of numerical simulation. To further analyze and explain this phenomenon, physical model verification was conducted.

4.2.1. Physical Model Verification Plan

The water model experiments are widely used in the validation of numerical simulation in molten bath fluid flow [31,33,34] and achieve better effect. Generally, the KCl solution was added simultaneously with bottom-blowing gas during the water model experiment process. The effect of bottom-blowing stirring was expressed by the mixing time, which was recorded when the conductivity difference between the two conductivity electrodes installed in the model bottom below 5%. However, the physical model verification, showed in Figure 6, was quite different from the above.

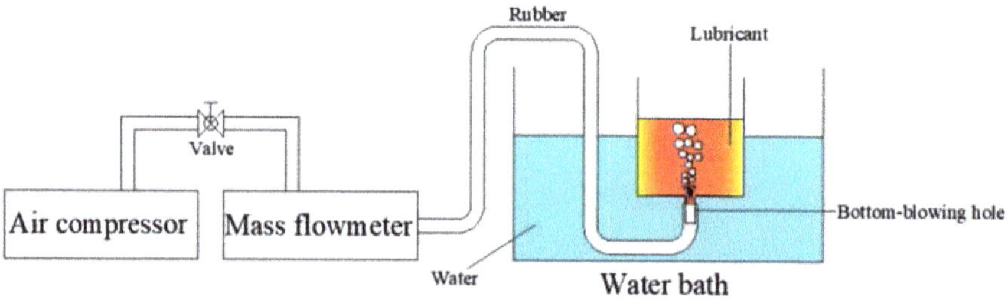

Figure 6. Connection diagram of the experimental device for physical model verification.

A transparent acrylic plexiglass tube mold, shown in Figure 7, with a cylinder at a diameter of 12 cm and a height of 10 cm in the upper part, and a circular bottom-blowing hole at a diameter of 1 cm and a height of 3 cm in the nether part, was used to simulate the effect of the EAF. The No. 11 lubricating oil inside the mold, produced by Mobil Glygoyle, was used to simulate the molten steel. It has stable performance and the relationship of the viscosity and density with temperature in the range of 0–100 °C was shown in Figure 8. The temperature characteristics of the different stages of smelting were simulated by adjusting the temperature of the water bath. The main instruments and parameters of the experiment are shown in Table 4.

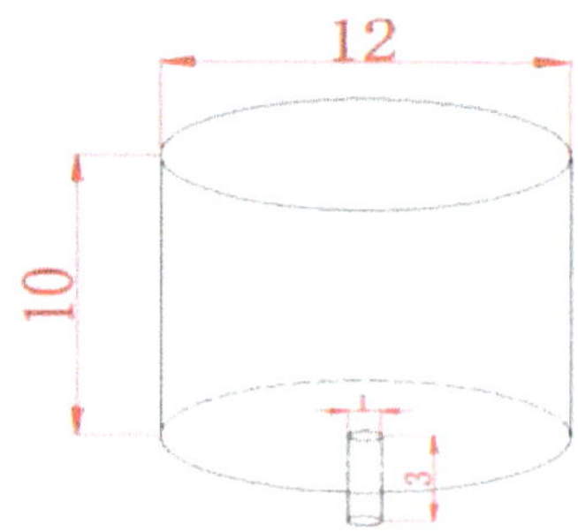

Figure 7. Schematic of the mold (cm).

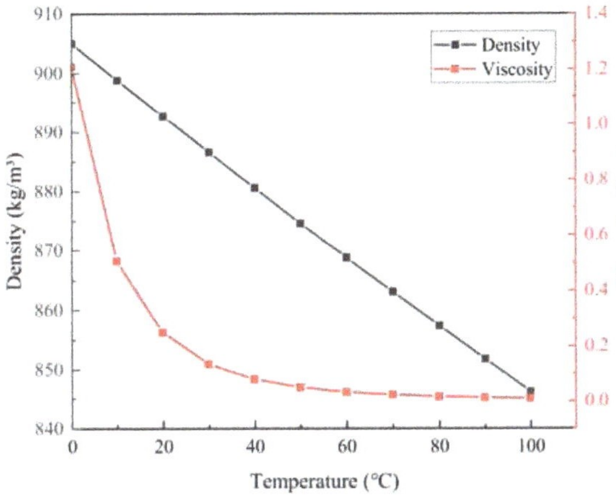

Figure 8. Changes in density and viscosity with the temperature of the No. 11 lubricating oil.

Table 4. Main instruments and parameters of physical model verification experiment.

Instruments	Parameters
HH-2J magnetic stirring water bath	Power supply and heating power: 220 V 50 HZ, 600 W; Constant temperature range: room temperature 40–90 °C; Temperature accuracy: $\leq \pm 0.5$ °C;
ACO (Air Compression Operation) series electromagnetic air pump	Power supply and heating power: 220 VAC/ 50 HZ, 35 W; Displacement: 40 L/min
Mass flowmeter	Variable flow adjustment, the flow rate of this experiment is constant at 1 L/min.
Transparent acrylic plexiglass tube mold	Wall thickness: 5 mm

To ensure the flows in physical model were similar to those in a real EAF, dynamic similarity between the two systems was determined based on the Froude number similarity criterion, shown as the following Equation (11) [15]:

$$\frac{Q_W}{Q_E} = \sqrt{\frac{d_W^2 D_W^2 H_W \rho_{lW} \rho_{gW}}{d_E^2 D_E^2 H_E \rho_{lE} \rho_{gE}}} \tag{11}$$

$$D_E = \frac{\pi\left(\frac{D_1}{2} \times \frac{A}{360} + \frac{D_2}{2} \times \frac{B}{360}\right) + 2L_1}{\pi} \tag{12}$$

where parameters with subscript W are those of physical model; parameters with E represent those of the EAF prototype; Q_W and Q_E are the gas flow rates of the physical model and the EAF prototype, m^3/h; d_W and d_E are the nozzle diameters of the physical model and the EAF prototype, m; H_W and H_E are the molten bath depth of the physical model and the EAF prototype, m, as 80 and 1650 mm, respectively; ρ_{lW} and ρ_{lE} are the liquid density of the physical model and the EAF prototype, kg/m^3; ρ_{gW} and ρ_{gE} are the gas density of the physical model and the EAF prototype, kg/m^3; D_W and D_E are the molten bath hydraulic diameter of the physical model and the EAF prototype, m. The value of D_W is 10 mm, while D_E can be calculated by Equation (12) [15]. Where A and B are the actual fan angles at the top of the EAF prototype, as 110° and 233°, respectively; D_1 and D_2 are the diameter of the reference circle on the top surface of the eccentric area and the diameter of the reference circle on the top surface of the main body of the EAF, as 1374 and 4072 mm, respectively; and L_1 is the length of the hypotenuse, as 2690 mm.

Before the experiment process, the flow rate of the mass flowmeter was set to 1 L/min, and it was confirmed that the connecting pipe was not leaking oil or gas. The water bath was warmed to 40 °C, and the temperature was maintained for 15 min. During the experiment process, an electromagnetic air pump was used to continuously blow the gas for 1 min. Bubbles were formed in the lubricating oil in the mold. The time required for the last bubble to rise to the liquid level was recorded. When the heat-transfer equilibrium reached the balanced state between the lubricating oil in the mold and the water in the magnetic stirring water bath, it can be considered that the temperature of the oil was equal to the water temperature (the external heat loss was ignored). The temperatures of the water bath pot set in the experiment were 40, 50, 60, 70, 80, and 90 °C, respectively. A group of bottom-blowing gas experiments was performed for each 10 °C, and each group was repeated seven times.

4.2.2. Physical Model Verification Results and Analysis

We recorded the time required from the start of the bottom-blowing operation to the rising of the last bubble in the lubricating oil to the liquid level in each test. The final data are shown in Table 5.

Table 5. Time of bubbles rising in the No. 11 lubricating oil at different temperatures.

Temperature	Number of Experiments							Average Value (s)
	1	2	3	4	5	6	7	
40 °C	66.40	66.55	66.63	64.46	65.26	64.72	65.94	65.77
50 °C	64.99	65.31	65.40	65.10	64.65	65.48	65.42	65.24
60 °C	65.49	64.31	64.53	64.24	64.46	65.36	64.23	64.58
70 °C	64.43	64.03	64.06	64.33	63.62	63.53	63.56	63.92
80 °C	63.05	63.07	63.25	62.86	63.44	63.17	60.28	63.08
90 °C	62.79	63.07	62.46	62.27	63.21	62.87	62.86	62.81

To ensure the accuracy and reliability of the data and improve the representative role of the average time in the sample data, the maximum and minimum values of the bubble escape time in each group were removed, and the rest of data were statistically verified.

The average value, variance between samples, and standard deviation of sample group were calculated by Equations (13)–(15), respectively.

$$\bar{t} = \frac{1}{n}\sum\nolimits_{i=1}^{n} t_i \tag{13}$$

$$D(t) = \frac{1}{n-1}\sum\nolimits_{i=1}^{n}(t_i - \bar{t})^2 \tag{14}$$

$$S(t) = \sqrt{\frac{1}{n-1}\sum\nolimits_{i=1}^{n}(t_i - \bar{t})^2} \tag{15}$$

where t is the bubble floating time, s; $\bar{t}$ is the average floating time of experimental bubbles in each group, s; n is the number of samples in each group, $n = 5$; $D(t)$ is the variance of each group; and $S(t)$ is the standard deviation of each group.

Under the condition that the confidence is 95%, the confidence interval of each group of samples is calculated by Equation (16).

$$\left(\bar{t} - \frac{S(t)}{\sqrt{n}}Z_{\frac{\alpha}{2}}, \bar{t} + \frac{S(t)}{\sqrt{n}}Z_{\frac{\alpha}{2}}\right)\alpha \tag{16}$$

In the formula: α is the area surrounded by the continuous random variable, and its distribution density function and the $\alpha-$ quantile of this distribution, $\alpha = 0.05$; $Z_{\frac{\alpha}{2}}$ is the $\alpha-$ subject to the normal distribution quantile, $Z_{\frac{\alpha}{2}} = 1.96$.

The statistically verified experimental results are listed in Table 6. From this table, it is known that there is a 95% confidence that the bubble escape time of the bottom-blowing gas was located in the confidence interval at the corresponding set temperature, and as the temperature increased, the confidence intervals of the upper and lower bubble escape time limits decreased. Thus, the bubble escape time data in the physical model verification are reliable.

Table 6. Statistical verification results.

Temperature	Variance	Standard Deviation	Confidence	Confidence Interval
40 °C	0.599	0.774	0.95	(65.096, 66.452)
50 °C	0.036	0.190	0.95	(65.077, 65.411)
60 °C	0.203	0.451	0.95	(64.185, 64.975)
70 °C	0.105	0.324	0.95	(63.636, 64.204)
80 °C	0.022	0.147	0.95	(62.951, 63.209)
90 °C	0.049	0.222	0.95	(62.616, 63.004)

According to the data in Table 5, the relationship between the bottom-blowing bubble floating time and the lubricating oil temperature is plotted in Figure 9. It can be seen from Figures 8 and 9 that under the total gas volume of 1 min bottom-blowing, as the temperature increased, the viscosity of the No. 11 lubricating oil decreased, and the bubble escape time was shortened accordingly. The experimental results confirm the conclusion of numerical simulation, that is, with temperature increase, under the same bottom-blowing stirring conditions, the viscosity of the molten fluid decreases, making it more difficult to be stirred [35], and the bottom-blowing efficiency continues to decrease with the smelting process.

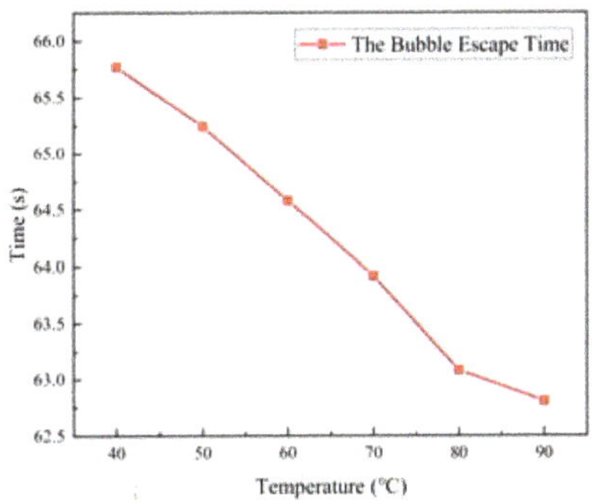

Figure 9. Relationship between bottom-blown bubble escape time and temperature.

To further explain this phenomenon, the relationship between argon and molten steel can be simplified as shown in Figure 10.

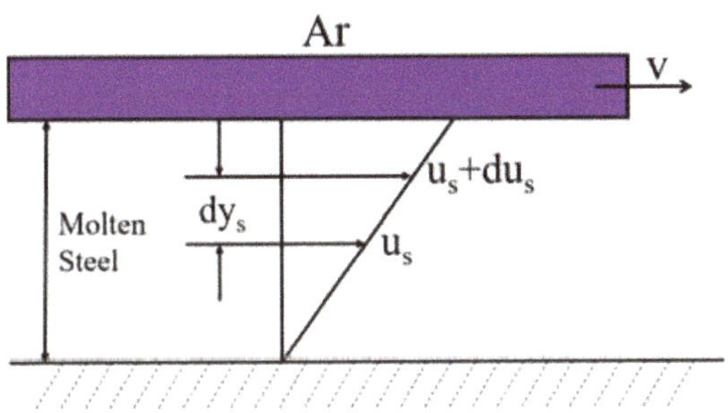

Figure 10. Schematic diagram of the interaction between argon and molten steel.

The bottom-blowing gas has a certain initial velocity and relies on the viscous force between the bottom-blowing gas and the molten steel to drive the flow of the molten steel. The argon gas close to the molten steel is in a laminar flow state, subject to Newton's viscosity theorem, as shown in Equation (17). According to Newton's viscosity theorem, the magnitude of the internal friction stress τ is proportional to the velocity difference du_s between the two fluid layers and inversely and the distance dy_s between the two fluid layers. Due to the same conditions used, du_s/dy_s is a constant, set to k_s.

$$\tau = \mu \frac{du_s}{dy_s} = \mu k_s \tag{17}$$

In the formula: τ is the internal friction shear stress between the two steel liquid layers, N; μ is the viscosity of the steel liquid, Pa·s;

The viscous force is the resistance for argon, and for the molten steel, the internally generated viscous force is the power of the molten steel and obeys Newton's second law. The relationship between speed, acceleration, and time is shown in Equation (20). Equation (20) can be obtained by combining Equations (17)−(19). The flow rate of molten steel is proportional to the viscosity and time.

$$\tau = ma \tag{18}$$

$$v = at \tag{19}$$

$$v = \frac{\mu k_s}{m} t \tag{20}$$

where a is the acceleration of the molten steel flow layer, m/s^2; m is the mass of the molten steel flow layer, kg; v is the flow rate of the molten steel, m/s; and t is the acceleration time of the molten steel, s.

From Equation (20), we can know that under the same bottom-blowing conditions, the farther away from the steel–slag contact surface, the longer the distance the molten steel is accelerated, and the greater the flow rate of the molten steel. In addition, with the progress of EAF smelting, the viscosity of the molten steel decreased, the acceleration effect of the bottom-blowing stirring on the molten steel worsened, and the flow rate of the molten steel decreased.

4.3. Operation System Discussion for Bottom-Blowing

Based on the simulation results and related analysis of the characteristic points of the four representative stages of EAF steelmaking, the four characteristic points selected can reflect the characteristics of the four representative stages. Combining simulation results with the physical model verification results, we suggest that the bottom-blowing flow control can be optimized according to bottom-blowing efficiency and smelting actual needs studied in this paper. The specific process can be controlled in four stages.

(1) Stage one represents the smelting period at the end of the smelting feed. In this stage, the melt pool has the highest viscosity, the bottom-blowing efficiency is high, and the power supply intensity is the largest. It is necessary to accelerate the melting of scrap steel. Therefore, using a higher bottom-blowing flow rate can accelerate the heat transfer of the melt pool without consuming a large amount of bottom-blowing gas and melting scrap;

(2) Stage two represents the early stage of decarburization in the smelting process. In this stage, the intensity of oxygen supply is large, and the internal chemical reaction in the molten pool is intensely exothermic. In addition to the stirring effect of the oxygen jet on the molten pool, CO produced by the decarburization reaction also has a larger mixing effect [36,37]. Combined with the fact that the molten steel has a higher viscosity and bottom-blowing efficiency, it is recommended to use a lower bottom-blowing flow rate to bring the molten steel to a better flow state at a lower flow rate;

(3) Stage three represents the mid-to-late stage of decarburization in the smelting process. In this stage, the oxygen supply intensity is relatively less, the chemical reaction

in the molten pool gradually weakens, and CO produced by the decarburization reaction weakens the stirring effect of the molten pool and reduces the amount of heat. The viscosity of the molten steel and the bottom-blowing efficiency is low. To ensure a good stirring effect, it is recommended to use a moderate bottom-blowing flow;

(4) Stage four represents the final stage of smelting. In this stage, the decarburization reaction in the furnace is close to the end point, and local oxidation is easy to cause iron loss. Although the viscosity and the bottom-blowing stirring efficiency are the lowest, to prevent local over-oxidation, reduce the loss of iron, and improve the metal yield, it is recommended to use a higher bottom-blowing flow rate.

5. Conclusions

In this paper, the industrial 100 t EAF was used as the reference model; the numerical simulation software was used to establish a three-dimensional grid model with 55,924 grids and 540,395 nodes; the fluid calculation software ANSYS Fluent 18.2 was used to calculate the velocity field of the molten steel under the same bottom-blowing gas flow rate at different EAF smelting periods; and the fluidity and bottom-blowing effect of the molten steel at different smelting stages were studied. The conclusions are summarized as follows:

(1) When the bottom-blowing gas flow rate was 100 L/min, the average flow rates of the four stages in EAF steelmaking were v_1 = 0.0081 m/s, v_2 = 0.0069 m/s, v_3 = 0.0063 m/s, and v_4 = 0.0053 m/s. Numerical simulation results showed that under the same bottom-blowing conditions, the longer the distance from the steel–slag contact surface, the greater the flow rate of the molten steel.

(2) The following relationship between the molten steel viscosity and flow rate was obtained: $v = -85048.18\eta^2 + 1808.75\eta - 1.473$. With the progress of the smelting stage, the viscosity of molten steel decreased, and flow velocity of the molten steel decrease.

(3) Physical model verification results indicated when the bottom-blowing stirring conditions remain unchanged, the temperature of the No. 11 lubricating oil increased, the viscosity of it decreased, and the bubble escape time would decrease. This confirmed the conclusion of numerical simulation, that is, with temperature increase, under the same bottom-blowing stirring conditions, the viscosity of the molten fluid decrease. The kinetic energy exchange between the bottom gas and the molten steel became worse, and the acceleration effect driven by bottom-blowing gas worsened, making the molten steel more difficult to be stirred, meaning the bottom-blowing efficiency continued to decrease with the smelting process.

(4) According to the numerical simulation and physical model verification results, relevant suggestions are made for the operation system of bottom-blowing in EAF steelmaking. The higher bottom blowing flow rates should be used during the smelting period and the end of smelting period while the lower bottom-blowing flow rates can be used in the early stage of decarburization and moderate bottom-blowing flow rates used in the middle and later stages of decarburization.

Author Contributions: Conceived of and designed the experiments, L.Y., Y.G., F.C., S.W., and H.H.; performed the experiments, H.H. and L.Y.; analyzed the data, H.H., F.Z.; searched the relevant literature and data, H.H. and B.L.; wrote the paper, H.H.; reviewed and contributed to the final manuscript, L.Y. and Y.G. All authors have read and agreed to the published version of the manuscript.

Funding: This research was funded by the National Natural Science Foundation of China (No. 51804345) and the Fundamental Research Funds for the Central Universities of Central South University (No. 2020zzts752).

Institutional Review Board Statement: Not applicable.

Informed Consent Statement: Not applicable.

Data Availability Statement: Restrictions apply to the availability of these datas. Datas were obtained from Central South University and are available from the Hang Hu with the permission of Central South University.

Acknowledgments: Financial support from the National Natural Science Foundation of China (No. 51804345) and the Fundamental Research Funds for the Central Universities of Central South University (No. 2020zzts752) is gratefully acknowledged.

Conflicts of Interest: The authors declare no conflict of interest.

References

1. Goodfellow, H.D.; Pozzi, M.; Maiolo, J. Dynamic process control and optimization for EAF steelmakers. *MPT* **2006**, *29*, 24–31.
2. Yin, R.Y. Thinking of Developing Modern Electric Furnace Route in 21st Century. *China Metall.* **2005**, *6*, 1–7.
3. Lv, M.; Zhu, R.; Yang, L.Z. High Efficiency Dephosphorization by Mixed Injection during Steelmaking Process. *Steel Res. Int.* **2018**, *90*, 1800454. [CrossRef]
4. Wei, G.S.; Peng, Y.H.; Zhu, R.; Yang, L.Z.; Wu, X.T. Fluid Dynamics Analysis of O_2–CaO Jet with a Shrouding Flame for EAF Steelmaking. *ISIJ Int.* **2020**, *60*, 481–491. [CrossRef]
5. Yang, L.Z.; Yang, Z.S.; Wei, G.S.; Guo, Y.F.; Chen, F.; Zheng, F.Q. Influence of Ambient and Oxygen Temperatures on Fluid Flow Characteristics Considering Swirl-type Supersonic Oxygen Jets. *ISIJ Int.* **2019**, *59*, 2272–2282. [CrossRef]
6. Jiang, M.F.; Li, L.F. Current Status of Development in Bottom Stirring Technique for Electric Arc Furnace. *Res. Iron Steel* **1994**, *5*, 46–49.
7. Wang, H.; Zhu, R.; Liu, R.Z.; Shou, D.; Xie, G.J.; Fan, S.L.; Gu, Y.L. Application Research of Carbon Dioxide in EAF Bottom Blowing. *Ind. Heat.* **2014**, *43*, 12–17.
8. Wei, G.S.; Zhu, R.; Dong, K.; Li, Z.Z.; Yang, L.Z.; Wu, X.T. Influence of bottom-blowing gas species on the nitrogen content in molten steel during the EAF steelmaking process. *Ironmak. Steelmak.* **2017**, *45*, 839–846. [CrossRef]
9. Wei, G.S.; Zhu, R.; Wu, X.T.; Dong, K.; Yang, L.Z.; Liu, R.Z. Technological Innovations of Carbon Dioxide Injection in EAF-LF Steelmaking. *JOM* **2018**, *70*, 969–976. [CrossRef]
10. Van Wijngaarden, M.J.U.T.; Pieterse, A.T. Bottom-stirring in an electric-arc furnace: Performance results at lscor Vereeniging Works. *J. S. Afr. Inst. Min. Metall.* **1994**, *1*, 27–34.
11. He, P.; Zhang, D.M.; Deng, K.W.; Zhang, R.S.; Wang, Q.K. Stirring Technology of Bottom Blowing Gas in EAF. *Iron Steel* **1992**, *9*, 65–70.
12. Ramirez, M.; Alexis, J.; Trapaga, G.; Jonsson, P.; Mckelliget, J. Modeling of a DC Electric Arc Furnace—Mixing in the Bath. *ISIJ Int.* **2001**, *41*, 1146–1155. [CrossRef]
13. Gu, Y.L.; Zhu, R.; Dong, K.; Bao, X.; Xie, G.J.; Liu, Z.Q.; Shou, D.; Ma, G.H. Optimization study on bottom blowing process in EAF. *Steelmaking* **2013**, *29*, 28–30.
14. Dong, K.; Zhu, R.; Liu, W.J. Bottom-blown Stirring Technology Practiced in Consteel EAF. *Adv. Mater. Res.* **2011**, *361*, 639–643. [CrossRef]
15. Wei, G.S.; Zhu, R.; Dong, K.; Ma, G.H.; Cheng, T. Research and Analysis on the Physical and Chemical Properties of Molten Bath with Bottom-Blowing in EAF Steelmaking Process. *Metall. Mater. Trans. B* **2016**, *47*, 3066–3079. [CrossRef]
16. Liu, F.H.; Zhu, R.; Dong, K.; Bao, X.; Fan, S.L. Simulation and Application of Bottom-Blowing in Electrical Arc Furnace Steelmaking Process. *ISIJ Int.* **2015**, *55*, 2365–2373. [CrossRef]
17. Ma, G.H.; Zhu, R.; Dong, K.; Li, Z.Z.; Liu, R.Z.; Yang, L.Z.; Wei, G.S. Development and application of electric arc furnace combined blowing technology. *Ironmak. Steelmak.* **2016**, *43*, 594–599. [CrossRef]
18. Suh, D.W.; Kim, N.J. Low-density steels. *Scr. Mater.* **2013**, *68*, 337–338. [CrossRef]
19. Zuazo, I.; Hallstedt, B.; Lindahl, B.; Selleby, M.; Soler, M.; Etienne, A.; Perlade, A.; Hasenpouth, D.; Massardier-Jourdan, V.; Cazottes, S.; et al. Low-Density Steels: Complex Metallurgy for Automotive Applications. *JOM* **2014**, *66*, 1747–1758. [CrossRef]
20. Feng, X.; Liang, F. Density of Liquid Steel over Temperature Range of 1803–1873 K. *J. Iron. Steel Res. Int.* **2004**, *11*, 37–40.
21. Zhao, S.; Wei, D.L.; Xu, J.H.; Chen, H.; Li, Z. Influence of Steel Strip-Feeding Process on Density and Segregation of Casting Ingot. *Mater. Sci. Forum* **2016**, *872*, 45–49. [CrossRef]
22. Hu, J.; Sun, B.L.; Li, C.W.; Xu, Y.; Wang, L. Numerical simulation of the influence of steel density change on the flow field in tundish. *Energ. Metall. Ind.* **2008**, *6*, 12–14.
23. Nakanishi, H.; Nakazato, K.; Abe, K. Temperature dependence of the density of molten germanium and silicon measured by a newly developed Archimedian technique. *J. Cryst. Growth* **1999**, *203*, 75–79. [CrossRef]
24. Huang, X.G. *Molten Metal. Ferrous Metallurgy Theory*, 4th ed.; Beijing Metallurgical Industry Press: Beijing, China, 2014; pp. 194–206.
25. Roscoe, R.; Bainbridge, W. Viscosity Determination by the Oscillating Vessel Method II: The Viscosity of Water at 20 °C. *Proc. Phys. Soc.* **1958**, *72*, 585. [CrossRef]
26. Hermann, S.; Frohberg, M.G.; Eberhard, S. Untersuchungen über wechselseitige Aktivitätseinflüsse in homogenen metallischen Mehrstofflösungen. *Steel Res. Int.* **1963**, *34*, 37–42.
27. Morita, Z.I.; Iida, T. Viscosity of molten iron and steel. *Iron Steel* **1982**, *2*, 54–61.
28. Wei, G.S.; Zhu, R.; Yang, L.Z.; Dong, K.; Liu, R.Z. Modeling on impact zone volume generated by coherent supersonic jet and conventional supersonic jet. *J. Iron. Steel Res. Int.* **2018**, *25*, 681–691. [CrossRef]
29. Guo, H.Z.; Zhao, P.; Wang, K.L.; Fu, J.; Ma, T.W. Numerical Calculation on Electromagnetic Stirring and Heat Transfer in DC-EAF Bath. *J. USTB* **1995**, *17*, 284–288.

30. Chen, J.X. *The Physical Properties of Iron and Steel. Manual of Chart Data for Steelmaking*, 2nd ed.; Beijing Metallurgical Industry Press: Beijing, China, 2010; pp. 355–393.
31. Yang, Z.S.; Yang, L.Z.; Cheng, T.; Chen, F.; Zheng, F.Q.; Wang, S.; Guo, Y.F. Fluid Flow Characteristic of EAF Molten Steel with Different Bottom-Blowing Gas Flow Rate Distributions. *ISIJ Int.* **2020**, *60*, 1957–1967. [CrossRef]
32. ANSYS Inc. ANSYS FLUENT 18.2 User' s Guide. 2018. Available online: https://www.ansys.com (accessed on 6 September 2018).
33. Alam, M.; Irons, G.; Brooks, G.; Fontana, A.; Naser, J. Inclined Jetting and Splashing in Electric Arc Furnace Steelmaking. *ISIJ Int.* **2011**, *51*, 1439–1447. [CrossRef]
34. Wei, G.S.; Zhu, R.; Wu, X.T.; Yang, L.Z.; Dong, K.; Cheng, T.; Tang, T.P. Study on the Fluid Flow Characteristics of Coherent Jets with CO_2 and O_2 Mixed Injection in Electric Arc Furnace Steelmaking Processes. *Metall. Mater. Trans. B* **2018**, *49*, 1405–1420. [CrossRef]
35. Barella, S.; Gruttadauria, A.; Mapelli, C.; Mombelli, D. Critical evaluation of role of viscosity and gas flowrate on slag foaming. *Ironmak. Steelmak.* **2010**, *39*, 463–469. [CrossRef]
36. Morales, R.D.; Rubén, L.G.; López, F. The Slag Foaming Practice in EAF and Its Influence on the Steelmaking Shop Productivity. *ISIJ Int.* **2007**, *35*, 1054–1062. [CrossRef]
37. Wei, G.S.; Zhu, R.; Yang, S.F.; Wu, X.T.; Dong, K. Simulation and application of submerged CO_2–O_2 injection in EAF steelmaking: Combined blowing equipment arrangement and industrial application. *Ironmak. Steelmak.* **2021**, *3*, 1–9. [CrossRef]

metals MDPI

Review

Water Electrolysis for the Production of Hydrogen to Be Employed in the Ironmaking and Steelmaking Industry

Pasquale Daniele Cavaliere [1,*], Angelo Perrone [1] and Alessio Silvello [2]

1 Department of Innovation Engineering, University of Salento, Via per Arnesano, 73100 Lecce, Italy; angelo.perrone@unisalento.it
2 Thermal Spray Center CPT, Universitat de Barcelona, 08028 Barcelona, Spain; asilvello@cptub.eu
* Correspondence: pasquale.cavaliere@unisalento.it

Abstract: The way to decarbonization will be characterized by the huge production of hydrogen through sustainable routes. Thus, the basic production way is water electrolysis sustained by renewable energy sources allowing for obtaining "green hydrogen". The present paper reviews the main available technologies for the water electrolysis finalized to the hydrogen production. We describe the fundamental of water electrolysis and the problems related to purification and/or desalinization of water before electrolysis. As a matter of fact, we describe the energy efficiency issues with particular attention to the potential application in the steel industry. The fundamental aspects related to the choice of high-temperature or low-temperature technologies are analyzed.

Keywords: water electrolysis; ironmaking; steelmaking; purification; desalinization; direct reduction; energy; renewables; high temperature; low temperature

Citation: Cavaliere, P.D.; Perrone, A.; Silvello, A. Water Electrolysis for the Production of Hydrogen to Be Employed in the Ironmaking and Steelmaking Industry. *Metals* **2021**, *11*, 1816. https://doi.org/10.3390/met11111816

Academic Editor: Thomas Gries

Received: 14 October 2021
Accepted: 11 November 2021
Published: 12 November 2021

1. Introduction

Ironmaking and Steelmaking are complex processing routes with many fundamental steps. The actual production routes are mainly based on carbon and gas for energy supply and to drive all the thermo-chemical reactions involved in the different iron oxides transformations. With the progress of technological advances, it is possible to reduce carbon dioxide emissions in different production steps. Obviously, the technological choices are strictly related to economic and local conditions [1]. Many gradual signs of progress are due to the gradual decarbonization of the steel industry with the partial replacement of coal with gas, oil, biomass, wastes, and electricity. Further improvements are due to the development of carbon capture and storage/utilization technologies. A big revolution is represented by the introduction of hydrogen in the ironmaking and steelmaking routes. As a matter of fact, and as the main difference with respect to coal utilization, hydrogen employment does not lead to carbon dioxide emissions. In addition, being an energy vercot, it can be used, transformed, and stored for various applications in a very versatile way. For all these reasons it is considered the best candidate in the transition of the steel industry toward decarbonization. Additionally, if hydrogen is produced through renewable energy sources, it allows for an emission-free scenario in those industries where it can be employed.

Among all the solutions developed for the decarbonization of the steel industry, the main routes are evolving toward the employment of hydrogen as an energy carrier (Figure 1).

At the present time, more than 90% of the produced hydrogen is obtained from fossil fuels through processes such as coal gasification or natural gas reforming; only 4% of hydrogen is produced via water electrolysis. The available main technologies are alkaline (AEL), Proton Exchange Membrane Electrolyzer (PEMEL), and Solid Oxide Electrolyzer (SOEL). PEMEL water electrolyzers can be further categorized into acidic PEMELs and alkaline PEMELs, but only the acidic PEM variant is widely commercially available [2]. Renewable hydrogen is believed to be essential for the restoring of industries "hard to abate"; in the future, it is believed that installed water electrolyzer capacities will range up to terawatts [3].

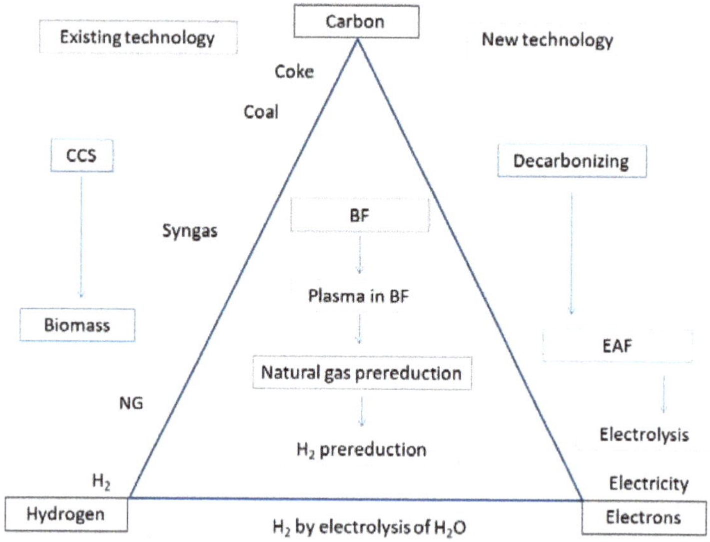

Figure 1. Pathways of technologies for greenhouse gases emissions abatement.

2. Water Electrolysis

Water electrolysis was developed in the last century mainly to produce hydrogen. Electricity allows for the splitting of water into hydrogen and oxygen then converting electrical energy into chemical energy. Being the hydrogen volume dependent on the electric current, the energy efficiency of the employed system is fundamental.

The AEM was the first to be developed. In the alkaline water electrolysis, the nickel-based anode and cathode are submerged in an aqueous KOH or NaOH solution as shown in Figure 2a.

Water is reduced at the cathode forming hydrogen gas and hydroxide ions OH^-. Hydroxide ions are then decomposed at the anode to form oxygen gas and water. The low-cost electrodes allow for the industrial scaling of this technology [4]. The separator normally is a Zirfon porous membrane of 500 μm in thickness [5]. The operating conditions are 70–90 °C at 30 bars with a current density <0.4 A/cm^2.

The PEMEL water electrolysis employs a solid polymer (sulfonated fluoropolymers) electrolyte that is a thin (250 μm thickness) proton-exchange membrane. The schematic of the process is given in Figure 2b.

Water is supplied to the anode, where water is first decomposed with sufficient electric potential to oxygen gas. H^+ protons cross the membrane, then, once arriving on the cathode surface, it recombines with electrons in order to produce gaseous hydrogen. Iridium and platinum are employed as a catalyst at the anode and cathode respectively. The current collector is made of titanium in order to increase the lifetime of the cell [6]. The designation 'PEMEL water electrolysis' refers to the commercial, traditional acidic PEMEL variant, and the alkaline PEMEL is referred to as anion exchange membrane (AEM) water electrolysis. Alkaline and PEMEL water electrolyzers are readily available, commercialized technologies, while SOEL is the least developed and not widely commercially available [7].

The schematic of the SOE electrolyzer is shown in Figure 2c.

AEL and PEMEL electrolysis technologies have operating temperatures below 100 °C. SOEL is based on the electrolysis of steam with operating temperatures in the range 700–1000 °C [8].

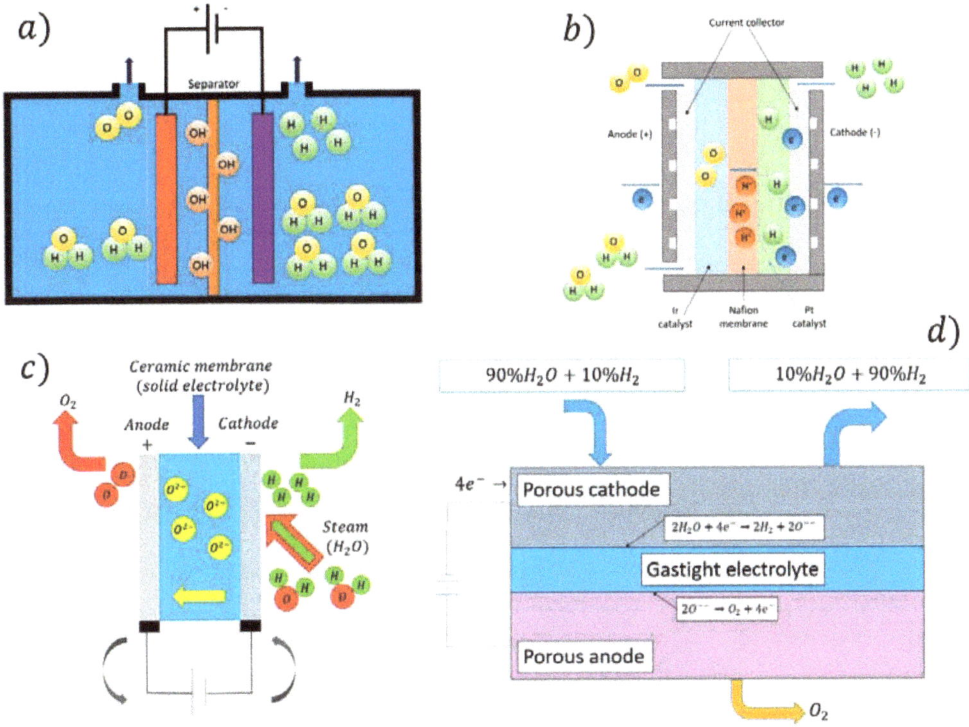

Figure 2. (**a**) Alkaline water electrolysis schematic; (**b**) PEMEL schematic; (**c**) SOEL schematic; (**d**) High-temperature steam electrolysis process schematic.

High-Temperature Steam Electrolysis (HTSE) is a new and high-potential instrument for hydrogen production in a clean way. It is based on a reverse fuel cell setting (Figure 2d).

In the case of steam employment, water dissociation is easier with respect to the case of AEM and PEMEL. This is because part of the required energy is provided by the high temperature of the steam. In general, this solution is suggested for those industrial applications where high-temperature sources are available. This allows HTSE to be very promising from an energy efficiency point of view. In general, as the temperature increases, the global energy efficiency of the process increases.

As a matter of fact, this technology requires less electricity consumption with respect to low-temperature water electrolysis. This is also due to the absolute less internal resistance of the high-temperature electrolysis cells. The cell has classical elements such as anodes, cathodes, and electrolytes. During electrolysis operations, steam and electricity are provided to the porous fuel electrode. This allows the reduction of water molecules for the oxygen ions (O^{2-}) and hydrogen at the cathode. The electrode, where oxidation reactions take place, is alimented with air. Other complex solutions foresee the employment of different gases or vacuum depending on the cell design. In SOEL, an electrolyte is a thin membrane allowing only oxygen ions to cross. This is a complex cell component that must stop all the hydrogen or different gases. As a function of the cell design, it can work in different ways such as thermo-neutral, endothermic, and exothermic depending on the steam temperature and on the provided electricity [9].

Not all the provided electricity is employed for the steam electrolysis, in fact, due to important losses, a part of the provided energy is wasted under the formation of heat. So, a large percentage of the global efficiency of the cell is dependent on the balance between the provided energy and the wasted one. The cell works in exothermic conditions once the provided electric energy exceeds the heat required for the water steam splitting. When the cell is working in endothermic conditions, the electrochemical reactions require more heat with respect to the one that is provided by electricity. So, heat is lost, and the cell temperature decreases. In these conditions, additional heat must be provided to the cell.

The different cell performances are traditionally evaluated as a function of the electric power that is consumed by the electrochemical reactions for a given hydrogen volume. A good summary is described in Figure 3 showing the difference among the different electrolyzers.

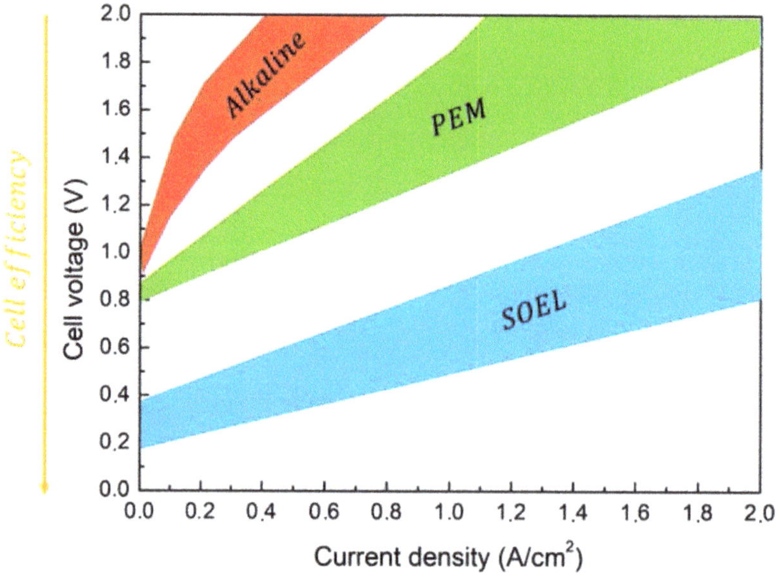

Figure 3. Electrolyzers comparison.

As a general behavior, water electrolysis for hydrogen production is a high energy-consuming process. This energy is provided mainly by electricity [10]. The order of electrical power consumption in AEL and PEMEL cells falls in the range 4.4–4.9 kWh/m^3 H_2. Given the heating provided to the HTSE cells, the electric power consumption falls in the range 3.8–3.9 kWh/m^3 H_2 [11].

As previously mentioned, hydrogen production through water electrolysis can be a completely fossil-free route once the electric power is produced through renewable sources, such as solar and wind. In these conditions, low-temperature electrolysis (LTE) is able to produce hydrogen with a specific electricity consumption (SEC) in the range 50–60 kWh/kg H_2.

The steam temperature in HTSE falls in the range 700–1000 °C. It is this condition that allows the SEC to be reduced at the order of 37 kWh/kg H_2 [3]. In each case, the main limitation to the technology development is the fast degradation of the cell components due to the high temperature and to the environment. Obviously, if the cell temperature is below 700 °C, the cell can reach a durability of around 25,000 h. Anyway, as the temperature decreases, the process efficiency decreases. So, an optimal balance must be done between the steam temperature and the materials consumption. In order for HTSE to become competitive with LTE, the performances should exhibit very high production rates in the order of ~40 mgH_2/cm^2/h or ~−1 A/cm^2. All this is in the case of a reasonable cost due to the fast electrodes and membranes consumption.

Very recently, a hydrogen production route very similar to electrolysis has been proposed [12]. This method is known as E-TAC. As with the other water-splitting technologies, the process employs electric power for splitting water into hydrogen and oxygen. The main difference with the previously described technologies is that hydrogen and oxygen gases are produced in two separate and different moments. These are known as an Electrochemical (E) step and a Thermally Activated Chemical (TAC) step (Figure 4).

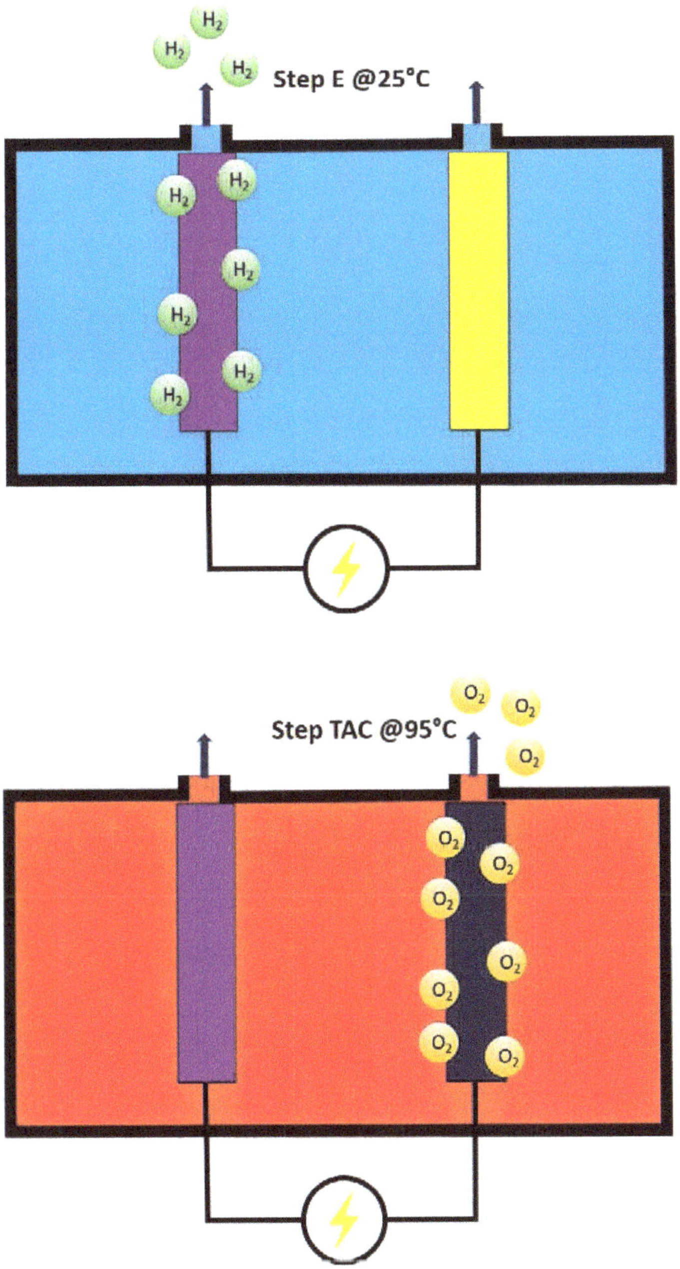

Figure 4. E-TAC hydrogen and oxygen production schematic.

Alkaline electrolysis is based on the contemporary oxygen and hydrogen evolution reactions taking place at high temperatures. In this solution, the membrane does not allow the O_2/H_2 crossover. In the case of E-TAC, the water-splitting process takes place in two separate steps. The first electrochemical step, acting at 25 °C, proceeds following the reactions (Equations (1) and (2)):

$$4H_2O + 4e^- \rightarrow 4OH^- + 2H_2 \quad (1)$$

$$\left(Ni(OH)_2 + OH^- \rightarrow NiOOH + H_2O + e^-\right) * 4 \quad (2)$$

During this step, only hydrogen bubbles are observed without oxygen production at the anode (in the supplementary material in [12], very interesting movies, describing the process, can be observed).

The second chemical step, acting at 90 °C, proceeds following the reaction (Equation (3)):

$$4NiOOH + 2H_2O \rightarrow 4Ni(OH)_2 + O_2 \quad (3)$$

The results belonging to pilot plants show an efficiency close to 95%. The potential for industrial scale-up is envisioned through a multi-cell design. In this design, cold and hot electrolytes are moved from one cell to another in order to regenerate the anode and let gases flow (Figure 5).

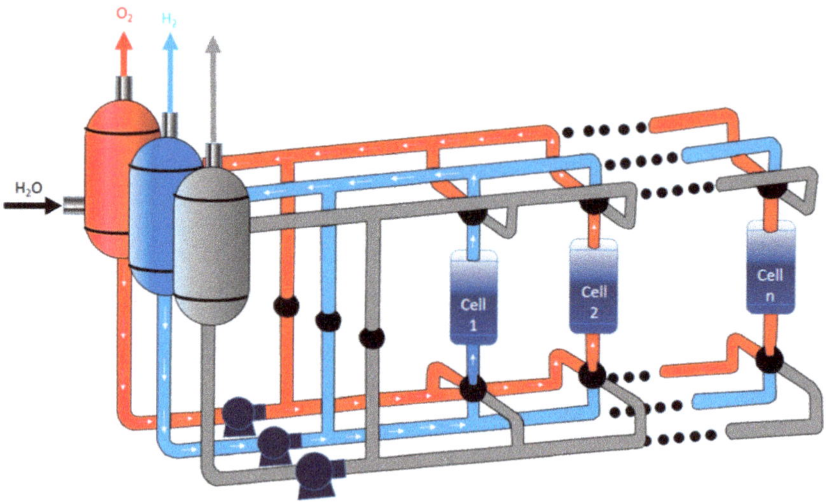

Figure 5. E-TAC multi-cell design.

The main advantage of this system is believed to be a membrane-free cell capable of producing hydrogen and oxygen in separate steps. This eliminates all the problems and costs related to membrane-based cells. The lack of a membrane enables high-pressure hydrogen production, potentially exceeding 100 bars. The possibility of operating with low anode potential leads this system to be very competitive in terms of energy balance.

3. Water Electrolysis Fundamentals

As already mentioned, the hydrogen production rate (mol/s) is linearly proportional to the current of the cell (Equation (4)):

$$\dot{n}_{H_2} = \eta_F \frac{i_{cell} A_{cell}}{zF} = \eta_F \frac{I_{cell}}{zF} \quad (4)$$

where $\dot{n}_{H_2}$ is the hydrogen production rate (mol/s), η_F is the Faraday efficiency, also known as the current efficiency, i_{cell} is the current density (A/cm^2), A_{cell} is the effective cell area (cm^2), z is the number of moles of electrons transferred in the reaction (for hydrogen, $z = 2$), F is the Faraday constant (9.6485 × 104 C/mol), and I_{cell} is the stack current (A).

As a general behavior, the anodic and cathodic reactions in acidic media is given by (Equations (5) and (6)):

$$\text{Anode}: 2H_2O \rightarrow 4H^+ + O_2 + 4e^- \tag{5}$$

$$\text{Cathode}: 2H^+ + 2e^- \rightarrow H_2 \tag{6}$$

While in alkaline media (Equations (7) and (8)):

$$\text{Anode}: 4OH^- \rightarrow O_2 + 2H_2O + 4e^- \tag{7}$$

$$\text{Cathode}: 2H_2O + 2e^- \rightarrow 2OH^- + H_2 \tag{8}$$

The electrolyte pH is the quantity mainly influencing the number of hydroxide ions and protons that drive the anodic and cathodic reactions. Here, in order to increase the ionic conductivity between the electrodes, strong basics or acids must be employed as electrolytes in order to increase the volume of energy carriers in terms of hydroxides ions and protons. For this reason, once the same electrolyte is used for the anode and for the cathode, the thermodynamic conditions of the water-splitting are driven by the electrolyte pH level.

The energy required for the water decomposition is the enthalpy change of the process, the enthalpy of formation of water, ΔH. The water electrolysis process is endothermic ($\Delta H > 0$). The free energy of the water-splitting reaction, called Gibbs free energy change ΔG, must be supplied to the electrodes as electrical energy. The remainder is the thermal energy Q, which is the product of the process temperature T and the entropy change ΔS. These thermodynamic quantities can be written as (Equation (9)):

$$\Delta H = \Delta G + Q = \Delta G + T\Delta S \tag{9}$$

H is the enthalpy, G is the Gibbs free energy, T is the temperature, S is the entropy, and Q is the required heat. In constant standard ambient conditions (298.15 K, one-atmosphere pressure), the required electrical work ΔG is equal to 237.2 kJ/mol (non-spontaneous reaction), and the amount of heat required Q is equal to 48.6 kJ/mol. Thus, the chemical reaction for water electrolysis can be expressed as (Equation (10)):

$$H_2O_{(l)} + 237.2\ \text{kJ}_{\text{el.}} * \text{mol}^{-1} + 48.6\ \text{kJ}_{\text{Heat}} * \text{mol}^{-1} \rightarrow H_{2(g)} + \frac{1}{2}O_{2(g)}\ \Delta H_0 = 285.8\ \text{kJ} * \text{mol}^{-1} \tag{10}$$

U_{rev} is the reversible voltage that is the minimum voltage necessary for the splitting of water. It is directly proportional to the Gibbs free energy change (Equation (11)):

$$U_{rev} = \frac{\Delta G}{zF} \tag{11}$$

Without the input of thermal energy, the minimum voltage required be-comes the thermoneutral voltage U_{tn} (Equation (12)):

$$U_{tn} = \frac{\Delta H}{zF} \tag{12}$$

In standard room conditions, the reversible voltage and the thermoneutral voltage result in 1.23 V and 1.48 V, respectively. For commercial water electrolyzers, all energy for the water electrolysis process is provided as electrical energy. Both the reversible voltage and the thermoneutral voltage are thermo-dynamic state functions. These quantities are a function of the cell pressure and temperature, albeit the thermoneutral voltage changes only slightly as a function of temperature and pressure [13]. The reversible voltage and the thermoneutral voltage are illustrated in Figure 6a as a function of cell temperature at standard room pressure.

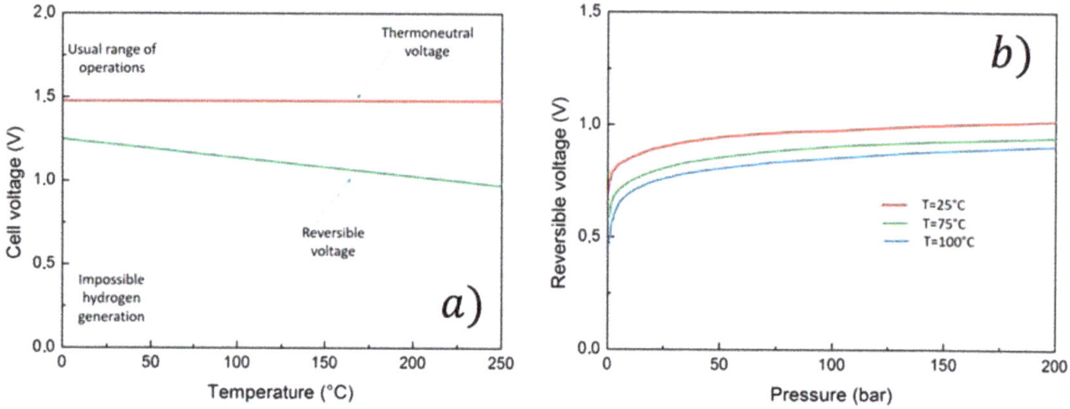

Figure 6. (**a**) Reversible and thermoneutral voltages as a function of cell temperature; (**b**) Reversible voltage as a function of pressure for different cell temperatures.

The effect of pressure on reversible voltage is exemplified in Figure 6b.

An increase in temperature will slightly reduce the overall energy demand ΔH of an ideal liquid water electrolysis process as the demand for electrical energy ΔG is more notably reduced than the demand for thermal energy $T\Delta S$ is increased [14]. Operation at higher temperatures is favorable as heat losses caused by overvoltages can be used to reduce the reversible voltage of water splitting. Thus, the utilization of thermal energy is an essential aspect of energy-efficient water electrolysis processes. The overall energy requirement ΔH will stay practically constant as a function of pressure in an ideal liquid water electrolysis process. However, a change in pressure will increase the demand for electrical energy ΔG; for instance, an increase from 0.1 MPa to 10 MPa at a cell temperature of 75 °C will increase the reversible voltage by 9%, but the demand for thermal energy $T\Delta S$ is correspondingly reduced.

Now, the electrolysis cell voltage results from the sum of the reversible voltage and all the overpotentials developing in the cell through the following equation (Equation (13)):

$$U_{cell} = U_{rev} + U_{ohm} + U_{act} + U_{con} \quad (13)$$

where U_{cell} is the cell voltage, U_{rev} is the open circuit. The reversible voltage results are a function of temperature and pressure. U_{ohm} is the overvoltage caused by Ohmic losses in the cell elements, U_{act} is the activation overvoltage, and U_{con} is the concentration overvoltage. The current–voltage characteristics of an electrolytic cell can be described by a polarization curve. An example of a polarization curve for AEL and PEMEL water electrolyzer cells is illustrated in Figure 7.

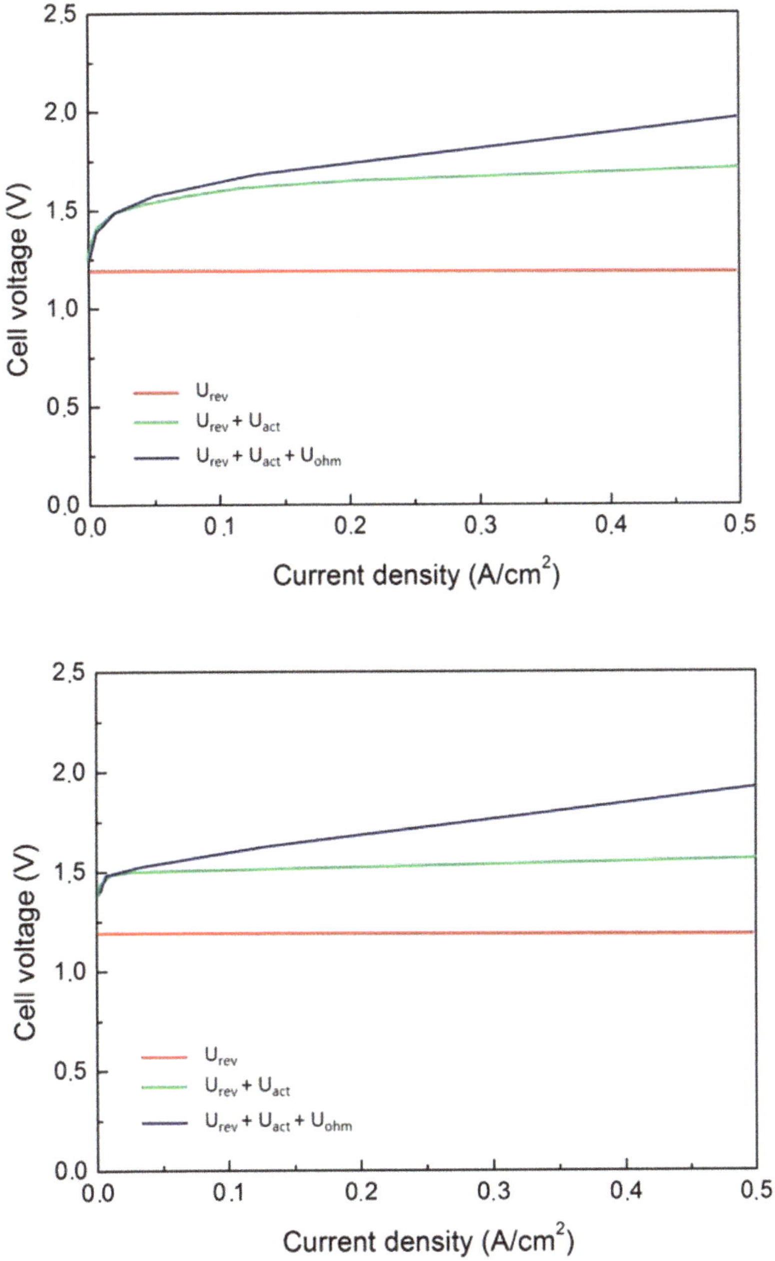

Figure 7. Overpotentials in AEL and PEMEL cells.

In the AEL water electrolysis, the Ohmic losses are mainly affected by the ionic conductivity of the liquid electrolyte, the thickness of the electrolyte layer, and the thickness and conductivity of the electrodes [15]. In the PEMEL water electrolysis, the ionic resistance of the polymeric membrane and the electrical resistance of the separator plates and current collectors are the main contributors to the Ohmic losses. The activation overpotential is caused by the anode and cathode reaction kinetics. The concentration overpotential is caused by mass transfer limitations at high current densities, where the supply of the reactant (water) is not sufficient to support the reaction rate of the production of hydrogen and oxygen gases at the electrode surfaces. The concentration losses are typically negligible for commercial water electrolyzers—especially for AEL electrolyzers—because of the relatively low current densities in the cells. Another non-linear region will appear in the cell polarization curve above the limiting (high) current density if mass transport losses occur. The evidence on the effect of temperature on the AEL performances underlines that all the overvoltages, as the anode activation, the cathode the Ohmic, and the supplied cell ones, increase in the case of the well-known bubble effect. So, as the bubble's dimensions and volume increase, the overvoltages increase. Now, the bubble's number, dimension, and volume are directly related to the current density; in addition, as the current density increases, the Ohmic overvoltage linearly increases. It is worth noting that the temperature increase due to an increase in the current density is more pronounced with respect to temperature increase due to the bubble effect. In addition, it should be considered that the power provided to the heater exponentially increases allow current densities levels while it linearly decreases at high current density levels. Power consumption decreases as the cell temperature increases [16].

The ideal cell behavior is shown in Figure 8.

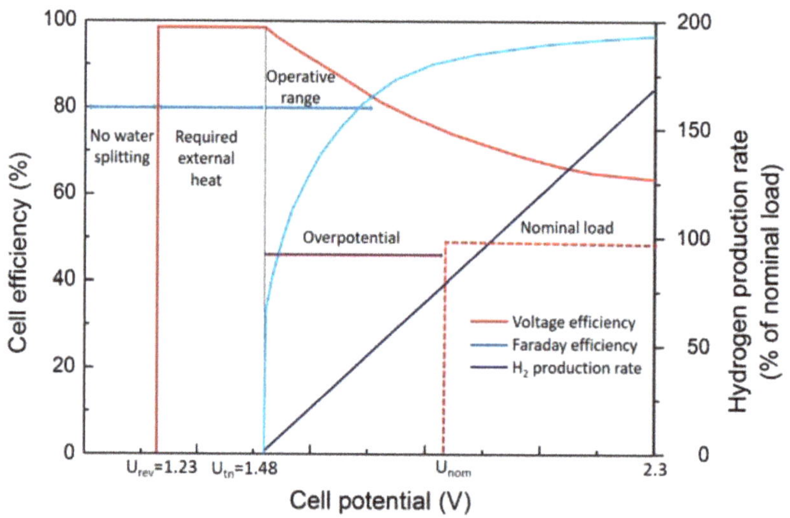

Figure 8. Ideal cell behavior as a function of the potential.

At room temperature and pressure, water decomposition starts at the thermoneutral voltage of 1.48 V. As the cell current increases, the hydrogen production increases. The Faraday efficiency typically approaches unity as cur-rent density is increased towards the nominal current density of the electrolyzer, but at low current densities, the Faraday efficiency may drop considerably.

The voltage efficiency is often used as an indicator of the energy efficiency of water electrolyzers. However, while the voltage efficiency represents the electrical losses in the cell or cell stack, it does not consider the possible stray current flow or gas crossovers over the anode and cathode compartments. The Faraday efficiency, or current efficiency, η_F is experimentally quantified as the ratio of the ideal hydrogen production rate to the actual hydrogen output from the electrolytic cell (or stack). To include both the voltage and

Faraday efficiencies in a single quantity, the definition-specific energy consumption should be used to assess the energy efficiency of a water electrolysis process (Equation (14)):

$$E_s = \frac{\int_0^{t_1} I_{stack} U_{stack} dt}{\int_0^{t_1} \dot{m}_{H_2} dt} \tag{14}$$

where E_s is the specific energy consumption, I_{stack} is the stack current, U_{stack} is the stack voltage, $\dot{m}_{H_2}$ is the hydrogen gas mass flow rate, and t_1 is the inspected time span. Hence, the specific energy consumption describes the amount of energy consumed to produce a mass unit of hydrogen gas. The energy efficiency of a water electrolysis process can be calculated from (Equation (15)):

$$\eta_e = \frac{HHV_{H_2}}{E_s} \tag{15}$$

where HHV_{H_2} is the higher heating value of hydrogen (39.4 kWh/kg or 3.54 kWh/Nm^3).

Alternatively, the lower heating value of hydrogen (LHV) can be used as a reference (33.3 kWh/kg or 3.00 kWh/Nm^3). The difference between the LHV and HHV is the latent heat of condensation. Typically, the HHV value is used as the reference for water electrolysis processes because liquid water is, in the case of alkaline and PEMEL water electrolyzers, usually supplied to the process, and the energy required for evaporation of water must be considered. The specific energy consumption of the water electrolysis process is further affected by the Faraday efficiency, which is non-linear with respect to the current density.

The selection between the currently commercially widely available alkaline and PEMEL technologies will set requirements for the required system components; alkaline technology needs a supply of liquid electrolyte, its controlled circulation, and separation from product gases. Furthermore, alkaline electrolyzer stacks are typically limited in their construction because of the liquid electrolyte supply and gas-liquid transport, which has made 200–300 V the typical stack voltage for industrial electrolyzers. The PEMEL technology avoids the construction limitations of the alkaline stacks, but the requirement for high electric currents (to produce more gases) and the exclusively bipolar construction of PEMEL stacks still set the stack DC currents relatively high compared with the stack DC voltage.

PEMEL water electrolyzers achieve comparable voltage efficiencies at higher current densities, in other words, they have lower cell impedances. Therefore, smaller variations in the instantaneous supply voltage cause greater fluctuations in the supplied current for PEMEL cells. As instantaneous, high variations in current density may have an adverse impact on cell degradation, actions may have to be taken to limit the current slew rate. Cell degradation increases the cell voltage over time, and the increased electrolyzer voltage resulting from degradation should be considered in the system design and operation.

Operating conditions, mainly cell temperature and pressure, affect the reversible voltage and impedance of the electrolytic cell and have an impact on the system efficiency. Increasing the cell temperature is generally beneficial to the cell voltage efficiency, but selected materials will limit the temperature. The cell reversible voltage is also affected by anode and cathode compartment pressures. The resulting change in voltage efficiency from an increase in pressure is comparable with the ideal isothermal compression of hydrogen gas. However, if the surrounding system, such as post-electrolysis synthesis processes or gaseous storage of hydrogen gas, requires elevated pressures, PEMEL electrolyzers may opt to operate at a differential pressure and output only hydrogen gas at elevated pressure, while alkaline electrolyzers are limited to balanced pressure operation. In the PEMEL water electrolysis, the change in hydrogen outlet pressure from 20 bar to 40 bar may result in un-changed electrical energy consumption [17]. However, an increase in pressure may compromise the control range of the electrolyzer and its specific energy consumption as the gas crossover rate is increased. Furthermore, operating the water electrolyzer in non-optimal conditions may risk the lifetime of the electrolytic cells.

As a matter of fact, the summarized properties of the described systems are listed in Table 1 [7].

Table 1. Electrolysis cells properties.

Technology	AEC	PEMEL	SOEC
Electrolyte	KOH	NAFION	YSZ
Cathode	Ni, Ni-Mo alloys	Pt, Pt-Pd	Ni, YSZ
Anode	Ni, Ni-Co alloys	RuO_2, IrO_2	LSM/YZS
Current density (A/cm^2)	0.2–0.4	0.6–2	0.3–2
Cell voltage (V)	1.8–2.4	1.8–2.2	0.7–1.5
Voltage efficiency (%)	62–82	67–82	~100
Cell area (m^2)	<4	<0.3	<0.01
Operating temperature (°C)	60–80	50–80	650–1000
Operating pressure (bar)	<30	<200	<25
Production rate (m^3/h)	<760	<40	<40
Stack energy (kWh/m^3)	4.2–5.9	4.2–5.5	>3.2
Lifetime (h)	<90,000	<60,000	<10,000
Maturity	Mature	Commercial	Demonstration
Capital cost (euro/kW)	<1200	<2300	>2000

Water supply and water purification are required to guarantee the normal operation of the water electrolysis process and to preserve the lifetime of the electrolytic cells. In the alkaline water electrolysis, water is consumed from the liquid electrolyte solution, whose concentration must be maintained by an inlet of deionized water. Meanwhile, the PEMEL water electrolysis is electrolysis of deionized water. The conductivity of the inlet water affects the operation of the electrolytic cell, its energy efficiency and aging, and the presence of alkaline electrolyte decreases the gas solubility, which has an impact on the gas crossover. The water supply can also be considered from the system integration point.

Alternative solutions are underlined such as in [18]. Here, energy-saving yet chlorine-free seawater electrolysis for efficient hydrogen production by a hybrid seawater splitting strategy is proposed. This chemistry consumes the seawater on the cathode to generate H_2 by hydrogen evolution reaction (HER); while the crossover of released OH^- to the anode side supply the hydrazine degradation to harmless H_2 and water with reduced salinity. Beyond the state-of-the-art seawater electrolysis, it enables hydrogen production at ultralow cell voltages but large current densities without chlorine hazards and limiting hydrogen-yielding efficiency. The hybrid seawater electrolyzer (HSE) using NiCo/MXene-based superaerophobichydrophilic and hydrazine-friendly electrodes requires a dramatically lower electricity expense of 2.75 kWh/m^3 H_2 than alkaline seawater electrolyzer (ASE) at industrial-scale current densities. This electrolyzer simultaneously allows fast hydrazine degradation to a rather lower residual while harvesting water with reduced salinity from seawater. On this basis, self-powered seawater electrolysis can be further realized by integrating the HSE into solar or hydrazine fuel cells for better cost-effectiveness and sustainability.

4. Decoupled Electrochemical Water Splitting

Interesting recent progress in water splitting technologies development is the so-called decoupled water splitting. It results in a highly flexible alternative to traditional water electrolysis technologies. During electrochemical water splitting, a redox mediator (M^-) is employed in an oxidation half-reaction that is coupled with the HER. On the other hand, the M half reduction is coupled with the OER (Equations (16)–(21)):

$$\text{Cathode}: \ 2H^+ + 2e^- \rightarrow H_2 \tag{16}$$

$$\text{Anode}: \ 2M^- \rightarrow 2M + 2e^- \tag{17}$$

$$\text{Overall}: \ 2H^+ + 2M^- \rightarrow H_2 + 2M \tag{18}$$

$$\text{Cathode}: \ 2M + 2e^- \rightarrow 2M^- \tag{19}$$

$$\text{Anode}: \ H_2O \rightarrow 2H^+ + 2e^- + \frac{1}{2}O_2 \tag{20}$$

$$\text{Overall}: \ 2M + H_2O \rightarrow 2M^- + 2H^+ + \frac{1}{2}O_2 \tag{21}$$

The overall reaction is exactly equivalent to the traditional water-splitting reaction previously described. The mediator (*M*) is continuously cycled through different oxidation states in a way similar to the well-known redox-flow battery (RFB) or a solid-state battery. Obviously, this behavior is dependent on its physical state.

As in traditional water-splitting technologies, inputs are water and electricity while the products are hydrogen and oxygen gases [19]. In the case of decoupled water splitting, pH is related to the electrolyte (as in the case of traditional electrolysis) but also to the current provided during the independent decoupled reactions.

The previous equations can be modified slightly for mediators that also up-take and release protons during operation (Equations (22)–(27)):

$$\text{Cathode}: \ 2H^+ + 2e^- \rightarrow H_2 \tag{22}$$

$$\text{Anode}: \ 2MH_2 \rightarrow 2M^* + 2H^+ + 2e^- \tag{23}$$

$$\text{Overall}: \ 2MH_2 \rightarrow H_2 + 2M^* \tag{24}$$

$$\text{Cathode}: \ 2M^* + 2H^+ + 2e^- \rightarrow 2MH_2 \tag{25}$$

$$\text{Anode}: \ H_2O \rightarrow 2H^+ + 2e^- + \frac{1}{2}O_2 \tag{26}$$

$$\text{Overall}: \ 2M^* + H_2O \rightarrow 2MH_2 + \frac{1}{2}O_2 \tag{27}$$

The overall reaction is the same as the overall water splitting reaction even if the system is proton balanced. In the case of alkaline electrolytes, the reactions are similar; the only difference is that balancing is performed by hydroxide ions.

The employment of the mediator allows for the separation of the hydrogen and oxygen evolution reactions in space and time. So, the benefits are mainly due to the possibility of controlling separately the steps with increased efficiency. Other positive aspects are represented by the possibility of employing low-cost electrolyzers; favoring the integration with renewable sources; improving the safety conditions of all the processes mainly related to the risks of explosion.

Given all those aspects, the decoupled water splitting systems are grouped into four classes depending on the physic conditions of the mediator, and on the input of the electricity (Figure 9).

The type 1 system is liquid mediated. This system consists of an RFB half-cell and an electrolysis half-cell. The mediator is contained in the electrolyte, it is reduced during step 1 (left) then, it is sent to the second cell. Here the mediator is re-oxidized. On the contrary, the electrode can be changed for the half-cell electrolysis by driving the reactions through a polarity opposite with respect to step 1.

In this kind of cell, the separator is employed in order to avoid the mixing of the mediator and the electrolysis products. Another kind of cell is represented by a situation where the second step evolves spontaneously (Type 2). Here the cell is composed of an RFB half-cell and an electrolysis half-cell with a catalyst bed. Here, HER acts spontaneously then, external input is not necessary.

Another type of decoupled water splitting is a solid-based technology based on two non-spontaneous steps (Type 3). This kind of cell employs only one electrode that is coupled alternatively to another electrode for the development of HER or OER in alternative ways. The basic construction of this solution is, for driving the two steps, an electrolysis half-cell and a battery half-cell.

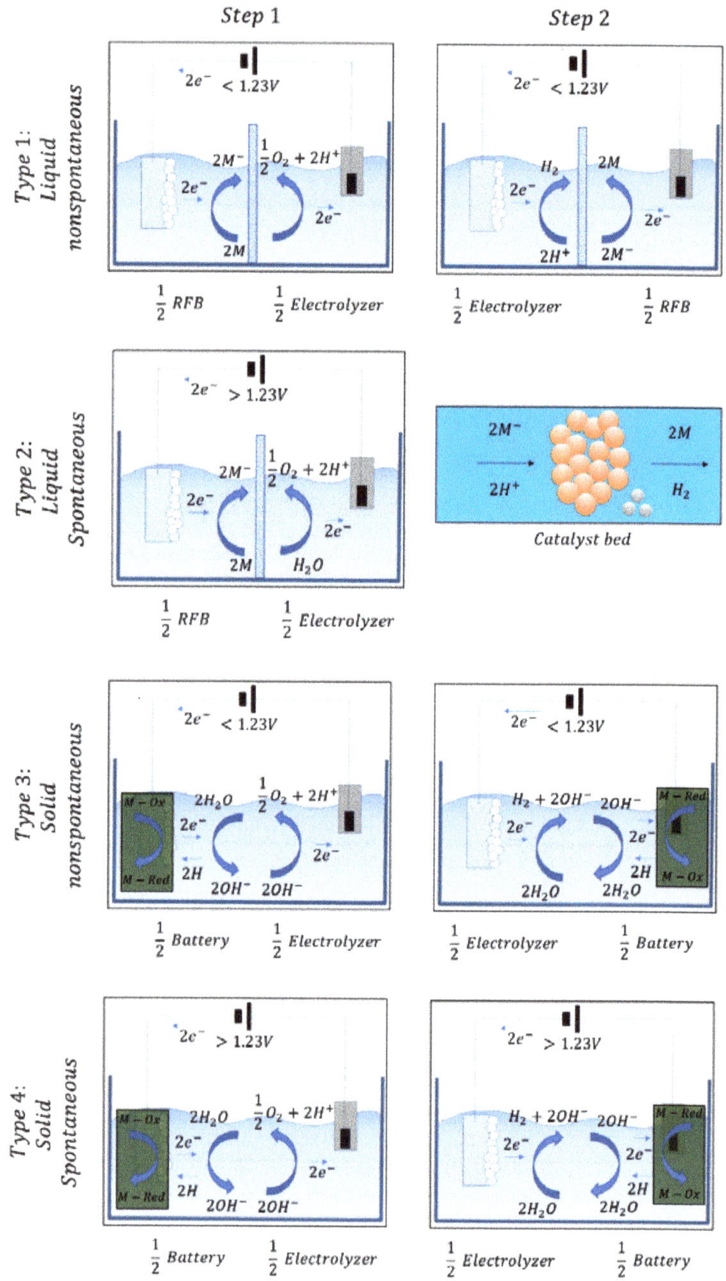

Figure 9. Different decoupled water splitting systems.

A different intermediate solution is the employment of a solid mediated system with only one spontaneous step (Type 4). The basic design is very similar to Type 3. In this solution, the second cell is not provided with electricity.

The continuous development of these systems led to the possibility of employing non-precious catalysts arriving to be competitive with the underdeveloped best PEM cells. It is also possible to employ low-cost membranes depending on the chosen configuration.

5. Water Purification

Water to be split during electrolysis must respond to many compositional requirements. These are standardized by the American Society for Testing and Materials (ASTM) Type II deionized (DI) water (resistivity 41 MΩ cm) while ASTM Type I DI water (410 MΩ cm) is preferred. ASTM defines Type II water, as required in commercial electrolyzers, as having a resistivity of 41 MΩ cm, sodium, and chloride content <5 μg L^{-1} and <50 ppb of total organic carbon (TOC).

The requirements needed by the water to be split in Alkaline electrolyzers are less stringent if compared to those needed for PEM. Anyway, as the water purity increases, the cell stability improves. Many techniques are available in order to reach such purity levels. The main applied ones are reverse osmosis (RO), multi-stage flash distillation (MSF), electrodialysis (ED), multiple-effect distillation (MED) to desalinate water. Sometimes they are coupled with an ion exchange or electrodeionization (EDI) for further improvement of the water quality.

In the case of seawater electrolysis, operating and capital costs must be seriously considered because of the equipment consumption depending on the starting water quality [20].

In the recent past, many developments have to reach toward the optimization of electrodes and catalysts for the OER in alkaline media of the seawater. The best conditions are underlined in the order of (4300 mA cm^{-2}) at the typical seawater pH (8). Now, seawater is very rich in borates and carbonites; the presence of these compounds limits high current densities. This limit can be overcome through the addition of additives such as KOH. All these additives led to an increase in the conductivity as the concentration increased. Obviously, it must be taken into account that the increase in additives concentration leads to a variation in the water pH. Strong variations in the pH impact the durability and stability of the electrodes. Other problems to be faced are eventual presences of other kinds of impurities such as various dimensions particles, ions, bacteria, microbes, all impacting on the system stability and long-term durability.

Much progress has been done in the recent past on desalinization through seawater reverse osmosis (SWRO). The improvements are mainly due to the development of highly resistant, stable, and efficient membranes, highly efficient energy recovery devices, and process optimization of reverse osmosis (RO) systems. All this progress led to the reduction in electricity consumption as well as to an important drop in operating costs. At the present time, the energy requirement of SWRO desalination plants has decreased from 9–10 kWh m^{-3} to <3 kWh m^{-3}.

A schematic of a typical desalinization plant coupled with a PEM electrolyzer is shown in Figure 10.

The PEM electrolysis plant consists of the electrolyzer stacks and the mechanical and electrical balance of plant (BoP) components. The electrical BoP consists of the AC to DC rectifier for converting grid electricity while the mechanical BoP consists of other auxiliary components such as pumps, heat exchangers, temperature swing adsorption (TSA) subsystem, and most importantly a deionizer (DI) system.

The SWRO plant contains the RO unit which uses a membrane barrier and pumping energy to separate salts from saline water. Using high-pressure pumps, water is forced through semi-permeable membranes that have a dense separation layer (thin-film composite membrane) allowing the passage of pure water molecules while rejecting dissolved salts and other impurities. In addition, in order to control RO membrane (bio)fouling and scaling, the SWRO system necessitates physical (e.g., dual media, sediment and carbon filters or low-pressure membranes, such as ultrafiltration) and chemical (e.g., coagulant polymer, antiscalant, acid, chlorination/dechlorination) pre-treatment steps with variable complexity depending on raw feed water quality. A combination of these filters provides a broad spectrum of reduction.

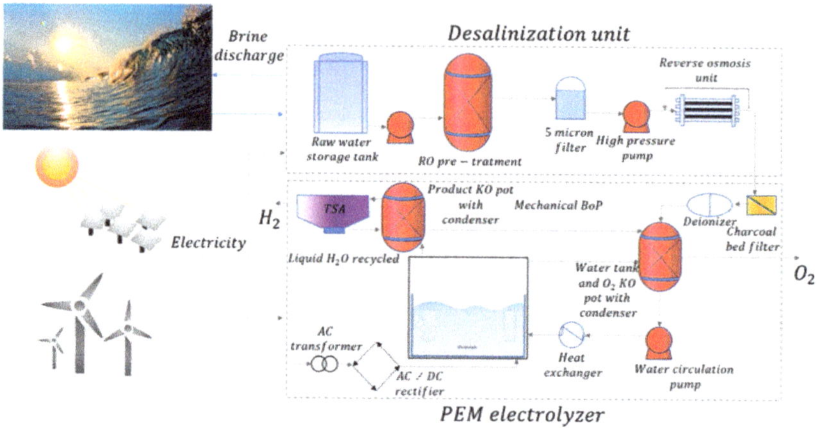

Figure 10. Seawater desalinization for hydrogen production.

There are several RO pre-treatment designs that could be adopted depending on the quality of water needed. The desalination unit consists of a double-pass RO system designed to attain the high purity of water required by the PEM electrolyzer. The SWRO-PEM coupled system could be located in coastal regions with intense solar irradiation and/or wind energy available to produce renewable electricity via photovoltaic cells, wind turbines, or even offshore structures if hydrogen supply for shipping for example was desired.

6. Efficient Hydrogen Production

The classically employed electrocatalysts for both the hydrogen evolution reaction (HER) and oxygen evolution reaction (OER) are precious metals such as platinum (Pt), and ruthenium (Ru) as well as their compounds. Obviously, these are very expensive materials; for this reason, electrolysis development limited in the past the large-scale industrial applications of these technologies.

For these reasons, technological and scientific research focused on the development of less expensive non-precious metals-based catalysts HER and OER.

The best alternatives have been recognized in metal-organic frameworks (MOFs) composites and transition-metal-based compounds such as metal oxides, hydroxides, sulfides, phosphides, nitrides, and selenides with high electrocatalytic properties for the development of HER and OER [21].

All the present efforts are devoted to increasing the efficiency of these non-precious materials catalysts in order to reduce the electric power required for hydrogen production.

Being electrochemical water splitting is one of the most environmentally friendly approaches used to produce hydrogen because it involves no carbon footprint. The half-cell reactions (HER and OER) occurring at cathode and anode, respectively; it must be fed by renewable sources. The final electrochemical decomposition of H_2O into H_2 and O_2 (by-product) is shown in (Equations (28) and (29)):

$$4H^+ + 4e^- \rightarrow 2H_2(\text{HER at cathode}) \quad (28)$$

$$2H_2O \rightarrow O_2 + 4H^+ + 4e^-(\text{OER at anode}) \quad (29)$$

so, the global reaction will be (Equation (30)):

$$2H_2O \rightarrow 2H_2 \uparrow + O_2 \uparrow \quad (30)$$

Hydrogen is produced during the development of HER representing one of the half-reactions in the overall electrochemical water splitting process. Obviously, only limited elements are favorable for efficient hydrogen production during HER. They are basically efficient for favoring the well-known fundamental mechanisms of the HER catalysis. These mechanisms are the Volmer-Heyrovsky and the Volmer-Tafel ones. The evolution of these mechanisms in acidic media is schematically shown in Figure 11a.

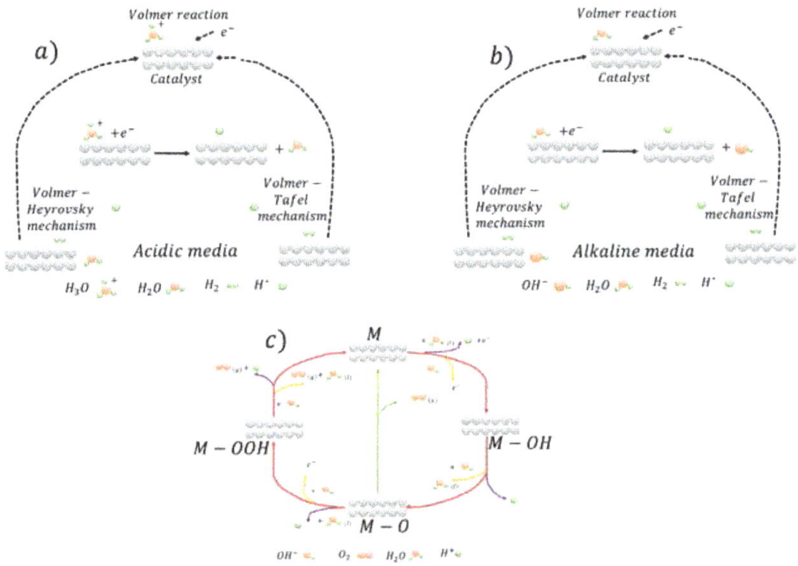

Figure 11. (**a**) HER mechanism in acidic media; (**b**) HER mechanism in alkaline media; (**c**) The oxygen evolution reaction (OER) mechanism in acidic solution (purple line) and alkaline solution in (yellow line).

The schematic of the evolution of the mechanisms in neutral and alkaline media is shown in Figure 11b.

In detail, the Volmer mechanism governs the primary step of HER during which the beginning of the transfer of the electrons takes place. Firstly, the H^+ (H_3O^+) combines with an electron to form H^* deposited/adsorbed on the electrode surface. The adsorbed H^* is then removed/desorbed by two paths to form H_2. At this stage, the first desorption step takes place. This is known as the Heyrovsky mechanism reaction during which the H^* atom in an adsorbed state merges with a free H^+/e^- couple from H_2. The final desorption step is governed by the Tafel reaction mechanism during which two adsorbed H^* combine to produce H_2.

By analyzing these fundamental mechanisms, in the case of high Tafel slopes, the Volmer reaction results in the rate-determining step. This indicates that the adsorption of hydrogen on the catalyst surface under low H coverage acts. Obviously, the electrode kinetics mechanisms during HER are strongly dependent on the solution pH. This is because, chemically, electron transfer is driven by the catalytic activity at the electrode surface. As a consequence, the focus must be on the evolution of the catalytic mechanism in different media.

The second half-reaction of water splitting is the oxygen evolution reaction (OER). It takes place at the anode electrode surface. The main difference with respect to the HER is that this reaction evolves through a four-electron transfer.

With respect to HER, OER requires higher overpotentials. This aspect is responsible for the strict matters related to the overall energy behavior and efficiency of the electrolysis processes. The main intermediate products of OER mechanisms are O*, OH*, and OOH*. Due to the nature of these products, the oxygen evolution follows different routes depending on the solution media. The main aspect of ORER is that, here, OH^- is absent in acidic solution. For this reason, the metal (M) reacts with water producing MOH. Then, it reacts continuously with electrons by transforming into MO.

MO and oxygen can be produced via two different mechanisms: in the first, two MO combine in order to produce oxygen directly; in the second, MO reacts continuously with oxygen by forming the intermediate species MOOH, later, MOOH forms O_2.

The decomposition of MOOH forming O_2 is shown in Figure 11c.

According to the Sabatier principle, the adsorption phenomenon is high pronounced to produce desorption; so, it is necessary to employ catalysts with the moderate binding capability to drive the whole process. In this view, when catalysts have weak bonding with O, the intermediate species OH* does not evolve easily into OOH*. It is believed that the best catalytic performance can be reached once the bonding of O acts in moderate conditions.

The best way to produce clean hydrogen is the employment of renewable power sources such as wind, solar, and biomass. Anyway, at the present time, only 4% of the consumed hydrogen is produced via this route. The remaining percentage is still related to technologies different from water electrolysis. The main problem is related to the origin of power and to the issues connected with the consumption of the electrolysis cells components.

Now, from a chemical point of view, the minimum energy needed for the water-splitting is the Gibbs free energy related to the needed voltage at a given temperature and pressure. In order to have efficient water splitting, the catalytic behavior during both OER and HER mast be kinetically increased for high overall electrolysis efficiency. At the present time, precious metals such as platinum are the best choices for efficient water splitting. However, much progress has been done in the recent past with respect to the development of catalysts based on less expensive materials. The main route is the replacement of precious metals with non-metal ones especially the transition materials.

Several transition metals compounds result in a very effective application in alkaline water electrolysis (AWE). They are mainly based on Fe, Mn, Co, and Ni-based, including their oxides resulting very efficiently during HER and OER half-cell reactions in water splitting. The main routes toward their large utilization are the improvement of electrical conductivity, the increased stability in different environments and media, the possibility of designing hybrid solutions by coupling such compounds with carbon-based materials.

More serious problems are faced during the development of water steam electrolysis. In fact, the very high temperature of the steam (in the order of 1000 °C) leads to fast degradation of the cell components, in particular, electrodes. In addition, such high-temperature conditions lead to the high instability of the electrocatalysts. In addition, critical issues are represented by ionic conductivity, electronic conductivity, and catalytic conductivity in SOEL. One of the primary solutions is the expansion of the surfaces of the electrode in order to reduce excessive heating. Costs can be reduced by deeply employing non-precious metals as catalysts. The cell duration can be increased by reducing the overpotential well as improving the stability of the catalysts.

Being the more recently developed, non-precious materials catalysts research for PEMEL cells is at the beginning. The main problem is that the stability of catalysts into acidic media is very crucial. The mainly employed materials for HER in acidic media are the oxides of transitional metals such as NbC, SnO_2, Ta_2O_5, TiO_2, WC, and TiC. The field is still very open because it is very difficult to substitute precious metals catalysts for the oxygen evolution reaction in PEMEL cells.

Other alternative solutions are represented by the employment of MOF electrocatalysts for OER and HER. They have large potentials because of their large surface area, optimal physiochemical properties, and porous structures.

7. Steel Manufacturing with Renewables Integration

During the direct reduction of iron oxides through hydrogen, water vapor is produced instead of carbon dioxide of the traditional ironmaking and steelmaking routes [22,23]. Then the water vapor can be separated and condensed or employed as input in the electrolyzers for water splitting. In this scenario, the only produced carbon dioxide is the one belonging to the power to be employed for the electrolysis operations. As above mentioned, H_2 is currently mainly produced from fossil sources (natural gas, coal, oil), by biomass gasification, and from non-carbon sources, such as water electrolysis [24]. Once fossil sources are employed, as in the case of natural gas or methane, the generated carbon dioxide must be captured and stored with largely increased costs. If green power is employed, the consumed power is in the order of ~5 kWh/m^3 of H_2. It is believed that further developments of the different technologies will lead to a remarkable reduction in consumption [25].

Hydrogen has large potentials for the direct reduction of iron ores. Obviously, the best solution is to reduce iron oxides through hydrogen produced via water electrolysis. In this way, it is estimated that the overall process can lead to carbon dioxide emissions in the order of 300 kg/t HRC.

The main employed direct reduction reactors in the world steel industry are based on the MIDREX or HYL processes. Here the dynamic control of the different transformations is optimal for the direct reduction through hydrogen. Obviously, many issues are related to the temperature and pressure in the reactor in terms of safety, product quality, and efficiency of the whole reduction process. Another fundamental aspect is represented by the metallization degree and by the presence of carbon in the reduced iron [1].

Generally, hydrogen can be employed as the only reductant gas in these king of plants as well as mixed with different percentages of natural gas. By considering the MIDREX® Plant, in the case of H_2 addition, one-third of the required natural gas can be substituted. For example, 60,000 Nm^3/h of H_2 can be substituted for approximately 20,000 Nm^3/h of natural gas in a 2.0 Mtpy plant, which represents approximately 30% of the total natural gas consumption. MIDREX® Plants generally employ three different ratios of H_2 and CO. Most use natural gas and a standard MIDREX® Reformer that produces a reducing gas with 55% H_2 and 36% CO (H_2/CO of 1.5). As above mentioned, hydrogen is very volatile, so the reactor pressure is fundamental. The equilibrium diagram for the reforming processes is shown in Figure 12.

The hematite reduction act through the following reactions (Equations (31)–(36)):

$$2Fe_2O_{3(s)} + H_{2(g)} \rightarrow 2Fe_3O_{4(s)} + H_2O_{(g)} \tag{31}$$

$$Fe_3O_{4(s)} + \frac{16}{19}H_{2(g)} \rightarrow \frac{60}{19}Fe_{0.95}O_{(s)} + \frac{16}{19}H_2O_{(g)} \tag{32}$$

$$Fe_{0.95}O_{(s)} + H_{2(g)} \rightarrow 0.95Fe_{(s)} + H_2O_{(g)} \tag{33}$$

$$2Fe_2O_{3(s)} + CO_{(g)} \rightarrow 2Fe_3O_{4(s)} + CO_{2(g)} \tag{34}$$

$$Fe_3O_{4(s)} + \frac{16}{19}CO_{(g)} \rightarrow \frac{60}{19}Fe_{0.95}O_{(s)} + \frac{16}{19}CO_{2(g)} \tag{35}$$

$$Fe_{0.95}O_{(s)} + CO_{(g)} \rightarrow 0.95Fe_{(s)} + CO_{2(g)} \tag{36}$$

The FMO MIDREX® Plant in Venezuela is designed with a steam reformer, and H_2/CO can vary from 3.3 to 3.8. There are six MIDREX® Modules that utilize gas made from coal, and these have hydrogen to CO ratios from 0.37 to 0.56. Thus, the MIDREX® Process has successfully produced DRI at H_2/CO ratios from 0.37 to 3.8. In these plants, 100% of pure hydrogen can also be employed. However, in the case of MIDREX, the reactor configuration does not change if natural gas or hydrogen is used as a reductant. H_2 input gas is generated external to the process and there is no reformer. With this design, a gas heater is employed to heat the gas to the required temperature. In the practice, the reducing gas H_2 content is about 90%, with the balance CO, CO_2, H_2O, and CH_4.

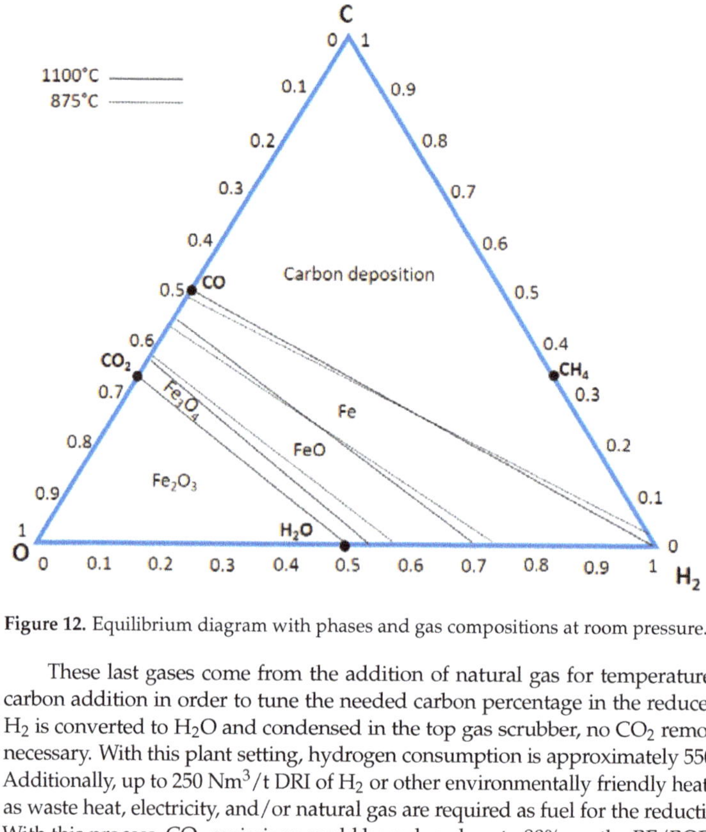

Figure 12. Equilibrium diagram with phases and gas compositions at room pressure.

These last gases come from the addition of natural gas for temperature control and carbon addition in order to tune the needed carbon percentage in the reduced iron. Since H_2 is converted to H_2O and condensed in the top gas scrubber, no CO_2 removal system is necessary. With this plant setting, hydrogen consumption is approximately 550 Nm^3/t DRI. Additionally, up to 250 Nm^3/t DRI of H_2 or other environmentally friendly heat sources such as waste heat, electricity, and/or natural gas are required as fuel for the reduction gas heater. With this process, CO_2 emissions could be reduced up to 80% vs. the BF/BOF steelmaking route. There are a number of considerations for the MIDREX H$_2$™ Process, the first of which is temperature. With these high hydrogen percentages, the DRI is deeply cooled. So, natural gas is necessary to sustain the needed temperature levels. According to Midrex indications, the addition of natural gas at a rate of 50 Nm^3/t DRI should accomplish this.

As mentioned above, the other crucial aspect is the iron carburizing. The vast majority of DRI is used in EAFs for further processing. EAF steelmaking practices today generally employs carbon added either in metallic charge materials such as DRI, HBI, and pig iron or as pure carbon. Burning this carbon with injected oxygen creates significant heat which reduces electricity consumption and enables faster melting. Since pig iron is made from BF hot metal that is saturated with carbon, it contains 4–4.5 percent carbon. DRI can have 1–4.5 percent carbon depending on the process, reducing gas used, and the way the DR plant is operated.

In the case of hydrogen produced via electrolysis, a crucial aspect is also represented by the location of electrolysis plants to produce hydrogen because this is the most electricity-consuming section of the integrated plant. To give an idea of electricity needed, for fossil-free electricity in the electric arc furnace, the amount of electricity is only about

0.5 MWh/ton liquid steel while the electrolysis consumes more than 2.5 MWh/ton liquid steel. For an energy system based on renewable and intermittent electricity, there is a need for balancing power, and hydrogen storage can play an important part.

The simplified schematic of the different sections of the analyses is shown in Figure 13.

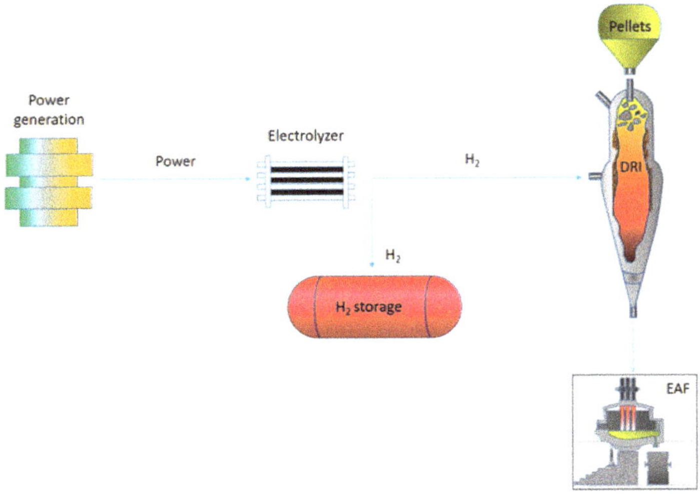

Figure 13. Schematic of integrated fossil-free primary steelmaking.

Energy is required by the DRI pre-heating, for the hydrogen pre-heating and compression before storage; this energy is very low if compared to the one required by the main sections of the plant.

In order to produce one ton of sponge iron through HDRI, 49 kg (545 Nm^3) of hydrogen is necessary.

Electrolysis is a modular technology where a big facility would consist of several parallel electrolysis stacks. The capacity of the current biggest stacks is approximately 600 Nm^3 H_2 per hour. The electricity consumption per Nm^3 hydrogen differs between 4.6–5.6 kWh/Nm^3, depending on the technology (alkaline electrolysis or PEM) with an expected decrease of approximately 0.2–0.4 kWh/Nm^3 for both technologies in the future due to technology and efficiency improvements (HYBRIT 2018). The 545 Nm^3 H_2 required for the production of one-ton sponge iron needs approximately 2.5–3 MWhe (de-pending on the electrolysis technology) for the hydrogen production.

Depending on the energy-saving requirements, electrolysis facility can either be placed close to the electricity generation e.g., close to a wind farm, and the hydrogen can then be transported in pipes to the DRI plant, or it could be located close to the plant, given that the conditions of the local energy system are right. One option is also to decentralize the system and locate the electrolyzers separately close to several different power production sites [26]. First of all, the power and electricity requirements depend on the DRI capacity. In order to produce one million tons of sponge iron, 2.7 TWh are required to correspond to 341 MW. These requirements increase to 8.2 TWh and 1022 MW for a three million-ton facility. At the moment this is the maximum size of the installed DRI plants. If a capacity of 5 million tons is reached, the requirement is 13.6 TWh and 1700 MW. In the case of hydrogen storage, the capacity of the electrolyzer must be increased. The optimal solution is believed to be the location of the electrolyzer plant directly connected with the renewables.

A very recent study investigates the integration of water electrolysis technologies in fossil-free steelmaking via the direct reduction of iron ore followed by processing in an electric arc furnace (EAF) with very broad and deep analyses of all the components. Hydrogen (H_2) production via low or high-temperature electrolysis (LTE and HTE) is considered for the production of carbon-free direct reduced iron (DRI) [3,27,28]. Producing carbon-free DRI with H_2 produced via LTE is probably the most straightforward route to fossil-free steel. HTE is attractive due to its high efficiency, given that steam is available, but is a much less proven technology than LTE [29]. Anyway, in all those situations where high-temperature sources and steam are available, THE is very efficient. Obviously, the performance and the durability of these high-temperature cells are fundamental for the large diffusion in the steel industry.

The schematic of the DRI plant integrated with LTE and HTE is shown in Figure 14.

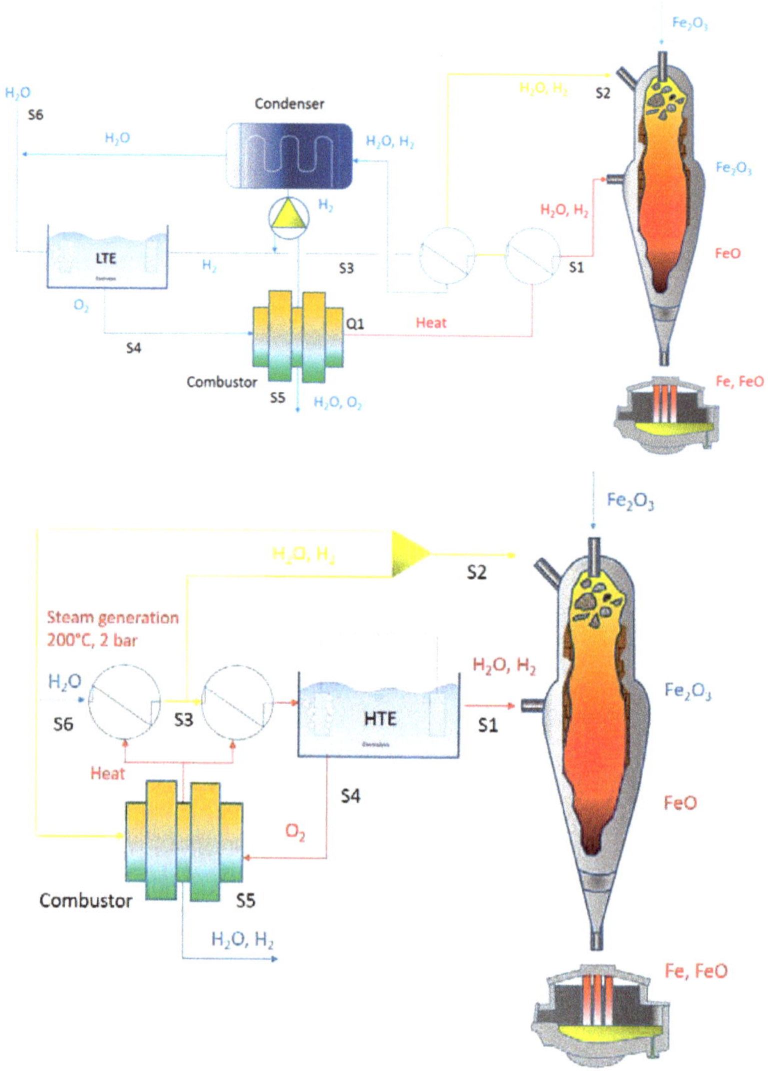

Figure 14. Schematic of the integration of DRI with LTE and HTE.

In all the described solutions, hydrogen is sent to the reactor at a temperature around 900 °C. By employing LTE, hydrogen must be pre-heated through heat exchange with the top gases, heating recovery from post-combustion of top gases, or traditional electricity facilities.

In order to avoid inert gas accumulation in the reactor, a top gas purge stream is employed. Recycled gas is normally purged at percentages around 10. This solution normally leads to increased hydrogen production in the case of low-temperature electrolysis. In this case, oxygen (produced in water splitting) is employed as fuel for the purge stream. In the case of high-temperature electrolysis facilities, top gas is directly sent to the electrolyzer. So, steam is produced through the heating provided by the top gas. In this solution, being the operating temperature of the electrolyzer in the order of 900–100 °C, it is not necessary the pre-heating of the produced hydrogen before the insertion in the reducing reactor. Of course, if the hydrogen is cooled, the extra heating will be provided to reach the required temperature levels of the reducing gas. In the case of HTE, a crucial aspect is represented by the sulfur removal before entering the DR shaft.

Generally, the top gas recycling stuck is a cyclone equipped with off-gas de-dusting and a condenser for heat recovery and hydrogen separation from the de-dusted off-gas. Pure hydrogen is produced via electrolysis based on renewables or nuclear energy. So, hydrogen from electrolysis is mixed with the separated hydrogen and then sent to the shaft furnace after compression (the pressure level depends on the type of plant; normally HYL reactor is designed for high-pressure hydrogen in order to increase the process efficiency).

The reducing gas is inserted from the bottom of the furnace, and it reduces the iron oxides at high temperature by encountering the iron-bearing pellets that are charged from the top of the furnace. The direct reduced iron is then collected at the bottom and sent to the electric arc furnace for further processing and melting.

The process is high heat consuming because of the endothermic nature of reactions involving pure hydrogen. So, in this view, it is fundamental to balance the employment of hydrogen from electrolysis and hydrogen from gas separation in order to sustain the process with constant and high efficiency.

So, the limiting aspect is those conditions related to the managing of hydrogen in terms of pressure and temperature.

The type of hydrogen also strongly influences the degree of metallization of the direct reduced iron. In the ideal case of 100% of metallization, the ratio recycled to electrolysis hydrogen is higher than 3. In this condition, the energy consumption is in the order of 13.7 GJ/t-DRI, where large energy consumption is due to the electricity required by the electrolysis plant. They provide heat that can be reduced if carbon monoxide is employed as an additional reductant. In this case, also carburization of iron is facilitated. Normally, Midrex plants employ hydrogen up to 75% and carbon monoxide up to 30%. With this carbon monoxide percentage, it is easy to take the temperature in the furnace under control.

As mentioned above, the carburization of iron is a fundamental aspect. In direct-reduced iron, carbon is contained in the form of cementite. Many studies are presented in the literature about the nature and the kinetics of cementite formation during the direct reduction of iron oxides. Different controlling mechanisms have been suggested for cementite formation such as the chemical reaction rates at the pore surfaces of reduced iron and the rate of mass transfer of carbon in iron.

In the case of sulfur presence, the stability of cementite is increased and the presence of free carbon in the iron is reduced. For this reason, sulfur presence is controlled with great precision. Again, the gas temperature has a large influence on the carbide's transformations and diffusion. In fact, the reduction rate is largely influenced by the temperature driving the chemical equilibrium of the reactions. At the same time, the cementite stability is governed by the carbon activity that is related to the temperature level. In addition, as the temperature is increased, the reduction rate increases because of a faster chemical reaction and mass transfer.

The general TENOVA-HYL direct reduction plant schematic is shown in Figure 15.

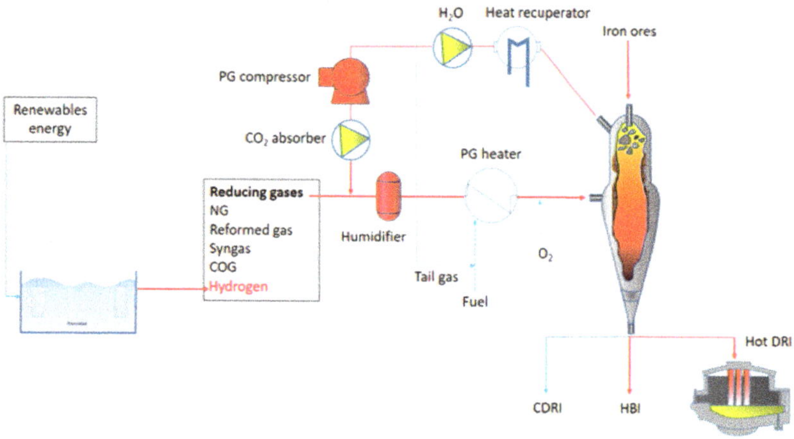

Figure 15. Schematic of HYL-DRI plant.

The operating temperature of the reactor is higher than 1050 °C, the reactor pressure falls in the range 6–8 bar. This pressure allows for the low volatility of the reducing gases with a remarkable increase in the overall efficiency. The plant is designed to employ various reducing gases including hydrogen in a broad range of percentages. The flexibility of the solution allows for the use of any energy source. The iron ores dimensions are in the range 3.2–18 mm. The plant allows for the selective removal of carbon dioxide demonstrating its intrinsic capability of CCSU.

The carbon dioxide capture corresponds to 45% of the total carbon input. The free CO_2 emissions to the atmosphere are 30% of the total carbon input. The carbon in the DRI is 25% of the total input.

For comparison, carbon input and output streams as well as the distribution of carbon flow within the MIDREX process are shown in Figure 16 (the figure shows the carbon balance for both NG and H_2 processes).

The main unit operations of the process comprise DR-shaft, reformer, top gas scrubber, product gas compressors, and heat exchangers. The reformer of the DR-NG schematic comprises one part for the reforming of NG providing the required energy for producing the syngas. The reforming process approaches a state of thermodynamic equilibrium in practice. The third main component is the top gas scrubber which has the goal to reduce the water content on the one hand and control the temperature levels of the two separated gas output streams on the other hand. Therefore, the temperature and the pressure level of these two streams—reducing gas input and top gas fuel—are set to fixed values. NG can be injected at different process stages: NG as input to the reforming gas, as an energy input for heating the reformer or reduction gas heater as well as directly to the bustle gas before entering the shaft furnace. The input of hydrogen in the DR-NG process is foreseen in the reforming gas before entering the reformer. In the hydrogen-based schematic, the reformer of the DR-NG process is replaced by a reduction gas heater representing a combination of a gas burner, mixer, and heat exchanger model. Hydrogen can be added either to the reduction gas and/or to the top gas fuel for heating purposes. The energy supplied to the heater can also be provided by NG. To achieve comparable results, the same solid input material compositions and amounts, as well as the same basic assumptions for the DR shaft (e.g., reduction degree, carburization behavior, temperature distribution, and so on), the top gas scrubber (temperatures and pressure levels) and the gas burner (excess air), is applied for both process models. In the proposed schematics, about 30% of the NG can be replaced by hydrogen without any process changes. The replacement of NG by hydrogen leads to an increased volume flow in the shaft furnace due to changing gas compositions. The gradual injection of hydrogen entails a corresponding decrease in carbon monoxide, whereas the content of CH4 is only slightly influenced by higher hydrogen shares. This behavior can be explained by the fact that even if 100% of NG is used, it is almost completely decomposed into carbon monoxide and hydrogen in the reformer before entering the shaft. Only a small and constant amount of NG is added for the enrichment of the bustle gas before entering the shaft. If hydrogen is added directly before the shaft furnace, it has to be preheated in a separate heating unit. The second schematic belongs to a design for the input of about 95% hydrogen. The remaining part is NG which is necessary to maintain process temperatures and the carbon content of the produced DRI. According to the DR-H_2 process setup, higher recirculating gas flows are necessary to maintain the required process temperatures. Residual amounts of CO and CO_2 are still present in the recycled gas and increase the total specific volume flow in the system. However, carbon-containing gas streams only play a minor role in the hydrogen-based DR process.

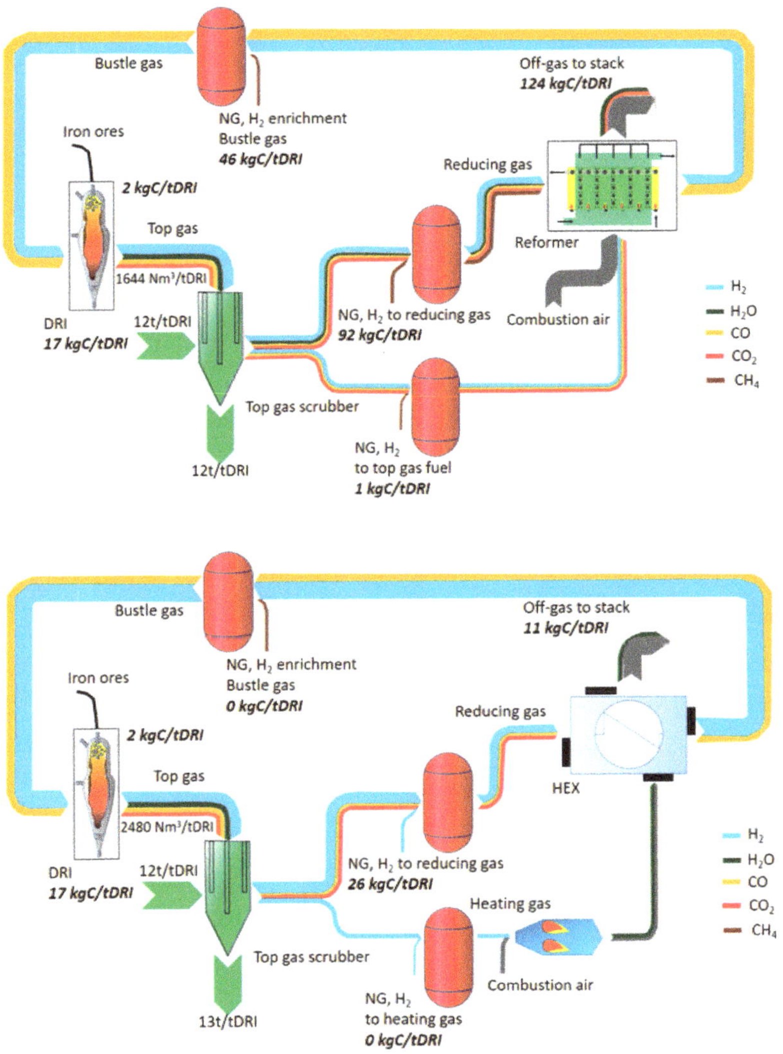

Figure 16. Schematic of the NG-and H2-based Midrex plants.

NG either used for reforming, enrichment, or heating purposes represents the main carbon input to the DR-NG reference process. In addition to the carbon output via the DRI product, the major emission source of carbon is the off-gas of the reformer. About 124 kg C/t DRI which are equivalent to about 453 kgCO_2/tDRI are emitted at this point. In comparison, the carbon output of the DR-H_2 process is almost equally distributed between the DRI and stack emissions (released by using the top gas as combustion gas for the heater), representing 17 and 11 kg C/tDRI, respectively. The main carbon source, in this case, is as well NG which is required for maintaining the carbon content of the DRI [30].

As mentioned, the TENOVA plant can work with different concentrations of hydrogen in the feeding gas. As the hydrogen content in the mixture increases, the total energy consumption in the reactor decreases. A strong decrease in electricity consumption is recorded as the hydrogen content increases.

8. Conclusions

Hydrogen is believed to be the best actor toward the decarbonization of industries hard to abate. It is considered the energy vector of the next future, capable of addressing multiple energy challenges and the cleanest solution to decarbonize industries. Hydrogen has the potential to be employed in the production of carbon steels, special metals, and semiconductors in the steel and electronics industries.

At the present time, more than 95% of the employed hydrogen is produced from fossil fuels. The most typical hydrogen production routes from fossil sources are coal gasification or natural gas reforming. At the present time, only 4% of hydrogen is produced via water electrolysis. Obviously, the decarbonization of human activities needs hydrogen to be produced through sustainable routes. One of the most promising ways is the electrolysis of water with the energy sources provided by renewables.

During electrolysis, water is split into hydrogen and oxygen through electricity. So, the electric energy is converted into chemical energy (though hydrogen) and thermal energy. At a given water volume, the produced hydrogen quantity directly depends on the electric current. The main water electrolysis technologies are: alkaline (AEL), Proton Exchange Membrane Electrolyzer (PEMEL), and Solid Oxide Electrolyzer (SOEL). The energy needed for polymer electrolyte membrane electrolyzers and alkaline water electrolysis is in the order of 4.4–4.9 kWh/m^3 H_2. Obviously, this energy demands decrease for high-temperature water electrolysis cells falling in the order of 3.8–3.9 kWh/m^3 H_2.

Water input represents a crucial aspect of these technologies, Alkaline electrolyzers are less stringent on water quality as compared to PEM, but it is recommended to employ high purity water in order to reach long-term stability and duration of the electrolysis plants. Such high purity water as required by water electrolysis systems is produced through a combination of either reverse osmosis (RO), multi-stage flash distillation (MSF), electrodialysis (ED), multiple-effect distillation (MED) to desalinate water, and commonly an additional technology such as ion exchange or electrodeionization (EDI).

The best solution for the steel industry seems to replace the traditional route based on coal and coke with the direct reduction reactors employing hydrogen as the main reductant. Midrex and HYL Energiron are the processes with the highest capacities in-stalled worldwide. H_2 can also be used as the reductant in conventional direct reduction reactors. In a route based on hydrogen and direct reduction, the output after the re-duction is the porous material of DRI, or sponge iron, which can possibly be transported to an EAF in a different location (or pressed to hot briquetted iron, HBI, which is favorable for transportation). The energy required for the melting in the EAF depends on the inlet temperature of the DRI, which will be different depending on if it has been transported or not.

If H_2 is produced by water electrolysis using hydro or nuclear electricity, then CO_2 emissions could be lowered to less than 300 kg/t HRC by saving almost 1700 kg/t HRC. Obviously, different gas mixings can be employed because of the temperature levels needed inside the reactor as well as because of the required metallization degree and cementite content in the reduced iron. In addition, the different gas mixings are responsible for the overall efficiency of the reduction process. The hydrogen pressure is also a fundamental parameter to be considered for the optimization of the new ironmaking and steelmaking routes.

Author Contributions: P.D.C.—Conceptualization, Supervision, Validation Writing; A.P.—Data curation; A.S.—Data curation. All authors have read and agreed to the published version of the manuscript.

Funding: This research received no external funding.

Institutional Review Board Statement: Not applicable.

Informed Consent Statement: Not applicable.

Conflicts of Interest: The authors declare no conflict of interest.

References

1. Cavaliere, P. Clean Ironmaking and Steelmaking Processes-Efficient Technologies for Greenhouse Emissions Abatement. In *Clean Ironmaking and Steelmaking Processes*; Springer: Cham, Switzerland, 2019. [CrossRef]
2. Schalenbach, M.; Tjarks, G.; Carmo, M.; Lueke, W.; Mueller, M.; Stolten, D. Acidic or Alkaline? Towards a New Perspective on the Efficiency of Water Electrolysis. *J. Electrochem. Soc.* **2016**, *163*, 3197–3208. [CrossRef]
3. Kruger, A.; Andersson, J.; Gronkvist, S.; Cornell, A. Integration of water electrolysis for fossil-free steel production. *Int. J. Hydrogen Energy* **2020**, *45*, 29966–29977. [CrossRef]
4. Shalenbach, M.; Zeradjanin, A.R.; Kasian, O. A Perspective on Low-Temperature Water Electrolysis –Challenges in Alkaline and Acidic Technology. *Int. J. Electrochem. Soc.* **2018**, *13*, 1173–1226. [CrossRef]
5. Lehner, M.; Tichler, R.; Steinm¨uller, H.; Koppe, M. *Power-to-Gas: Technology and Business Models*; Springer International Publishing: New York, NY, USA, 2014. [CrossRef]
6. Carmo, M.; Fritz, D.; Mergel, J.; Stolten, D. A comprehensive review on PEM water electrolysis. *Int. J. Hydrogen Energy* **2013**, *38*, 4901–4934. [CrossRef]
7. Schmidt, O.; Gambhir, A.; Staffell, I.; Hawkes, A.; Nelson, J.; Few, S. Future cost and performance of water electrolysis: An expert elicitation study. *Int. J. Hydrogen Energy* **2017**, *42*, 30470–30492. [CrossRef]
8. Herz, G.; Muller, N.; Adam, P. High Temperature Co-Electrolysis as a Key Technology for CO2 Emission Mitigation—A Model-Based Assessment of CDA and CCU. *Chem. Ing. Tech.* **2020**, *92*, 1044–1058. [CrossRef]
9. Motylinski, K.; Wierzbicki, M.; Kupecki, J.; Jagielski, S. Investigation of off-design characteristics of solid oxide electrolyser (SOE) operating in endothermic conditions. *Renew Energy* **2021**, *170*, 277–285. [CrossRef]
10. Koj, J.C.; Wulf, C.; Schreiber, A.; Zapp, P. Site-Dependent Environmental Impacts of Industrial Hydrogen Production by Alkaline Water Electrolysis. *Energies* **2017**, *10*, 860. [CrossRef]
11. Smolinka, T.; Wiebe, N.; Sterchele, P.; Palzer, A.; Lehner, F.; Jansen, M.; Kiemel, S.; Miehe, R.; Wahren, S.; Zimmermann, F. *Studie IndWEDe: Industrial-Isierung der Wasserelektrolyse in Deutschland: Chancen und Herausforderungen für nachhaltigen Wasserstoff für Verkehr, Strom und Wärme*; Fraunhofer-Institut für Produktionstechnik und Automatisierung: Stuttgard, Germany, 2018.
12. Dotan, H.; Landman, A.; Sheehan, S.W.; Malviya, K.D.; Shter, G.E.; Grave, D.A.; Arzi, Z.; Yehudai, N.; Halabi, M.; Gal, N.; et al. Decoupled hydrogen and oxygen evolution by a two-step electrochemical–chemical cycle for efficient overall water splitting. *Nat. Energy* **2019**, *4*, 786–795. [CrossRef]
13. Ulleberg, Ø. Modeling of advanced alkaline electrolyzers: A system simulation approach. *Int. J. Hydrogen Energy* **2003**, *28*, 21–33. [CrossRef]
14. Ursua, A.; Gandia, L.M.; Sanchis, P. Hydrogen production from water electrolysis: Current status and future trends. *Proc. IEEE* **2012**, *100*, 410–426. [CrossRef]
15. Milewski, J.; Guandalini, G.; Campanari, S. Modeling an alkaline electrolysis cell through reduced-order and loss-estimate approaches. *J. Power Sources* **2014**, *269*, 203–211. [CrossRef]
16. Jang, D.; Choi, W.; Cho, H.-S.; Cho, W.C.; Kim, C.H.; Kang, S. Numerical modeling and analysis of the temperature effect on the performance of an alkaline water electrolysis system. *J. Power Sources* **2021**, *506*, 230106. [CrossRef]
17. Koponen, J.; Kosonen, A.; Ruuskanen, V.; Huoman, K.; Niemelä, M.; Ahola, J. Control and energy efficiency of PEM water electrolyzers in renewable energy systems. *Int. J. Hydrogen Energy* **2017**, *42*, 29648–29660. [CrossRef]
18. Sun, F.; Qin, J.; Wang, Z.; Yu, M.; Wu, X.; Sun, X.; Qiu, J. Energy-saving hydrogen production by chlorine-free hybrid seawater splitting coupling hydrazine degradation. *Nat. Commun.* **2021**, *12*, 1–11. [CrossRef] [PubMed]
19. Ifkovits, Z.P.; Evans, J.M.; Meier, M.C.; Papadantonakis, K.M.; Lewis, N.S. Decoupled electrochemical water-splitting systems: A review and perspective. *Energy Environ. Sci.* **2021**, *14*, 4740–4759. [CrossRef]
20. Khan, M.A.; Al-Attas, T.; Roy, S.; Rahman, M.M.; Ghaffour, N.; Thangadurai, V.; Larter, S.; Hu, J.; Ajayan, P.M.; Kibria, G. Seawater electrolysis for hydrogen production: A solution looking for a problem? *Energy Environ. Sci.* **2021**, *14*, 4831–4839. [CrossRef]
21. Anwar, S.; Khan, F.; Zhang, Y.; Djire, A. Recent development in electrocatalysts for hydrogen production through water electrolysis. *Int. J. Hydrogen Energy* **2021**, *46*, 32284–32317. [CrossRef]

22. Chevrier, V. Midrex H2 TM: Ultimate low-CO2 ironmaking and its place in the new hydrogen economy. In Proceedings of the AISTech-Iron and Steel Technology Conference Proceedings, Philadelphiam, PA, USA, 7–10 May 2018; pp. 725–729.
23. Ghadi, A.Z.; Valipour, M.S.; Biglari, M. CFD simulation of two-phase gas-particle flow in the Midrex shaft furnace: The effect of twin gas injection system on the performance of the reactor. *Int. J. Hydrogen Energy* **2017**, *42*, 103–118. [CrossRef]
24. Prammer, J. The actual problems of current decarbonization. *Chernye Met.* **2019**, *1*, 55–59.
25. Lei, Q.; Wang, B.; Wang, P.; Liu, S. Hydrogen generation with acid/alkaline amphoteric water electrolysis. *J. Energy Chem.* **2019**, *38*, 162–169. [CrossRef]
26. Cavaliere, P. Direct Reduced Iron: Most Efficient Technologies for Greenhouse Emissions Abatement. In *Clean Ironmaking and Steelmaking Processes*; Springer: Cham, Switzerland, 2019. [CrossRef]
27. Müller, N.; Herz, G.; Reichelt, E.; Jahn, M.; Michaelis, A. Assessment of fossil-free steelmaking based on direct reduction applying high-temperature electrolysis. *Clean. Eng. Technol.* **2021**, *4*, 100158. [CrossRef]
28. Cavaliere, P. *Ironmaking and Steelmaking Processes-Greenhouse Emissions, Control, and Reduction*; Springer: Cham, Switzerland, 2016. [CrossRef]
29. Andersson, J.; Grönkvist, S. A comparison of two hydrogen storages in a fossil-free direct reduced iron process. *Int. J. Hydrogen Energy* **2021**, *46*, 28657–28674. [CrossRef]
30. Rechberger, K.; Spanlang, A.; Conde, A.S.; Wolfmeir, H.; Harris, C. Green Hydrogen-Based Direct Reduction for Low-Carbon Steelmaking. *Steel Res. Int.* **2020**, *91*, 2000110. [CrossRef]

MDPI
St. Alban-Anlage 66
4052 Basel
Switzerland
Tel. +41 61 683 77 34
Fax +41 61 302 89 18
www.mdpi.com

Metals Editorial Office
E-mail: metals@mdpi.com
www.mdpi.com/journal/metals

www.ingramcontent.com/pod-product-compliance
Lightning Source LLC
Chambersburg PA
CBHW061440210926
51850CB00008B/1560